University Initiatives in Climate Change Mitigation and Adaptation

AF181700

Walter Leal Filho · Rafael Leal-Arcas
Editors

University Initiatives in Climate Change Mitigation and Adaptation

Springer

Editors
Walter Leal Filho
Faculty of Life Sciences
Hamburg University of Applied Sciences
Hamburg
Germany

Rafael Leal-Arcas
School of Law
Queen Mary University of London
London
UK

ISBN 978-3-030-07809-6 ISBN 978-3-319-89590-1 (eBook)
https://doi.org/10.1007/978-3-319-89590-1

© Springer International Publishing AG, part of Springer Nature 2019
Softcover reprint of the Hardcover 1st edition 2019
This work is subject to copyright. All rights are reserved by the Publisher, whether the whole or part of the material is concerned, specifically the rights of translation, reprinting, reuse of illustrations, recitation, broadcasting, reproduction on microfilms or in any other physical way, and transmission or information storage and retrieval, electronic adaptation, computer software, or by similar or dissimilar methodology now known or hereafter developed.
The use of general descriptive names, registered names, trademarks, service marks, etc. in this publication does not imply, even in the absence of a specific statement, that such names are exempt from the relevant protective laws and regulations and therefore free for general use.
The publisher, the authors and the editors are safe to assume that the advice and information in this book are believed to be true and accurate at the date of publication. Neither the publisher nor the authors or the editors give a warranty, express or implied, with respect to the material contained herein or for any errors or omissions that may have been made. The publisher remains neutral with regard to jurisdictional claims in published maps and institutional affiliations.

Printed on acid-free paper

This Springer imprint is published by the registered company Springer International Publishing AG part of Springer Nature
The registered company address is: Gewerbestrasse 11, 6330 Cham, Switzerland

Preface

Much research on climate change is being performed by many universities around the world. But despite this positive trend, there are only a few publications where an interdisciplinary set of researchers may be able to tackle climate change issues from a variety of perspectives: social sciences, natural sciences, economics, etc. This book is an attempt to fill in this gap.

This publication involves researchers in the field of climate change in the widest sense, not only from traditional climate science, but also from the fields of environment, human geography, business and economics, arts, administration, and media studies.

The aims of this book are twofold:

I. To provide researchers at universities from across the world performing research on issues pertaining climate change with an opportunity to present their works and research projects and also educational initiatives;
II. To introduce innovative methodological approaches and projects which aim to offer a better understanding of climate change across society and economic sectors.

Moreover, a further aim of this book, consistent with the philosophy of the "Climate Change Management Series," is to document and disseminate the wealth of experiences on climate change research at universities taking place today.

This book is divided into two parts:

- Part I contains papers which describe experiences from climate change research, education, and studies.
- Part II describes experiences and lessons from climate change and related projects.

We thank the authors for their willingness to share their knowledge, know-how, and experiences, as well as the many peer reviewers, which have helped us to ensure the quality of the manuscripts.

Enjoy your reading!

Hamburg, Germany
London, UK
Summer 2018

Walter Leal Filho
Rafael Leal-Arcas

Acknowledgements

The co-editor of this book, Prof. Dr. Rafael Leal-Arcas, gratefully acknowledges the financial help from two European Union grants: Jean Monnet Chair in EU International Economic Law (project number 575061-EPP-1-2016-1-UK-EPPJMO-CHAIR) and the WiseGRID project (number 731205), funded by the European Commission's Horizon 2020. Both grants have been awarded to Prof. Dr. Rafael Leal-Arcas.

Summer 2018 Prof. Dr. Rafael Leal-Arcas

Contents

Part I Experiences From Climate Change Research, Education and Studies

Introducing the International Climate Change Information Programme (ICCIP) 3
Walter Leal Filho

Educating Students and Their Future Employers to Minimise Environmental and Climate Impacts Through Cost-Effective Environmental Management Strategies 13
Kay Emblen-Perry and Les Duckers

Carbon Management Planning in UK Universities: A Journey to Low Carbon Built Environment 33
Muhammad Usman Mazhar, Richard Bull, Mark Lemon and Saad Bin Saleem Ahmad

Understanding the Role of Networks in Stimulating Adaptation Actions on the Ground: Examples from Two African Case Studies 57
Gina Ziervogel, Lorena Pasquini and Jessica Lee

Climate Resilience Planning and Organizational Learning on Campuses and Beyond: A Comparative Study of Three Higher Education Institutions 77
Camille Washington-Ottombre, Sarah E. Brylinsky, Dennis B. Carlberg and Dano Weisbord

Language and Climate Change: Towards Language of Sustainability in Promoting Climate Change Mitigation in Malawi 95
Peter Mayeso Jiyajiya

Behavioural Approaches of Rural Women Farmers to Mitigation and Adaptation Measures of Climate Change in Abia State, Nigeria 111
Ogechi Jubilant Umeh and Ike Nwachukwu

Climate Change Impacts and Research in the Caribbean: Constraints, Opportunities and the Role of Tertiary Institutions 131
Lisa Benjamin and Adelle Thomas

Greening Theatre Landscapes: Developing Sustainable Practice Futures in Theatre Graduates 143
Linda Hassall and Stephen Rowan

Integrated Rainwater Harvesting Practices for Poverty Reduction Under Climate Change: Micro-Evidence from Ethiopia 159
Anteneh Girma, Menale Kassie, Siegfried Bauer and Walter Leal Filho

Impact of Renewable Technology on Lignocellulosic Material of Palm Fruit Fibre: Strategy for Climate Change and Adaptation 175
Onifade Tawakalitu Bola, Wandiga Shem Oyoo, Bello Isah Adewale, Jekayinfa Simeon Olatayo and Harvey J. Patricia

Maize Yields in Varying Rainfall Regimes and Cropping Systems Across Southern Africa: A Modelling Assessment 203
Siyabusa Mkuhlani, Walter Mupangwa and Isaiah Nyagumbo

Climate Change Education Trends in Canadian Post-secondary Educational Institutions (PSIs) 229
Ana Maria Martinez and Steve Alsop

Part II Experiences and Lessons From Climate Change and Related Projects

Geospatial Analysis of Rainfall and Temperature Variations Effect on Maize (Zea Mays) Yield 247
Olumuyiwa Idowu Ojo, Johnson Olawale Olonnu and Masengo Francois Ilunga

Climate Change Adaptation for Russian Cities: A Case Study of the Thermal Comfort Assessment 265
Natalia Shartova and Pavel Konstantinov

Integrating Biofertilizers with Conservation Agriculture Can Enhance Its Capacity to Mitigate Climate Change: Examples from Southern Africa 277
Isaiah I. C. Wakindiki, Mashapa E. Malobane and Adornis D. Nciizah

Exergy-Based Responsibility Allocation of Climate Change 291
Hossein Khajehpour, Yadollah Saboohi and George Tsatsaronis

Geo-Web Services and New Exchange Formats to Develop Future Services Supporting Climate Change Adaptation Measures 317
Martin Scheinert, Hardy Pundt and Andrea Heilmann

Interdisciplinary Research on Energy Efficiency Standards and Climate Change Mitigation: Methods, Results, and Communication 333
Carl J. Dalhammar and Jessika Luth Richter

***Counting the Hot Air*: Greenhouse Gas (GHG) Accounting Principles for National GHG Emission Inventories (NEIs)** 351
Sooksiri Chamsuk

Climate Change "Conditionality": The Case for Bundling the Fate of International Trade Legal Obligations and Climate-Change-Relevant Legal Obligations 369
Eduardo Alvarez Armas

Peruvian Amazonian Natives and Climate Change: Minorities Facing a Global Problem 385
Guillermo Mandelli Constant, Erika Sänger Do Amaral, Alberto Paucar-Caceres, Silvia Quispe Prieto and Manuel Caipa Ramos

Part I
Experiences From Climate Change Research, Education and Studies

Introducing the International Climate Change Information Programme (ICCIP)

Walter Leal Filho

Abstract This final chapter introduces the International Climate Change Information Programme (ICCIP) and describes some of its activities, outlining what the Hamburg University of Applied Sciences is doing in respect of climate change adaptation. It also outlines various opportunities for cooperation with universities interested on matters related to climate change.

Keywords Climate change · Adaptation · Training · Information-research

1 Introduction

Over the past 15 years, a noticeable increase in the level of attention given to climate issues has been observed. Finding practical, workable and cost-efficient solutions to the problems posed by climate change has become a priority to many countries. Also, even though the engagement of the private sector on climate matters is not as high and it could be, non-governmental organisations as well as the general public are interested on climate matters in a way not seen before.

But even though climate change is a matter of great scientific relevance and of broad general interest, there are many problems related to its communication (e.g. Moser 2007). There is a need to consider the importance and difficulties inherent in talking about climate change to different types of publics using various types of communication tools and strategies (Nerlich et al. 2010). For instance, climate change is often regarded as too broad in scope, as too abstract in respect of its implementation, too complex and therefore too difficult to understand. Yet, much could be gained by ensuring matters related to climate change are better understood

W. Leal Filho (✉)
Research and Transfer Centre "Sustainable Development and Climate Change Management", Faculty of Life Sciences, Hamburg University of Applied Sciences, Lohbrügger Kirchstraße 65, 21033 Hamburg, Germany
e-mail: info@iccip.net

© Springer International Publishing AG, part of Springer Nature 2019

W. Leal Filho and R. Leal-Arcas (eds.), *University Initiatives in Climate Change Mitigation and Adaptation*, https://doi.org/10.1007/978-3-319-89590-1_1

and if people—especially the youth (Leal Filho et al. 2010) are motivated to engage in the global efforts to address the challenges posed by climate change.

There is thus a perceived need for concrete action in order to address the problems inherent to the communication of climate change and to undertake a set of information, communication, education and awareness-raising initiatives which may allow it to better understood. It is on the basis of this reality that the "**International Climate Change Information Programme**" (ICCIP) has been created.

The need for the "**International Climate Change Information Programme**" was identified during "Climate 2008", the world's first scientific conference on climate change held online on 3–7 November 2008. The effectiveness of Climate 2008—which was followed by various other online climate conferences since (Leal Filho et al. 2015) and is an initiative led by the Hamburg University of Applied Sciences in Germany, allowed it to extend it further, in partnership with a wide range of national and international organisations such as UNEP, UNESCO, WMO, IPCC, FAO and many others agencies.

Since the governments of 195 nations endorsed the Paris Agreement in 2015, a new momentum was provided towards efforts aimed at reducing global greenhouse gas emissions. The Paris Agreement establishes the principle that future national plans will be no less ambitious than existing ones, which means that national climate action plans will provide a firm floor and foundation for higher ambition. In addition, countries will submit updated climate plans—called nationally determined contributions (NDCs)—every five years, thereby steadily increasing their ambition in the long term. Climate action, according to the Paris Agreement, will also be taken forward in the period before 2020. Countries will continue to engage in a process on mitigation opportunities and will put added focus on adaptation opportunities. This, in turn, means that the global demand for information on climate change has increased, and so has the need for mechanisms which allow a broader understanding of what it means and how it affects people's lives.

Due to its scope, it is necessary to consider climate change as a process influenced by various variables, as outlined in Fig. 1. Therefore, the search for solutions to the problems caused by climate change cannot be uni-dimensional: it needs to be pursued in an integrated way.

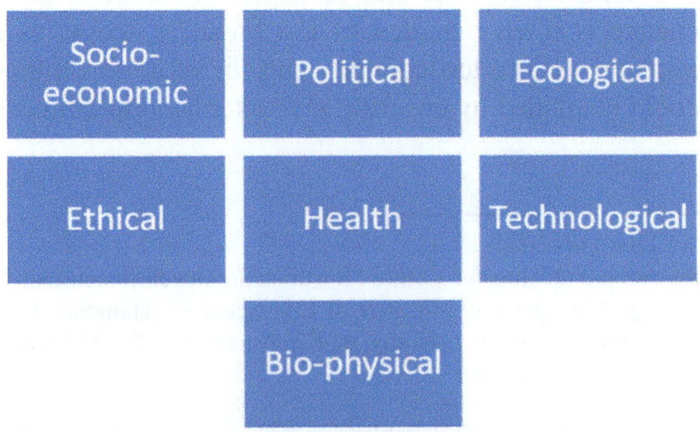

Fig. 1 Some of the variables that influence climate change (Leal Filho 2009)

The creation of the "**International Climate Change Information Programme**" is a concrete step towards the goal of "climate change understanding for all", supporting the ongoing efforts towards the search for solutions for the problems associated with climate change, an issue which is global in nature, but which needs to be supported by concrete regional and local efforts.

2 Aims of the "International Climate Change Information Programme"

The aims of the "**International Climate Change Information Programme**" are:

- to disseminate the latest findings from scientific research on climate change, including elements related to its environmental, social, economic and policy aspects in a way that allow them to be understood by the non-specialist audience. This will take place by means of books, chapters, journal articles and information via the media;
- to undertake education, communication and awareness-raising projects on matters related to climate change in both industrialised and developing countries in cooperation with UN agencies, universities, scientific institutions, government bodies, NGOs and other stakeholders;
- to network people and organisations ways to discuss the problems, barriers, challenges and chances and potentials related to communication on climate change.

Based on the fact that personal interactions are important in order to foster a dialogue and the search for new solutions, ICCIP also organises specialist events round the world. These events have over the years encouraged more networking and information exchange and have catalysed many new cooperation initiatives and projects.

3 Target Groups

The ICCIP aims to reach a broad audience which consists of:

- Scientists,
- Decision-makers,
- Enterprises,
- NGOs,
- Universities,
- Schools,
- Local communities and
- Interested individuals

All these groups have benefitted from the activities undertaken as part of the ICCIP, since its creation in 2008.

4 Activities

Some of the most common misconceptions related to climate change (Leal Filho 2009) are:

i. *Climate change is too abstract an issue*
ii. *Climate change is too broad a topic*
iii. *Climate change is mostly a technical matter where calculations and forecasts are made*
iv. *There are no trained people to handle the approach of climate change topics in an understandable way*
v. *The amount of resources needed to communicate climate change do not justify it*
vi. *Climate change has too wide a scientific basis*

If one carefully examines them, the above outlined misconceptions have quite deep roots. It is thus important to understand them so as to allow misconceptions to be overcome. The following activities have been performed by the "**International Climate Change Information Programme**" since its creation in 2008:

- Organisation of information events on different aspects of climate change, including environmental, social, economic and policy aspects, which were attended by over 5.000 delegates to date. Some of the events in 2018/2019 are:

  7–9 February 2018: World Symposium on Climate Change Communication, Graz, Austria
  3–5 April 2018: World Symposium on Climate Change and Biodiversity (WSCCB-2018), Manchester, UK
  10–12 April 2018: International Symposium on Climate Change and Museums: critical approaches to engagement and management, Manchester, United Kingdom
  14–15 May 2018: Symposium on Climate Change Adaptation, University of Ibadan, Nigeria
  19–21 June 2018: Symposium on Climate Change and Coastal Zone Management, University of Prince Edward Island, Canada
  3–5 July 2018: International Scientific Conference on Climate Change Adaptation in Eastern Europe, Banja Luka, Republika Srpska, Bosnia and Herzegovina
  13–15 February 2019: World Symposium on Climate Change and Tourism, Bariloche, Argentina

20–21 February 2019: Symposium on Climate Change Adaptation in Latin America, Lima, Peru
11–13 September 2019: 3rd World Symposium on Climate Change Adaptation, Akure, Nigeria

- Production of books, chapters, on climate change, tackles its various ramifications. The book series "Climate Change Management" was initiated with Springer in 2009 http://www.springer.com/series/8740 which has published over 30 volumes to date, including ground-breaking publications such as:

  Universities and Climate Change (2010)
  The Political, Social and Economic Aspects of Climate Change (2011)
  Climate Changes and Disasters Risk Management (2012)
  Handbook of Climate Change Adaptation (2014)
  Climate Change Research at Universities (2017)
  Handbook of Climate Change Communication (2018)

- The publication of scientific papers in peer-reviewed journals. Some of the recent papers published by the ICCIP team and associates are:

  Leal Filho et al. (2016) An assessment of smallholder soil and water conservation practices and perceptions in contrasting agro-ecological regions in Zimbabwe. In Water Resources and Rural Development, https://doi.org/10.1016/j.wrr.2016.09.001
  Leal Filho, W., Ayal, D. (2017) Farmers' perceptions of climate variability and its adverse impacts on crop and livestock production in Ethiopia. In Journal of Arid Environments, https://doi.org/10.1016/j.jaridenv.2017.01.007—https://authors.elsevier.com/a/1UPwBVu7-iGLd
  Leal Filho, W., Modesto, F., Nagy, G., Saroar, M., Toaukum (2017) Fostering coastal resilience to climate change vulnerability in Bangladesh, Brazil, Cameroon and Uruguay: a cross-country comparison. In Mitig Adapt Strateg Glob Change (2017) https://doi.org/10.1007/s11027-017-9750-3
  Leal Filho, W., Nzengya, D., Muasya, G., Wanzuu, J. (2017) Climate change responses among the Maasai Community in Kenya. Climatic Change, Volume 145, Issue 1–2, pp 71–83|https://doi.org/10.1007/s10584-017-2087-9
  Leal Filho, W., Chérif, S., Azeiteiro, U. (2017) The role of farmers' perceptions in coping with climate change in Sub-Saharan Africa. Journal of Global Warming, Vol. 12, Nos. 3/4, 2017. pp. 483–498. https://doi.org/10.1504/ijgw.2017.10005907.

- The "International Journal of Climate Change Strategies and Management", created in 2009 http://www.emeraldgrouppublishing.com/ijccsm.htm, is a fully indexed journal and is since 2017 an open-access periodical;
- Execution of education, communication and awareness-raising projects on matters related to climate change in both industrialised and developing countries;

- Organisation of joint activities in cooperation with UN agencies, universities, scientific institutions, government bodies, NGOs and other stakeholders.

The "**International Climate Change Information Programme**" has also a mailing list, which networks people and organisations ways to discuss the problems, barriers, challenges and chances and potentials related to communication on climate change.

Ultimately, the "**International Climate Change Information Programme**" breaks the barriers seen when one tries to communicate climate change and suggest measures to address the existing deficiencies.

5 Partnership

The "**International Climate Change Information Programme**" is an initiative led by the Hamburg University of Applied Sciences in Germany, working in cooperation with Manchester Metropolitan University, UK and a wide number of organisations. These are:

- United Nations Environment Programme (UNEP)
- World Meteorological Organisation (WMO)
- Intergovernmental Panel on Climate Change (IPCC)
- United Nations Educational, Scientific and Cultural Organisation (UNESCO)
- European Space Agency (ESA)
- Food and Agriculture Organisations of the United Nations (FAO)
- Global Environment Facility (GEF)
- Caribbean Community Climate Change Centre (CCCCC)
- Emerald Group Publishing Ltd.
- Sahara and Sahel Observatory
- Information Board of Climate Change Communication (IOCCC)

In addition, ICCP has worked with various **media partners** which have reported on its activities:

- WELT Gruppe
- NDR Info
- TIME Magazine
- The Economist
- European Sustainability Review
- Baltic Sea Magazine

The fact that climate change is now seen and perceived as being a major challenge to both industrialised and developing countries, means that substantial efforts are now being seen, in order to address the problem and its various ramifications.

The "**International Climate Change Information Programme**" has over the years engaged on a variety of projects, one of which is the project "ACP-EU

Technology-Transfer Network on Rainwater Harvesting Irrigation Management for Sustainable Dryland Agriculture, Food Security and Poverty Alleviation in sub-Saharan Africa" or AFRHINET http://afrhinet.eu/.

Through AFRHINET, the implementation of integrated theoretical and practical capacity-building, and the development of technology-transfer and demonstration projects in the field of rainwater harvesting irrigation (RWHI) took place in a sample of African countries (Ethiopia, Kenya, Mozambique and Zimbabwe). In addition, the knowledge and use of RWHI management for small-scale irrigation in rural dryland areas of sub-Saharan Africa were enhanced.

Moreover, AFRHINET set in motion the development of research and technology-transfer centres, and a transnational network, as platforms for cooperation and the exchange of experience in RWHI management. The network comprises micro-enterprises, non-governmental and public actors, academic/scientific institutions, and rural dryland local communities, especially farmers, women and youth groups.

Parallel to the international efforts being undertaken in respect of the mitigation of climate change and of its impacts, there is a perceived need for sound climate change adaptation strategies, which may be implemented by means of concrete projects, all of which need to have measurable and tangible goals, so as to yield the expected outputs. **Climate change adaptation projects** are important for two main reasons:

- firstly, they offer the possibility to cope with the impacts or consequences of climate change in the **short or medium term**, hence alleviating the pressure on people and on ecosystems suffering from it, especially in developing nations;
- secondly, climate change adaptation projects—especially if implemented in the context of adaptation strategies at the macrolevel—can serve the purpose of mobilising public and private stakeholders, engaging them in the problem-solving process.

Since there is a perceived need to develop evaluation techniques for climate change projects, ICCIP also started the project **Evaluating Climate Change Adaptation**, whose structure is described in Box 1.

Box 1 The Project Evaluating Climate Change Adaptation

Worldwide, there are in excess of 300 internationally funded climate change adaptation projects being implemented at present, whose combined budget is well in excess of US$1 billion. All of them have specific targets and goals, but a few of them make provisions for external assessments of their effectiveness, cater for proper quality control procedures, or have—apart from periodical reports—systems in place which allow them to fully address current and future problems in a pre-emptive way. This states of affairs illustrates the need for a research project which looks at the evaluation of the effectiveness and impacts of climate change projects in a dynamic way. It is for this purpose that the project **Evaluating Climate Change Adaptation** is being undertaken.

Aims of the project

The project **Evaluating Climate Change Adaptation** aims to:

(a) investigate the elements which prevent climate change adaptation projects from being fully successful, including structural and logistical problems, as well as other implementation issues;
(b) review, by means of a survey of individual projects, the extent to which quality control is embedded into projects and the effectiveness of current mechanisms aimed at improving the quality of their delivery;
(c) provide recommendations for further improvements, identifying needs and shortcomings, as well as factors that currently hinder—or may in the future endanger—the achievement of a project's targets or from achieving its expected results.

Thanks to its scope and approach, the project **Evaluating Climate Change Adaptation** will provide project teams with an opportunity enhance the delivery of their projects, hence honouring the commitments in respect of funding and staff time.

Expected Outputs

The project **Evaluating Climate Change Adaptation** will produce a report which will outline the main problems, issues and barriers seen in implementing climate change adaptation projects. This state-of-the-art document will be useful in offering guidance and supporting the delivery of the hundreds of climate change adaptation projects currently being undertaken, worldwide.

6 Conclusions and Opportunities for Collaboration

Since its creation in 2008, ICCIP has evolved to become the world's largest non-government funded information, communication and education programme on climate change. Thanks to the inclusive approach and diversity of activities, it is highly attractive and engages thousands of people around the world, who benefit from its works, publications and events.

ICCIP is keen to cooperate with academic and research organisations, in the execution of the research projects, events and preparation of joint publications. Please contact the ICCIP Team in Hamburg to discuss possible cooperation opportunities: info@iccip.net.

References

Leal Filho, W. (2009). Communicating climate change: challenges ahead and action needed. *International Journal of Climate Change Strategies and Management, 1*(1), 6–18.

Leal Filho, W., Grima, J., & Pace, P. (2010). Perceived frameworks of young people on global warming and ozone depletion. *Journal of Baltic Science Education, 9*(1), 35–49.

Leal Filho, W., Mannke, F., Manolas, E., & Al-Amin, A. Q. (2015). The effectiveness of climate change communication and information dissemination via the internet: experiences from the online climate conference series. *International Journal of Global Warming, 8*(1), 70–85.

Moser, S. C. (2007). More bad news: The risk of neglecting emotional responses to climate change information. In S. C. Moser & L. Dilling (Eds.), *Creating a climate for change: Communicating climate change and facilitating social change* (pp. 64–80). https://doi.org/10.1017/CBO9780511535871.006.

Nerlich, B., Koteyko, N., & Brown, B. (2010). Theory and language of climate change communication. *WIREs Clim Change, 1,* 97–110. https://doi.org/10.1002/wcc.2.

Educating Students and Their Future Employers to Minimise Environmental and Climate Impacts Through Cost-Effective Environmental Management Strategies

Kay Emblen-Perry and Les Duckers

Abstract Traditionally, internal views from UK businesses expect the maximisation of profits and delivery of value for the customer; the external view expects realistic profits and provision of employment. However, this view is changing, with customers increasingly demanding products and services that also demonstrate environmental responsibility and minimise climate impacts. Although the cause–effect relationship between business operations, negative environmental impacts and climate change is well established, there is now an enhanced appreciation that environmental challenges are systemic, interlinked and cannot be addressed in isolation. Despite the proliferation of 'low-cost' or 'no-cost' technological and behavioural opportunities, businesses struggle to realise opportunities that address these interlinked challenges, demonstrate environmental responsibility and minimise climate impacts, as they are embedded in economic systems in which improvement equals investment. Environmental improvement interventions have become synonymous with cost consumption rather than cost saving—frequently at odds with corporate financial strategies. In an attempt to change this view, support the mitigation of climate change through the reduction in environmental impacts and develop successful employment-ready graduates skilled in effective environmental improvement techniques, an innovative Environmental Strategy Module engaging postgraduate students in environmental management strategy design is taught at Coventry University. This offers students a more financially accessible approach to environmental improvement: a self-funding environmental management strategy created through the Environmental Value for Money Framework. This paper presents a conceptual study of the Environmental Value for Money Framework and its engagement of students as future employees in creating self-funded, economically

K. Emblen-Perry (✉)
Worcester Business School, University of Worcester, Worcester WR1 3AS, UK
e-mail: k.emblenperry@worc.ac.uk

L. Duckers
Coventry University, Coventry CV1 5FB, UK
e-mail: L.Duckers@coventry.ac.uk

© Springer International Publishing AG, part of Springer Nature 2019
W. Leal Filho and R. Leal-Arcas (eds.), *University Initiatives in Climate Change Mitigation and Adaptation*, https://doi.org/10.1007/978-3-319-89590-1_2

viable environmental management strategies. It also offers this framework as a mechanism to encourage businesses to engage in carefully planned and economically viable strategic environmental improvements.

Keywords Environmental management strategy · Environmental responsibility Environmental impacts · Environmental Value for Money Framework

1 Introduction

Over the last 3 decades, there has been a growing recognition from pressure groups, trade organisations, politicians and the public of the need to rethink businesses' role on creating sustainable futures. Firstly, there is an acknowledgement that organisations can severely affect climate change through their day-to-day operations and therefore need to minimise their environmental impacts (Finke et al. 2016; Carbon Trust 2015). Secondly, as highlighted within the Sustainable Development Goals (SDGs) 4 and 12, there is a growing expectation that educational systems should contribute to developing a sustainable society (United Nations 2017). UNESCO (2017) argues that the momentum for Education for Sustainability (EfS) has never been stronger but to achieve the development of skills, values and attitudes required to develop sustainable futures, education systems must introduce pedagogies that empower learners to transform the way they think and act.

The understanding of the cause–effect relationships of environmental challenges from the use of natural resources, generation of waste and creation of environmental pollution (which are key environmental impacts of businesses) and climate change has evolved in recent years. It is now generally accepted that these challenges are systemic in nature and cannot be tackled in isolation (European Environment Agency 2015). This paper therefore supports the mitigation of climate change by provoking environmental impact reduction.

Although there is widespread external pressure for organisations to demonstrate environmental responsibility, a positive, proactive response has not been forthcoming across all UK commercial sectors. Whilst it has become normal for large UK organisations to recognise their actual and/or potential environmental impacts and implement a strategic response, it is still uncommon for SMEs to adopt formal environmental management strategies (Graafland and Smid 2016). Rather, where environmental responsibility is accepted, it is more usual for SMEs to adopt an ad hoc approach to reduce and mitigate environmental impacts (Panwar et al. 2016).

During the authors' attempts to recruit smaller companies for a previous student-led, live environmental audit project, it was recognised that whilst many businesses expressed an interest in minimising their impact on the environment very few were familiar with environmental management practices or how to implement environmental improvement processes. Cassells and Lewis (2017) found a similar lack of engagement with environmental management strategies despite growing expectations from stakeholders to behave environmentally responsibly and adopt principles and practices of environmental management. Larger organisations generally recognise that management systems, auditing processes and certification schemes (e.g. ISO14001, ISO50001, EMAS) offer tested approaches for improving environmental sustainability performance that can give them a badge of environmental responsibility.

However, there is a lack of appropriately trained staff to drive the sustainability agenda within organisations, which is reducing their ability to contribute to environmental protection and minimise climate impacts. EfS has lagged behind the sustainability interests of businesses (Benn and Dunphy 2009; Lambrechts and Ceulemans 2013) so that management curricula have not adequately prepared students to deal with sustainability issues (Waddock 2007). This growing sustainability skills gap has left only 13% of UK organisations possessing the skills required to compete successfully in a sustainable economy (IEMA 2014). Universities are now playing an increasingly important role in closing this skills gap, with increasing resource invested in enhancing graduate employability (Cashian et al. 2015) and adopting real-world settings within learning and teaching (Wiek et al. 2014) to develop employment-ready graduates.

In response to this demand for environmentally literate students and to potentially reduce the environmental impacts of businesses, the authors have designed an innovative postgraduate Environmental Strategy Module that is an integral part of the Environmental Management M.Sc. at Coventry University. It focuses learning, teaching and assessment (LTA) around the creation of a proactive, self-funded five-year environmental management strategy for a simulated business utilising the Environmental Value for Money Framework (EVFM Framework). This engages students in a real-world scenario to enhance EfS and feed forward practical skills that are able to reduce environmental impacts through cost-effective environmental management strategies to their future employers.

This paper presents a conceptual study of a practical approach to EfS that contributes to the minimisation of environmental impacts that are linked to climate change, energy consumption, pollution and use of natural resources. It explores the Environmental Management Strategy Module, with a particular focus on the EVFM Framework as a dual-purpose tool. Firstly, a learning, teaching and assessment approach generates academic success through the engagement of students in environmental strategy design in a simulated real-world setting. Secondly, it facilitates an innovative methodology for environmental management strategy design for organisations seeking a cost-effective approach to environmental improvements. This self-funded approach to environmental strategy design is particularly valuable for organisations that do not currently benefit from the value for money environmental improvement interventions that are widely available as they expect them to require initial financial investment. The authors have linked these two aspects of the EVFM Framework to provide a 'feed forward' experience for students who can take this knowledge and academic experience into their business careers.

This paper adds to the discourse on LTA for EfS and offers experience-based guidance to other educators implementing active learning in real-world settings. It also offers a tool that may help mitigate climate change by provoking environmental impact reduction.

2 Environmental Management Strategy in an Organisational Context

An Environmental management strategy is important as it allows an organisation to contribute to sustainable development through the application of environmental management processes (Shrivastava and Shrivastava 1995). Benefits from adopting an environmental management strategy include cost savings, reduction of risk, increased stakeholder engagement and demonstration of responsibility (Fothergill et al. 2017), regulatory compliance (Khalili and Duecker 2013) and more market-focused benefits such as client demands and image benefits (Potoski and Prakash 2004; Campos 2012).

Environmental audits are a fundamental part of an environmental management strategy. These voluntary, essential management procedures allow organisations to detect problems before they effect operations (Beckett and Murray 2000), develop systematic approaches to improving environmental performance (Hillary 2004), avoid or manage environmental harm whilst improving economic performance (Viegas et al. 2013) and provide a benchmark from where to measure subsequent change (Clark 1998).

Whilst it is normal for large organisations to have an environmental management strategy, conduct environmental audits and possibly obtain accreditation, most smaller companies in the UK appear reluctant to adopt such strategic approaches so that many environmental issues remain peripheral to the day-to-day running of the business (Studer et al. 2008). Khalili and Duecker (2013) consider financial constraints the key to this disengagement and suggest the biggest challenge facing organisations is access to resources to support the development of an environmental management strategy. They highlight the limiting effects of the conflict of maximising financial performance whilst minimising negative environmental impacts. The perception of intervention costs, rather than their reality, may also act as a barrier to formal environmental management (Ervin et al. 2012). Reluctance to engage formally in environmental management may also involve cultural barriers and challenges. Large (2012) suggests accreditation forces a specific approach to environmental improvement that may not suit a number of businesses, particularly SMEs.

An environmental management strategy can contribute to the achievement of Sustainable Development Goals 9 and 12 that promote the adoption of sustainable practices so that organisations do more and better with less (United Nations 2017). SDG 12 advocates awareness raising, education and the provision of adequate information to encourage sustainable production whilst SDG 9 supports the achievement of environmental objectives through technological progress (United Nations 2017). An environmental audit as part of an environmental management strategy also supports the adoption of sustainability reporting promoted by SDG 12. However, the lack of a standardised framework to integrate environmental excellence into a business strategy discourages the adoption of strategic interventions to minimise climate impacts.

3 Environmental Management Strategy in a Learning, Teaching and Assessment Context

Both students and their potential employers now expect academics to promote employment skills within LTA (Pegg et al. 2012) so that employment-ready graduates and postgraduates with environmental sustainability knowledge and appropriate employability skills are able to join organisations after completing their studies. This contributes to the achievement of SDG 4, which targets all learners to have relevant skills for employment and the knowledge and skills needed to promote sustainable development (United Nations 2017). The promotion of employment skills within Higher Education (HE) is particularly important in the UK, as 85% of graduate roles now require environmental sustainability knowledge (Drayson 2014). However, an environmental skills gap exists, and in many cases, the environmental knowledge of the graduates does not meet business needs (Laurinkari and Tarvainen 2017).

Alongside the need for environmentally literate employees, students' preferences for interactive, experiential learning have also reshaped the practice of Education for Sustainability (EfS) in HE (HEFCE 2013; Higher Education Academy 2016). Together, these are replacing traditional instructivist approaches to learning, teaching and assessment with participatory and collaborative user interactions (Conole and Alevizou 2010).

The need for sustainability advocates and drive for sustainable futures requires more integrated and practical solutions that engage future sustainability professionals in proactive actions rather than reactive resolutions. However, EfS programmes frequently exclude such proactive approaches to the inherently complex environmental sustainability (Viegas et al. 2016). Ferreira et al. (2006) suggest this proactive, integrated learning develops by combining environmental management with project-based learning to provide a holistic view of reality.

The inclusion of an environmental management strategy project as a LTA methodology provides a significant opportunity to develop students' theoretical and practical work. This stimulates self-directed learning (Moalosi et al. 2012), increases integrated thinking which creates knowledge through collating and synthesising information (Nonaka 1994) and enhances softer employment skills such as commitment and responsibility (Ferreira et al. 2006). Crosthwaite et al. (2006) suggest these generic and transferable employability skills are more likely to develop when students engage with realistic and relevant experiences in contexts that they find meaningful.

Oblinger and Oblinger (2005) and Wiek et al. (2014) emphasise the importance of educating students in real-world settings. This learning, teaching and assessment approach is intended to trigger students' thinking, which develops learning for insight (Beech and MacIntosh 2012), and challenge preferences for just-in-time learning to achieve the long-term transformation rather than short-term victories advocated by Sharp (2012). The authors consider that utilising a company brief-driven environmental management strategy assignment immerses students in real-world project-based and solution-orientated learning and supports the positive outcomes of 'learning by doing'—engaging students (Dewey 1916) and complex problem-solving (Wiek et al. 2014). Corcoran and Wals (2004) recognise that an audit project is both an outcome and a process of learning.

4 The Design of the Study

This conceptual study focuses on the authors' experiences of the creation and application of an innovative, cost-effective environmental management strategy that has been adopted as the LTA methodology for the postgraduate Environmental Strategy Module. It makes use of observations of students' in-class and post-study applications of the EVFM Framework and analysis tools provided to support in-class and assignment activities. Secondary research has been undertaken to provide the evidence base for the strategic interventions presented in the EVFM Framework example. The authors have undertaken this research to offer other educators in the sustainability community a practical tool for EfS that may encourage students' long-term transformation to successful environmental practitioners and sustainability advocates and provide practical support to organisations seeking cost-efficient environmental improvement opportunities.

5 Environmental Management Strategy in the Module Context

The Environmental Strategy Module forms a fundamental part of the Environmental Management M.Sc., providing vital sustainability knowledge, skills and values and practical employment skills to promote environmental responsibility within individuals and organisations. The authors aim to encourage students to develop an individual and collective sense of responsibility that Burgess (2006) and Ellison and Wu (2008) consider able to motivate learning for good practice. The authors designed the module to introduce students to the opportunities available to organisations that can minimise climate impacts through cost-effective environmental management strategies. This learning, teaching and assessment approach aligns with the SDGs, which advocate utilising resources more effectively and doing more and better with less (United Nations 2017). It is the authors' intention to train students in the use of tools and techniques and develop environmental knowledge, skills and values that can feed forward into their future workplaces to provoke environmental responsibility and climate action.

The module runs in six, 4-h sessions held weekly. Each session includes the foundation knowledge of business sustainability appropriate for postgraduate students, practical activities to promote the integration and synthesis of this information and exploration of potential solutions to emerging issues. This enables students to explore the effect of sustainable and unsustainable behaviours on businesses, particularly their internal and external pressures from and impacts on stakeholders. The authors also provide audit process and skills training throughout the module in formal and informal training sessions.

The module assignment comprises of two elements; firstly, students are required to undertake an environmental audit of the simulated, real-life company presented in the form of a mixed-media case study, and secondly, they utilise the audit findings to design a five-year self-funded environmental management strategy utilising the EVFM Framework. Within this, the students are required to generate an Environmental Fund to create a budget to pay for larger interventions that can generate more significant mitigation impacts and financial savings. Whilst students create a theoretical environmental management strategy, the approach is equally valid for real-world organisations.

In preparation for the module's taught sessions and assignment, each student receives a company briefing document that contains a detailed profile of a simulated real-life engineering company based in the UK's West Midlands region. This company brief includes mixed-media information (text, photographs, data sheets, etc.) on the company's size, history and operational activities, including supply chain transport and material handling. In order to complete an effective environmental audit, the results of which provide baseline data for the EVFM Framework, the student is able to request additional information in an 'audit meeting' in which one of the authors acts as the company's auditee. Additional company data requested typically include site plan, energy consumption and utility bills, raw materials purchased, waste quantities, environmental incidents, stakeholder complaints and chemical storage. The authors note that they do not provide a split of energy data between heating and process energy to engage students in tools and techniques for energy consumption analysis such as Degree Day Analysis.

The student undertakes a detailed environmental audit of the simulated company utilising the company brief and information obtained from the audit meeting. Their audit can identify areas of good practice, detect problems and provide the benchmark from which to recommend subsequent strategic change. This audit, the initial part of the assignment, then feeds forward into the assessed five-year improvement strategy for the case study company.

As part of the in-class support, the authors present additional environmental performance analysis tools and techniques to assist students to design the 'no-cost' foundation year of their strategy. Examples of these tools are presented within the interventions explored below. Undertaking and understanding the value of more detailed performance evaluation can raise awareness, educate and encourage both students and organisations to measure and analyse their own performance. This can, in turn, enhance organisational performance improvement and provoke sustainability reporting in line with the aims of SDG 12. The incorporation of investment in environmentally sustainable technologies in the environmental management strategy supports SDG 9.

The authors take this self-funded approach to strategic environmental management to demonstrate that environmental performance can be improved without relying on fixed and variable asset budgets that could be used elsewhere within an organisation to improve competitiveness. The 'no-cost' starting point is incorporated to encourage students and organisations to rethink assumptions that environmental improvement is costly and research simple starting interventions that are able to be implemented in all sizes of organisation to extend and provoke environmental responsibility within silent and ad hoc environmental actors.

6 The EVFM Framework

The EVFM Framework provides the students and educators a LTA tool that comprises the core of the module assignment and facilitates in-class activities and independent study that can enhance formative and summative feedback. It also, perhaps most importantly, provides a tool that students can use to demonstrate their environmental skills and knowledge within their future workplaces and promote a new, value for money, 'no-cost' approach to environmental management and minimising climate impacts.

The framework captures the current costs of facilities and operations that have an environmental impact in the 'Environmental Activity' column. Evidence-based improvement interventions are then proposed and recorded as a 'Saving Measure' along with their financial benefits (shown as a positive cost) and implementation costs (shown as a negative cost). Over the five-year period of the strategy, the interventions and actions combine to form a self-funded environmental management strategy.

The EVFM Framework captures and evaluates improvement opportunities against the benchmark of the current activity cost to demonstrate the cost saving available from each improvement intervention and the long-term cost and environmental benefits of the initiative implemented. Savings achieved against the current operating cost (for students this is calculated from information provided in their case study brief) during each year are identified and balanced to create the total for the year. This savings total is then transferred to the following year's 'Environmental Fund at the start of the year', to create the investment budget for costlier and more impactful improvement interventions. The savings raised in each year do not have to be spent in the following year; budget surpluses can accumulate over the current strategy period. In practice, a planned surplus could also be deliberately built up and carried forward into future strategy periods to allow high-cost interventions to be funded. As the EVFM Framework generates a self-funded environmental strategy, the Environmental Fund starting balance is zero by design.

Although the EFVM Framework captures the benefits of environmental improvement interventions and efficient utilisation of resources as financial savings, the authors recognise that these could also convert into environmental impact savings such as reduced carbon emissions or water consumption reduction. Financial parameters are adopted, as organisations are most likely to change practices if they can benefit financially (Hillary and Burr 2011). The financial savings achieved in the environmental management strategy example presented below highlight the value offered by the self-funding approach to environmental performance improvement through the EVFM Framework, particularly for financially constrained organisations.

7 Cost-Effective Environmental Management Strategy

Figure 1 contains an example of a self-funded, cost-effective environmental management strategy generated through the EVFM Framework. This example, taken from a student's assignment, utilises the material provided in the company brief and the additional information obtained through participating in the audit meeting.

Environmental performance improvement interventions are included in the example presented to demonstrate the process and positive outcomes of this self-funded approach to minimisation of climate impacts. The EVFM Framework in Fig. 1 includes short descriptions of the actions proposed; additional details are included below, along with benefit calculations and some analysis tools presented in class. These provide students a sense of agency in their strategy design and are of use to organisations adopting this strategic approach to environmental performance improvement. The environmental management strategy presented here highlights the learning available from this innovative approach to EfS and the potential financial benefits emanating from environmental impact minimisation for real-world organisations.

8 Self-funded Environmental Management Strategy in Practice

1. Energy consumption reduction: Good housekeeping

Implementing good housekeeping is an effective intervention that can introduce long-term 'no-cost' behavioural change. Switching off unwanted heating, lights, production equipment and IT, closing windows, etc., could reduce consumption of electricity by 30% (Carbon Trust 2011). In addition, ensuring effective control of

		Year 1		Year 2		Year 3		Year 4		Year 5	
Environmental Fund balance at start of year	→		£0		£67,200		£23,460		£167,435		£329,840
Environmental Activity	**Current Cost**	**Saving Measure**	**Saving**	**Saving Measure**	**Saving**	**Saving Measure**	**Saving**	**Saving Measure**	**Saving**	**Saving Measure**	**Saving**
Lighting	£126,000	Good housekeeping measures	£6,300	Continued good housekeeping measures	£6,000	Continued good housekeeping measures	£5,500	Continued good housekeeping measures	£4,800	Continued good housekeeping measures	£4,000
Lighting		Programme of installing LED lighting as conventional bulbs fail	£12,600	Continued saving	£18,900	Continued saving	£25,200	Continued saving	£31,500	Continued saving	£37,800
Motors	£83,000			Replace low efficiency motor with high efficiency motor and variable speed drive	-£10,790	Continued saving	£14,210	Continued saving	£14,210	Continued saving	£14,210
Gas Space Heating	£200,000	Good housekeeping and Degree Day analysis	£8,000	Continued saving	£8,000	Continued saving	£8,000	Continued saving	£8,000	Continued saving	£8,000
Gas Process Heating	£600,000	Good housekeeping	£10,000	Replace with process efficient equipment (50% funded by Local Authority energy efficiency grant)	-£30,000	Initial saving	£264,000	Continued saving	£264,000	Continued saving	£264,000
Water		Good housekeeping measures	£8,000	Continued saving	£8,000	Continued saving	£8,000	Continued saving	£8,000	Continued saving	£8,000
Water	£100,000					Install lower flow water equipment	-£30,000	Continued saving	£22,500	Continued saving	£15,600
Water								Rainwater harvesting/reuse system	-£50,000	Initial saving	£19,000
Waste	£50,000	Good housekeeping measures	£2,000	Good housekeeping measures	£1,700	Good housekeeping measures	£1,450	Good housekeeping measures	£1,230	Good housekeeping measures	£1,045
Waste		Establish contract for sale of waste metal	£20,000	Solvent consumption reduction	£22,000	Continued saving	£21,800	Continued saving	£21,600	Continued saving	£21,400
IT	£10,000	Good housekeeping measures	£300	Continued saving	£300	Energy efficient IT equipment	-£20,725	Continued saving	£3,000	Continued saving	£1,000
Material handling	£4,800			Replace leased LPG forklift truck with electric	-£650	Fuel saving	£1,000	Fuel saving	£1,000	Continued saving	£1,000
Drain System and Storage Bund						Install new drains with interceptor and new storage bund	-£131,000				
Power system										Install Combined Heat and Power Plant	-£320,000
Balance at year end	→		£67,200		£23,460		£167,435		£329,840		£75,055

Fig. 1 Environmental management strategy created through the EVFM framework

the heating system (the heating is gas powered in the simulated company) can reduce energy consumption. For example, reducing building temperatures by 1 °C can reduce heating costs by 8% (Carbon Trust 2011). Good housekeeping continues in all years within this strategy. The financial benefit of energy-efficient lighting reduces year-on-year as the lighting gradually changes to LEDs (see subsequent intervention).

2. Energy consumption reduction: Lighting

Traditional energy-inefficient lighting within the workplace represents approximately 20% of total electricity consumption (Carbon Trust 2015). Conventional light bulbs are extremely energy intensive compared to fluorescent tubes and Compact Fluorescent Bulbs (CFBs). European Directive 2005/32/EC is phasing them out. Despite being costlier, lighting systems integrating Light Emitting Diodes (LEDs) offer significant longer-term financial benefits; LED lighting can use 50% less energy than fluorescent tubes or CFBs (Carbon Trust 2015). However, as they are up to 300% more expensive than fluorescent bulbs and CFBs (Carbon Trust 2015) and require different light fittings, a cost-efficient short-term strategy would be to replace CFBs as they fail. The strategic priority would be to generate the investment required to install LED light fittings through the good housekeeping measures identified above. Since CFBs and fluorescent tubes have a long life, this process may continue for several years and this is computed in the spreadsheet.

3. Energy consumption reduction: Gas

The simulated company is energy intensive and consumes high levels of gas for space heating and within manufacturing operations. However, the company brief does not specify individual gas consumption of these two applications. To investigate gas utilisation, a Degree Day Analysis can be conducted which apportions total gas consumption to heating and operational processes. 'Degree Days' is a measure of the difference between the base heating temperature and the actual outdoor temperature multiplied by the number of days. In the UK, the base temperature used to calculate Degree Days in the UK is 15.5 °C, because at this temperature most UK buildings do not need supplementary heating. If the outside air temperature drops below the heating base temperature, the building needs heating. The findings of the Degree Day Analysis can promote the prioritisation of improvement actions against the largest area of expenditure. To explore this further, and assist other educators and businesses to introduce the self-funded environmental management strategy approach, a Degree Day Analysis utilising the figures from the company brief, along with its strategic implication, is presented in Fig. 2.

Figure 2 shows the gas consumption and Degree Day figures for the simulated company over a 12-month period. When displayed graphically in a plot of gas consumption against Degree Days and a trendline added (Fig. 3), the intercept value on the *Y*-axis identifies the process energy requirement. In this example, operational processes consume 853,891 kWh of gas per month or 10,247,000 kWh of over a 12-month period as annually. Subtracting the process requirement from the total energy consumption establishes the energy required for the company's

Fig. 2 Gas consumption and Degree Days

Consumption period	Gas consumption kWh	Degree Day Value
January	1315433	402
February	1372222	415
March	1351111	362
April	1123111	230
May	1013333	140
June	823333	15
July	844233	0
August	848877	0
September	962666	35
October	1076667	160
November	1245556	280
December	1387000	325
Total	**13363544**	**2364**

space heating; in this example, 3,117,000 kWh of gas is required annually for space heating and 10,247,000 kWh for operational processes.

Within a cost-effective, self-funded environmental management strategy, a Degree Day Analysis can highlight where improvement actions should be prioritised. In this example, the operational processes consume over three times the quantity of gas used for space heating; therefore, improvement actions should focus here initially to reduce consumption and maximise contributions to the Environmental Fund. In years 2–4 when funding may be available for additional potential improvements, the Degree Day Analysis suggests that it would be more cost-effective, and obtain greater returns, to invest in process efficiency prior to improving space heating.

The authors recognise that whilst this analysis tool only provides an estimate of the consumption of energy they recommend utilising it as a strategy development tool to create a baseline of organisational energy use from which to benchmark the impacts of energy-efficient interventions on energy consumption.

4. Motors

Electric motors consume up to 70% of industrial electricity (Gynther et al. 2016), yet many of those utilised remain relatively inefficient despite European Directive 640/2009 implementing mandatory efficiency levels for new electric motors. The application of a motor also determines its' energy consumption; many are wastefully operated at full output, rather than matched to the load.

In this strategy, the Environmental Fund generated within Year 1 finances the replacement of the ten standard motors identified in the company brief with high-efficiency units. These are to be operated via variable speed drives which helps to match the output to the load to maximise efficiency (Saidur 2010). Subsequent savings achieved from operating the new motors efficiently can reduce energy consumption and generate cost savings of c. 20% (Chan and Kantamaneni 2015).

5. Water consumption reduction

The company brief indicates there is a high operational water consumption, so that simple housekeeping measures make a small but nonetheless valuable contribution

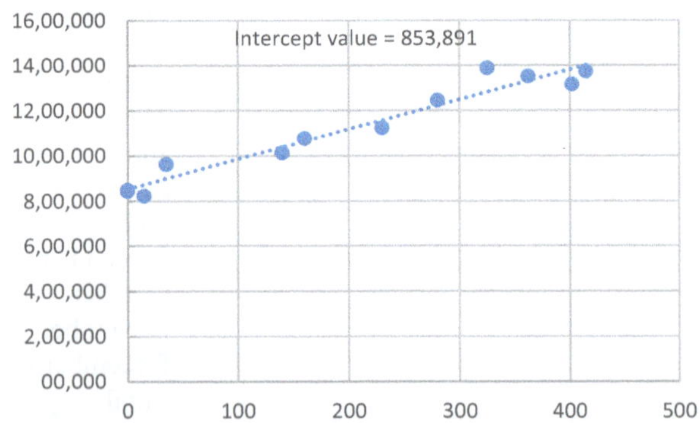

Fig. 3 Correlation between gas consumption and Degree Days

to financial savings. Carbon Trust (2015) suggests savings of 15% are possible through no-cost interventions. However, once there is sufficient investment available in the Environmental Fund, the strategy proposes the installation of low-flow equipment and a rainwater harvesting system. Efficient use of these high-cost interventions will deliver a significant reduction in water consumption and generate financial savings of up to 30% (Envirowise 2016).

6. Waste reduction

The company brief indicates the simulated company sends 40 tonnes of mixed waste to landfill per year. Incorporating the current landfill tax of £86/tonne for general waste and £200/tonne for hazardous waste (Gov.uk 2017) with waste removal costs can generate the baseline cost of waste produced. The suggested strategic priorities are the sale of waste metal and adoption of good housekeeping measures to maximise reuse and waste reduction. These simple ‘no-cost’ housekeeping and behaviour changes can reduce waste by at least 15% (DEFRA 2011). Annual reductions in savings apportioned over the life of the strategy are included to reflect the reduction in volumes of waste generated.

7. Material handling

The environmental improvements to material-handling equipment, replacing the LPG-powered forklift truck identified in the company brief with an electric alternative, increase the rental costs by c. 25% (Anonymous 2017). However, this intervention generates overall savings as the reduced running costs of c. 40% offset the increase in rental charges. These savings prime the Environmental Fund and are used to invest in larger water consumption reduction interventions.

8. Power generation system

Within the environmental management strategy, investment in interventions that make operational processes’ more energy efficient is prioritised as these processes consume a higher amount of gas than space heating. In Year 5, when the Environmental Fund has sufficient budget, investment in a combined heat and power (CHP) plant is suggested. This would supply both heat and power to the organisation and offers considerably more efficient overall use of fuel than generating them individually. Despite the capital cost of £310,000 (Renewable Energy Hub 2017), energy consumption reduction could generate savings of c. £60,000 per year, as a CHP system requires c. 33% less fuel to generate the same amount of heat and power (Northern Ireland Environment Agency 2017). This can therefore generate financial savings of c. 20% in the future strategy period.

9. Drainage system and storage bund

Improvements to the drainage system and hazardous materials storage are examples of measures that can help to avoid environmental damage, but not directly save money. A clearly recorded drainage system, incorporating an interceptor to remove oils and sediments from storm water and spills, will protect water resources. In addition, the strategy suggests bunds are employed to offer secondary containment during the storage of hazardous materials and so further protect the environment. Whilst these interventions do not directly generate cost savings, they will reduce environmental risk and the consequent threat of financial penalties and provide evidence to stakeholders of environmental responsibility. They also contribute to potential future opportunities to recycle and reuse wastewater in the next strategy period.

9 Impact and Implications of the Strategic Approach to Environmental Management Utilising the EVFM Framework

The authors believe that organisational environmental performance improvement does not need to be costly or divert scarce resources from alternative operational requirements. This belief has led to the creation of the self-funded approach to environmental management strategy development presented in this paper, which has the EVFM Framework at its heart. Adopting this as a LTA methodology may be of use to other educators looking to incorporate active learning into business strategy or a practical approach into EfS. It may also be valuable for organisations seeking 'no-cost' environmental improvement opportunities or a self-funded environmental strategy.

The EFVM Framework offers an environmental improvement tool that promotes environmental sustainability literacy, develops employment skills within the experiential, interactive learning environment advocated by HEFCE (2013) and Higher Education Academy (2016) and has re-fashioned EfS within Coventry University's Environmental Strategy Module. This can promote relevant employment skills for all learners, which contributes to the achievement of SDG 4 and provokes awareness raising and education advocated by SDG 12.

The framework can engage students in generative sustainability through their preferred experiential, active learning style in a real-world setting advocated by Oblinger and Oblinger (2005) and Wiek et al. (2014). Students appear to find this learning, teaching and assessment approach challenging, as they must engage in generative sustainability to develop a creative innovative 5-year environmental management strategy. This assessed environmental management strategy is valuable, as it requires students to analyse the company brief and audit meeting information, undertake independent research and develop a customised strategic approach that is not available from the internet or within previous research papers. This level of challenge is, however, considered appropriate for postgraduate students.

The LTA approach presented in this paper also offers many opportunities for educators to provide ongoing formative feedback and support. The in-class activities such as audit meetings, quizzes and informal presentations link theory and practice to the audit and environmental strategy to develop sustainability literacy and challenge students' thinking, which can develop learning for insight, considered valuable by Beech and MacIntosh (2012).

The authors suggest that incorporating an environmental management strategy into LTA disguises their intention to generate a long-term transformation of students' knowledge, skills and values, which is required to achieve sustainable futures and develop students' awareness and confidence (Harvey and Green 1993). As overt transformative learning appears unpopular with students, concealing this within a series of preferred short-term successes (Sharp 2012) through the audit, development of environmental management strategy components and self-directed learning can covertly develop knowledge and skills. Utilising a cost-effective environmental management strategy as a LTA tool can also enhance the integrated thinking required for sustainability literacy, advocacy and employment skills (Ferreira et al. 2006) and contribute to an individual and collective sense of responsibility (Burgess 2006; Ellison and Wu 2008).

The EVFM Framework promotes both hard and soft employment skills demanded by students and their future employers (Pegg et al. 2012). Incorporating the framework into both teaching and assessment allows educators to encourage the development of student's personal values and support the development of their confidence in the inherent complexity of environmental management problems and solutions. This can contribute to closing the environmental skills gap identified by IEMA (2014) and Viegas et al. (2016).

Undertaking an audit and designing an environmental management strategy allows students to embed awareness of environmental sustainability issues, impacts and potential solutions by collating and synthesising information obtained in class and through individual assignment research to create knowledge (Nonaka 1994). This engages students in understanding organisations' environmental performance and develops strategic responses that are vital for sustainability advocacy and successful future careers rather than simply engaging in just-in-time learning. The environmental audit of a simulated real-life company also presents an opportunity to develop students' knowledge and experience of developing integrated solutions, which Ferreira et al. (2006) suggest will assist them to appreciate the different dimensions and complexity of environmental problems.

Students participating in the Environmental Strategy Module are exposed to an environmental management technique that can feed forward into their future (or current) workplaces. This contributes to the achievement of SDG 4 as students leave Coventry University with relevant skills for employment and the knowledge and skills needed to promote sustainable development advocated by United Nations (2017). The authors' self-funded strategic approach to minimising climate impacts has already proven to be successful; one part-time student utilised the EVFM Framework in his employer's car component manufacturing company and generated a saving of £300,000.

Although the perception of investment may act as a barrier to formal environmental management (Ervin et al. 2012), the authors suggest that a self-funded environmental management strategy utilising initial 'no-cost' interventions can overcome this barrier and create sufficient funds to allow costlier improvements to be financed. The EVFM Framework offers this 'no-cost' route into formal, strategic environmental management for organisations that have previously resisted environmental performance improvement, as they perceived they lacked financial resources to fund investments. The concept of an environmental management strategy based on a dedicated Environmental Fund is valuable as it demonstrates to organisations that they can make substantial improvement to satisfy stakeholders' demands for environmental responsibility (Cassells and Lewis 2017) without incurring upfront costs or funding interventions from existing budgets allocated for operational priorities. The incorporation of an environmental audit to provide the baseline performance appraisal, from which to develop an environmental management strategy, may also encourage sustainability reporting within organisations as advocated by SDG 12 (United Nations 2017).

Stimulating environmental responsibility through the EFVM Framework, which provides the structure for a management strategy and offers a systematic approach to improving sustainability performance whilst improving economic performance (Viegas et al. 2013), contributes to the achievement of SDGs 9 and 12 at local, corporate, regional, national and international levels. As students come from Coventry, the wider West Midlands, UK and Europe as well as Africa, Asia and the Americas to study at Coventry University, the EVFM Framework and the self-funded approach to promoting environmental awareness and enhancing environmental responsibility can have worldwide reach.

10 Conclusion

The drive for sustainable futures requires more integrated and practical solutions that engage future sustainability professionals in constructive actions rather than reactive resolutions. The cost-effective environmental management strategy presented offers a self-funded, proactive and accessible response to environmental challenges that can contribute to the development of environmental responsibility within individuals and businesses and the mitigation of climate change through the reduction in energy consumption, prevention of pollution and prudent use of natural resources.

This paper has presented a conceptual study of the authors' innovative cost-effective and self-funded approach to environmental management strategy development that positions the EVFM Framework at the heart of the development process. This environmental strategy development approach has two purposes: firstly, it provides a LTA methodology that can develop students' sustainability literacy, skills and values in a real-world setting that can contribute to the closure of current environmental skills gaps whilst engaging students' in experiential learning. Secondly, the carefully planned, economically viable strategic approach to environmental performance improvement can offer businesses a route to develop cost-effective environmental responsibility. Evidence from a former student suggests this framework can feed forward to provoke the development of an environmental management strategy within students' workplaces. This environmental management strategy approach and framework may therefore be of value for both educators in the sustainability community seeking a practical methodology for EfS that encourages students' long-term transformation to successful environmental practitioner and sustainability advocate, rather than simply delivering just-in-time learning supporting (thus SDGS 4 and 12) and organisations seeking 'no-cost' environmental improvement opportunities.

The authors recognise that this paper presents a conceptual study of the innovative strategic, self-funded environmental management approach and framework for both LTA and organisational development. The EVFM Framework's current focus on financial benefits of environmental interventions offers a single motivator for business improvements. However, incorporating environmental impact savings measured as reductions in carbon emissions, water use or waste generated may provide an additional driver for change.

Additional research into its applications and outcomes can validate its effectiveness for EfS and ability to promote businesses' environmental improvement practices that will contribute to the mitigation of climate change. The authors suggest undertaking an exploration of students' engagement with the EVFM Framework to test whether it can generate environmental insight and deeper learning to develop the sustainability literacy, knowledge, skills and values that are required to promote sustainable futures. Testing the EVFM Framework within businesses could validate commercial applicability and overcome perceptions that environmental interventions are cost consuming rather than cost saving.

References

Anonymous. (2017). *Fork lift truck hire*. http://www.bendigomitchell.com/forklifts-for-hire/. Last Accessed September 19, 2017.

Beckett, R., & Murray, P. (2000). Learning by auditing: A knowledge creating approach. *The TQM Magazine, 12*(2), 125–136.

Beech, N., & MacIntosh, R. (2012). *Managing change: Enquiry and action*. Cambridge, United Kingdom: Cambridge University Press.

Benn, S., & Dunphy, D. (2009). Action research as an approach to integrating sustainability into MBA programs: An exploratory study. *Journal of Management Education, 33*(3), 276–295.

Burgess, J. (2006). Blogging to learn, learning to blog. In A. Bruns & J. Jacobs (Eds.), *Use of blogs* (pp. 91–103). New York: Peter Lang Publishers.

Campos, L. (2012). Environmental management systems (EMS) for small companies: A study in Southern Brazil. *Journal of Cleaner Production, 32,* 141–148.

Carbon Trust. (2011). *Heating control: Maximising comfort, minimising energy consumption*. https://www.carbontrust.com/media/10361/ctg065_heating_control.pdf. Last Accessed September 14, 2017.

Carbon Trust. (2015). *Guidance Note 10: Lighting—Energy technology list*. https://www.gov.uk/government/uploads/system/uploads/attachment_data/file/474456/Lighting_Guidance_Note_November_2015.pdf. Last Accessed September 14, 2017.

Cashian, P., Clarke, J., & Richardson, M. (2015) *Perspectives on: Employability. Is it time to move the employability debate on?* Chartered Association of Business Schools: https://charteredabs.org/wp-content/uploads/2015/06/Employability-Debate1.pdf. Last Accessed September 13, 2017.

Cassells, S., & Lewis, K. (2017). Environmental management training for micro and small enterprises: The missing link? *Journal of Small Business and Enterprise Development, 24*(2), 297–312.

Chan, Y., & Kantamaneni, R. (2015). *Study on energy efficiency and energy saving potential in industry and on possible policy mechanisms*. Brussels: ICF International.

Clark, G. (1998). Maximising the benefits from work-based learning: The effectiveness of environmental audits. *Journal of Geography in Higher Education, 22*(3), 325–334.

Conole, G., & Alevizou, P. (2010). *A literature review of the use of Web 2.0 tools in higher education*. York: Higher Education Academy. http://www.heacademy.ac.uk/.

Corcoran, P., & Wals, A. (2004). *Higher education and the challenge of sustainability: Problematics, promise, and practice*. Boston: Kluwer Academic Publishers.

Crosthwaite, C., Cameron, I., Lant, P., & Litster, J. (2006). Balancing curriculum processes and content in a project centred curriculum: In pursuit of graduate attributes. *Education for Chemical Engineers, 1*(1), 39–48.

DEFRA. (2011). *The economics of waste and waste policy*. https://www.gov.uk/government/uploads/system/uploads/attachment_data/file/69500/pb13548-economic-principles-wr110613.pdf. Last Accessed September 19, 2017.

Dewey, J. (1916). *Democracy and education: An introduction to the philosophy of education*. New York: Macmillan.

Drayson, R. (2014). *Employer attitudes towards, and skills for, sustainable development higher education academy*. https://www.heacademy.ac.uk/system/files/executive-summary-employers.pdf. Last Accessed September 14, 2017.

Ellison, N., & Wu, Y. (2008). Blogging in the classroom: A preliminary exploration of student attitudes and impact on comprehension. *Journal of Educational Multimedia and Hypermedia, 17*(1), 99–122.

Envirowise. (2016). *Reduce your water use and collect the savings*. http://www.wrap.org.uk/sites/files/wrap/EN847_final.pdf. Last Accessed September 18, 2017.

Ervin, D., Wu, J., Khanna, M., Jones, C., & Wirkkalas, T. (2012). Motivations and barriers to corporate environmental management. *Business Strategy and the Environment, 22*, 390–409.

European Environment Agency. (2015). *The European environment—State and outlook 2015*. https://www.eea.europa.eu/soer. Last accessed October 28, 2017.

Ferreira, A., Lopes, M., & Morais, J. (2006). Environmental management and audit schemes implementation as an educational tool for sustainability. *Journal of Cleaner Production, 14*, 973–982.

Finke, T., Gilchrist, A., & Mouzas, S. (2016). Why companies fail to respond to climate change: Collective inaction as an outcome of barriers to interaction. *Industrial Marketing Management, 58*, 94–101.

Fothergill, J., Smith, D., & Payne, R. (2017). *Driving sustainable resource management through ISO 14001*. https://www.iema.net. Last Accessed September 15, 2017.

Gov.uk. (2017). *Waste costs*. https://www.gov.uk/green-taxes-and-reliefs/landfill-tax. Last Accessed September 18, 2017.

Graafland, J., & Smid, H. (2016). Environmental Impacts of SMEs and the effects of formal management tools: Evidence from EU's largest survey. *Corporate Social Responsibility and Environmental Management, 23*(5), 297–307.

Gynther, L., Mustonen, S., & Saarivirta, E. (2016). *Policies and measures for promoting efficient electric motors in industry*. https://wec-policies.enerdata.net/Documents/cases-studies/WEC-case-study-Electric-motor.pdf. Last Accessed September 19, 2017.

Harvey, L., & Green, D. (1993). Defining quality. *Assessment & Evaluation in Higher Education, 18*(1), 9–34.

Higher Education Academy. (2016). *Higher education academy frame WORKS, 04 Essential frameworks for enhancing student success*. http://www.heacademy.ac.uk/transform. Last Accessed September 16, 2017.

HEFCE. (2013). *Sustainable development in higher education: Consultation on a framework for HEFCE*. Higher Education Funding Council for England. http://www.hefce.ac.uk. Last Accessed September 16, 2017.

Hillary, R. (2004). Environmental management systems and the smaller enterprise. *Journal of Cleaner Production, 12*(6), 561–569.

Hillary, R., & Burr, P. (2011). *Evidence-based Study into the Benefits of EMSs for SMEs*. http://www.randd.defra.gov.uk/Document.aspx?Document=FINALEV0440report.pdf. Last Accessed September 19, 2017.

IEMA. (2014). *Preparing for the perfect storm*. https://www.iema.net/. Last Accessed September 15, 2017.

Khalili, N., & Duecker, S. (2013). Application of multi-criteria decision analysis in design of sustainable environmental management system framework. *Journal of Cleaner Production, 47,* 188–198.

Lambrechts, W., & Ceulemans, K. (2013). Sustainability assessment in higher education. Evaluating the use of the auditing instrument for sustainability in higher education (AISHE) in Belgium. In S. Caeiro, W. Leal Filho, C. Jabbour, & U. Azeiteiro (Eds.), *Sustainability assessment tools in higher education institutions. Mapping trends and good practice around the world*. Switzerland: Springer International Publishing.

Large, C. (2012). *ISO14001: Loved by procurement, hated by SMEs*. https://www.theguardian.com. Last Accessed September 15, 2017.

Laurinkari, J., & Tarvainen, M. (2017). *The policies of inclusion*. London: EHV Academic Press.

Moalosi, R., Molokwane, S., & Mothibedi, G. (2012). Using a design-orientated project to attain graduate attributes. *Design and Technology Education, 17*(1), 30–43.

Nonaka, I. (1994). A dynamic theory of organizational knowledge creation. *Organization Science, 5*(1), 14–37.

Northern Ireland Environment Agency. (2017). *Combined heat and power (CHP) benefits and opportunities for business*. http://www.netregs.org.uk/environmental-topics/carbon-reduction-and-efficiency/combined-heat-and-power-chp-benefits-and-opportunities-for-business/. Last Accessed September 15, 2017.

Oblinger, D., & Oblinger, J. (2005) *Educating the net generation*. https://www.educause.edu/ir/library/pdf/pub7101.pdf. Last Accessed September 15, 2017.

Panwar, R., Nybakk, E., Hansen, E., & Pinkse, J. (2016). The effect of small firms' competitive strategies on their community and environmental engagement. *Journal of Cleaner Production, 129,* 578–585.

Pegg, A., Waldock, J., Hendy-Isaac, S., & Lawton, R. (2012). *Pedagogy for employability*. https://www.heacademy.ac.uk/system/files/pedagogy_for_employability_update_2012.pdf. Last Accessed September 15, 2017.

Potoski, M., & Prakash, A. (2004). The regulation dilemma: Cooperation and conflict in environmental governance. *Public Administration Review, 64*(2), 152–163.

Renewable Energy Hub. (2017). *Micro combined heat and power microchip CHP information*. https://www.renewableenergyhub.co.uk/. Last Accessed September 19, 2017.

Saidur, R. (2010). A review on electrical motors energy use and energy savings. *Renewable and Sustainable Energy Reviews, 14*(3), 877–898.

Sharp, L. (2012). Green campuses: the road from little victories to systemic transformation. *International Journal of Sustainability in Higher Education, 3*(2), 128–145.

Shrivastava, P., & Shrivastava, P. (1995). The role of corporations in achieving sustainability. *Academy of Management Review, 4,* 936–960.

Studer, S., Tsang, S., Welford, R., & Hills, P. (2008). SMEs and voluntary environmental initiatives: A study of stakeholders' perspectives in Hong Kong. *Environment and Planning B: Planning and Design, 51*(2), 285–301.

UNESCO. (2017). *Education for sustainable development goals learning objectives*. http://unesdoc.unesco.org/images/0024/002474/247444e.pdf. Last Accessed September 22, 2017.

United Nations. (2017). *Sustainable development knowledge platform: Sustainable development goals*. https://sustainabledevelopment.un.org/sdgs. Last Accessed September 22, 2017.

Viegas, C., Bond, A., Duarte Ribeiro, J., & Selig, P. (2013). A review of environmental monitoring and auditing in the context of risk: Unveiling the extent of a confused relationship. *Journal of Cleaner Production, 47,* 165–173.

Viegas, C., Bond, A., Vaz, C., Borchardt, M., Pereira, G., Selig, P., et al. (2016). Critical attributes of sustainability in higher education: A categorisation from literature review. *Journal of Cleaner Production, 126,* 260–276.

Waddock, S. (2007). Leadership integrity in a fractured knowledge world. *Academy of Management Learning & Education, 6*(4), 543–557.

Wiek, A., Xiong, A., Brundiers, K., & van de Leeuw, S. (2014). Integrating problem-and project-based learning into sustainability programs. *International Journal of Sustainability in Higher Education, 15*(4), 431–449.

Dr. Kay Emblen-Perry has several years of senior environmental and ecology consultancy experience delivering consultancy projects in renewable energy technologies, contaminated land remediation, biodiversity offsetting and ecological assessment for UK organisations. She is qualified as an environmental and quality lead auditor, has implemented environmental management systems for both UK and multinational organisations and has trained environmental and quality assessors. In previous roles, Kay gained senior project management and purchasing management experience in international automotive companies. Her project managed the implementation of sustainable supply chain strategies, new vehicle projects and EU REACH Regulations. Kay's specialisation is in Sustainable Management including Environmental Management and Justice, Social Responsibility and Economic Sustainability.

Dr. Les Duckers has been lecturing at Coventry University since 1975, initially in Physics and Engineering, later going on to establish the department of Environmental Sciences in 1992. Since then, he has primarily taught masters course and supervised Ph.D. students across a wide range of topics, reflecting his major research interest in environment and renewable energy and especially wave energy. He is a visiting professor in Japan and Sri Lanka and visiting lecturer to Reading and Loughborough Universities in the UK. He has experience of about 100 industrial environmental audits, is associate editor for The Renewable Energy Journal and has acted as consultant to a number of companies.

Carbon Management Planning in UK Universities: A Journey to Low Carbon Built Environment

Muhammad Usman Mazhar, Richard Bull, Mark Lemon and Saad Bin Saleem Ahmad

Abstract Climate change and increasing carbon emissions are the biggest challenges for the modern world. Organisations are facing increasing pressure from governments and stakeholders to reduce carbon emissions. The Higher Education (HE) sector has a huge environmental, social and economic impact. In 2012–13, Higher Education Institutions (HEIs) consumed 7.9 billion kWh of energy and emitted 2.3 million tonnes of carbon emissions, which strengthens the role of universities in implementing carbon management for a low carbon built environment. The HE sector is not exempt from implementing carbon management strategies and respond to the UK government's Climate Change Act by developing its own targets in England, which are in line with the national targets—80% reduction by 2050 and 34% by 2020 from the 1990 baseline. This indicates the scale of the challenge to implement carbon management through effective planning procedures. The aim of this paper is to explore the key elements of the carbon management planning process in UK universities and identify potential areas of improvements. This exploratory study adopted a qualitative and inductive research approach. The data were collected through the content analysis of eighteen universities' carbon management plans (CMPs). The study found that key elements of carbon management planning are senior management leadership, carbon footprinting, carbon reduction targets,

---

M. U. Mazhar (✉)
Department of Management, Nottingham Business School, Burton Street, Nottingham NG1 4BU, UK
e-mail: muhammadusman.mazhar@ntu.ac.uk

R. Bull · M. Lemon
Institute of Energy and Sustainable Development, De Montfort University, The Gateway, Leicester LE1 9BH, UK
e-mail: rbull@dmu.ac.uk

M. Lemon
e-mail: mlemon@dmu.ac.uk

S. B. S. Ahmad
Department of Civil and Environmental Engineering, Norwegian University of Sciences and Technology, Trondheim, Norway
e-mail: saad.ahmad@ntnu.no

© Springer International Publishing AG, part of Springer Nature 2019

W. Leal Filho and R. Leal-Arcas (eds.), *University Initiatives in Climate Change Mitigation and Adaptation*, https://doi.org/10.1007/978-3-319-89590-1_3

stakeholder engagement, funding and resources, governance and evaluation and reporting. Universities have shown policy commitment and developed CMPs for implementation, but the performance of universities varies significantly. There is also a disconnect between planning and delivery. The findings of this research show that CMPs can be valuable tools to assist universities in their carbon management journey. However, weaknesses are identified in the current design of CMPs, for example, overly focusing on the technical issues of carbon management (to the detriment of socio-technical factors), unsupportive of stakeholder engagement, not aligned with core policies and strategies and being static documents. CMPs are not comprehensive with regard to the operational boundary of carbon emissions and need standard approach for measuring, targeting and reporting. This study will be useful to academics and practitioners aiming to improve carbon management planning in universities and other organisations.

Keywords Carbon management · Planning · Universities · Low carbon built environment

1 Introduction

Carbon management is now high on the global and UK policy landscape. In 2006, the Stern Review indicated that a 25% carbon emissions reduction below current levels is required to stabilise global CO_2 emissions at levels that will not have harmful impacts and states that the cost of not acting on climate change is greater than the cost of acting now (Stern 2006). In response to global and the European climate change policies, the UK became the first country to introduce the Climate Change Act 2008 aiming for 80% carbon emissions reduction by 2050 and 34% by 2020 as compared to a 1990 baseline (Her Majesty's Stationery Office 2008). The public sector (including the higher education) recognises the benefits of carbon management and sees the short-term as well as long-term advantages of its implementation (The Carbon Trust 2012).

The Higher Education Funding Council for England (HEFCE) adopted the same targets, which against a 2005 baseline are equivalent to a reduction of 43% by 2020 and 83% by 2050 (HEFCE 2010a). Universities in the UK had to develop CMPs to respond to their funding bodies such as universities in England developed institutional CMPs for policy compliance on the advice of HEFCE. At that time, the success in meeting the targets in CMPs was a contributory factor in future capital funding allocations for universities (HEFCE 2008a). This was a strong policy driver to move universities forward with regard to carbon management and get senior management leadership on board. Higher Education Institutions (HEIs) including universities have recognised themselves that they are well placed to take on a leadership role in implementing carbon management (Klein-Banai and Theis 2013). In 2012–13, HEIs consumed 7.9 billion kWh of energy and produced 2.3 million tonnes of emissions (Higher Education Statistics Agency 2014). This indicates the scale of the problem in universities. The HE sector emits these carbon emissions due to the nature of its operational business activities. The increasing carbon emissions in universities need to be managed through effective planning.

CMPs present a 'public' position of universities on how they are planning to implement carbon management strategies to meet their targets. Brite Green (2015) found that the majority of the English universities are further behind from achieving their 2020 carbon reduction target. Therefore, universities need to take a strategic approach to develop and manage CMPs and address the necessary gaps. This research aims to explore the key elements of the carbon management planning process in UK universities and identify potential areas of improvements. This helps assess universities' level of compliance and adoption of national and HE sector carbon reduction policies and strategies by developing CMPs which is a major carbon management tool. The literature is reviewed around carbon management planning in organisations and universities. Then, the research methodology is presented before discussing the data analysis and results. Finally, the paper includes a discussion of the main research findings with conclusions and recommendations presented.

2 Organisations and Carbon Management Planning

Organisations are beginning to understand the need to implement carbon management strategies (Carbon Disclosure Project 2010). There is a relationship between an organisation's carbon management strategy, the sector it operates in and the size of the organisation (Lee 2012). Research into the strategic response to climate change in public and private sector organisations found that carbon management was dependent upon how those organisations 'thought' about carbon emissions (Bebbington and Barter 2011). Frameworks underpinning corporate climate change strategies are developed to reduce carbon emissions in energy intensive companies within the manufacturing and process industries (Kolk and Pinkse 2004). Some previous studies are focused on organisations' efforts to reduce emissions from the production processes and their participation in carbon emissions trading systems. Some of the studies have stressed the need to focus on improving the product and supply chain measures (e.g. Weinhofer and Hoffman 2010; Hoffman 2007). Cadez and Czerny (2016) explore corporate climate change strategies of carbon-intensive firms to identify configurations of firms pursuing similar strategies and appraise the relationships between nineteen carbon reduction practices and their underlying strategies. The public sector has an opportunity to lead by example in carbon management and influence the private sector organisations (Wehrmeyer et al. 2009). Ball et al. (2009) recommend considering research into 'carbon neutrality' within public sector organisations, as there has not yet been any analysis on how these organisations address emissions. There is a considerable grey literature on public sector carbon management strategies, focussed on higher education, local authorities and National Health Services (NHS) trusts. This suggests that, despite much good practice

(The Carbon Trust 2013), proactive actions are required by public sector organisations to reduce carbon emissions.

The Higher Education (HE) sector has significant potential to play a key role in supporting the transition to a low carbon economy (Davies and Dunk 2015; Mazhar et al. 2014). The UK HE sector is significant in terms of population, economic contribution and societal influence and is an important sector for implementing carbon management (Robinson et al. 2017). Universities are increasingly reporting their carbon emissions as part of their sustainability performance measurement (Klein-Banai and Theis 2013). Much of the focus on carbon emissions reductions and sustainability strategies in universities has been on the energy consumption of buildings; this is largely because they are major contributors to an institution's carbon emissions and they are usually under their direct control (Klein-Banai and Theis 2013). HE institutions should consider how a carbon management strategy will fit into the wider context of their business operations, integrates with existing policies and strategies and can contribute to the delivery of regulatory obligations (HEFCE 2010b). However, the case of HEIs' energy and carbon emissions reduction is considered more complex than other organisations due to the heterogeneity of the sector (Altan 2010). This is due to the fact that many universities have complex built environment, different stakeholders and competing core business priorities in a rapidly changing HE environment.

There is a lack of academic research on issues around carbon management in universities (Robinson et al. 2015; Mazhar et al. 2012, 2014). Much of the existing literature and related research is focused on wider sustainability or green issues rather than planning approaches. Wittneben et al. (2009) state that the theoretical developments in organisational studies and corporate management practices for climate change mitigation have been obsolete. There is literature available in the form of CMPs and HEFCE strategy documents. The focus on research into carbon management planning and its main elements remains limited. Robinson et al. (2015) provided a reality check on carbon management in universities with a focus on English Russell Group universities. Mazhar et al. (2017) explored the critical success factors for embedding carbon management in universities and offered lessons to be learnt for organisations. These two studies are among a few carbon management studies that have looked into existing CMPs of universities. Robinson et al. (2015) found that current CMPs are not a good indicator of future performance and the HE sector in England has underestimated the challenge of carbon management.

3 Research Methodology

The study adopted a qualitative and inductive research approach to produce insights into universities' carbon management planning approaches. The data were collected through the content analysis of eighteen UK universities' carbon management plans (CMPs) and strategies. Most of the CMPs were publicly available on university websites due to the HEFCE requirements. Carbon management plans and strategies were chosen for the analysis, as this is the only document that addresses the issue of carbon emissions in universities. The rationale is that it may be difficult to access university management to gain comprehensive insights into carbon management planning process. Therefore, carbon management strategies and plans were used to indicate the perspectives and approaches of different universities. The sample of 18 universities is representative of different types of universities across the UK based on the foundation year (pre-1992 and post-1992). While these universities have other carbon-related strategies and policies such as sustainability strategies, environmental strategies, travel plans, energy policies, waste management policies and procurement policies, detailed analysis of these documents falls outside the scope of this study. Out of the eighteen CMPs studied, sixteen were available on websites. There were two universities that do not have CMPs publicly available on the website. One of them has put a summary of the CMP and the other has placed it on the corporate website, but it is only available on staff web pages. Table 1 presents the names of the eighteen universities with the titles of their carbon management documents. Two universities have given different names to documents for their own branding.

Themes drawn from the literature were used to underpin a systematic analysis of the CMPs which were then analysed. Additional themes and sub-themes also emerged and are discussed in the ensuing paper. The qualitative data analysis software NVivo 10 was used for the thematic analysis. All of the CMPs were systematically reviewed to develop these themes, as presented in Fig. 1. The data

Table 1 Carbon management documents analysed for the study

No.	Name of the University	Name of the Document	Year
1	De Montfort University	Carbon Management Plan	2011
2	University of Leicester	Strategy and Implementation Plan	2007
3	Loughborough University	Carbon Management Plan	2010
4	Nottingham Trent University	Strategy and Implementation Plan	2008
5	University of Derby	Carbon Management Plan	2009
6	The University of Northampton	Carbon Management Plan	2011
7	University of East Anglia	Carbon Reduction Plan	2012
8	University of Cambridge	Carbon Management Plan	2010
9	Leeds Beckett University	Carbon Management Strategy	2012
10	University of Lincoln	Carbon Management Plan	2011
11	University of Nottingham	Carbon Management Plan	2010
12	University of Birmingham	Carbon Management Implementation Plan	2010
13	University of Bradford	Ecoversity-One Planet Strategy	2011
14	The University of Edinburgh	Climate Action Plan	2010
15	Heriot-Watt University	Carbon Management Plan	2009
16	Cardiff University	Carbon Management Plan	2013
17	Aberystwyth University	Implementation Plan	2007
18	Queen's University Belfast	Carbon Management Plan	2013

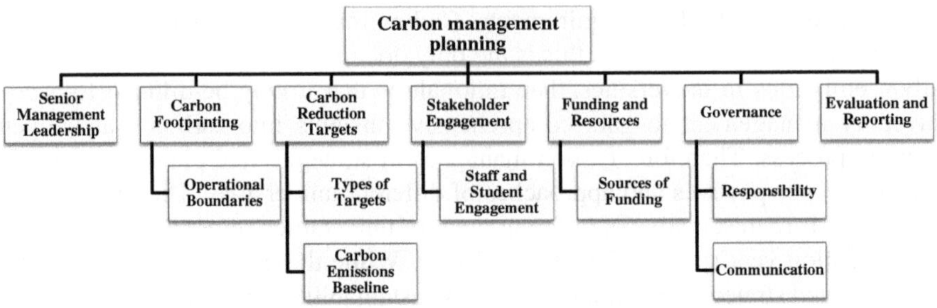

Fig. 1 Key elements of carbon management planning

were then coded to produce findings. The themes and sub-themes were recorded based on the frequency of their occurrence and their usefulness for the carbon management process.

4 Analysis and Results

Figure 1 presents the key elements of carbon management planning drawn out during the content analysis of CMPs of universities.

4.1 Senior Management Leadership

Senior management leadership is the starting point for carbon management planning in universities. This theme explores the role of senior management leadership and the extent to which senior management is engaged in carbon management. The carbon management plans (CMPs) of eight, out of the eighteen, universities have mentioned senior management commitment and their increasing role in carbon management. All of the CMPs officially get signed off by one of the members of senior management team. However, the content analysis suggests that CMPs do not provide evidence for the level of their true engagement, as plan/strategy and implementation may differ. The Nottingham Trent University's Strategy and Implementation Plan describes the importance of senior management support:

> In order to achieve the greatest possible success, it is essential to have visible top level support for the Carbon Management Programme. This top level support will in turn open up avenues such that the programme can filter through each level of the organisation and as such enable Nottingham Trent University to achieve and exceed its carbon reduction targets as identified in the Carbon Trust Higher Education Carbon Management Programme and the University's Strategic Plan. (Nottingham Trent University Strategy and Implementation Plan, p. 59)

Due to the importance of senior management leadership, five out of the eighteen universities are trying to ensure the involvement of senior management. Only three universities mentioned in CMPs that their senior management have already bought into the carbon management process, which suggests a lack of senior management leadership in majority of the universities in reality. The analysis indicates that universities have awareness and an agreement on the role of senior management leadership, but they have not secured their full engagement yet. However, CMPs do not clearly mention a lack of senior management leadership and do not develop strategies or action plans to achieve their engagement. The CMP of the University of Lincoln states the role of senior management leadership for embedding carbon management in the university:

> Carbon management will be led by senior management and will be embedded across the institution through the alignment of policies, management practices and procedures in support of the low carbon vision. (University of Lincoln Carbon Management Plan, p. 8)

In contrast, seventeen out of the eighteen universities' CMPs have a 'foreword' from either a member of senior management team or Vice Chancellor (VC) with the signature to demonstrate that carbon management is backed up by senior leadership. The sign-off does not provide evidence for the real commitment and engagement of senior management, but universities have started to at least consider it in strategic management agenda, if not implement effectively. The University of Leicester's Environment Action Programme (ULeap) supports this:

> The Higher Education Carbon Management Programme (HECMP) has helped to bring environmental issues to the top of the University's agenda. (University of Leicester Environment Action Programme (ULeap), p. 7)

Senior management leadership can potentially lead to carbon management being embedded in corporate strategy of the university with a clear strategic vision and support for implementation.

4.2 Carbon Footprinting

Carbon footprinting is the measurement of carbon emissions from an organisation's business operations. The content analysis found that carbon emissions measurement is an important element of the carbon management process. The majority of the universities have done carbon footprinting or in the process of doing so. The measurement of carbon emissions is the first step to start managing and reducing emissions.

> Calculating an emissions baseline is the first step in enabling the university to quantify its carbon footprint and to gain a better understanding of its overall carbon contribution. (Loughborough University Carbon Management Plan, p. 15)

Universities are using different methods for carbon footprinting, and there is no specific international standard for HE carbon footprinting. HEFCE provided guidelines to measure carbon emissions, and many universities have followed that approach. There are other internationally recognised carbon footprinting guidelines published by the Global Reporting Initiative (GRI), UK government's Department for Environment, Food and Rural Affairs (DEFRA), the Greenhouse Gas Protocol and the International Standardisation Organisation (ISO).

4.2.1 Operational Boundaries

The majority of the universities have measured scope 1 and scope 2 emissions. The World Resources Institute (WRI) and World Business Council for Sustainable Development (WBCSD) classified scope 1 as direct emissions that occur from sources owned or controlled by the organisation and scope 2 accounts for emissions from the generation of purchased electricity. Scope 3 is all other indirect emissions that arise as a consequence of various organisational activities, but occur from sources not owned or controlled by the organisation (WRI and WBCSD 2004). Nine out of the eighteen universities (50%) have measured scope 1 and 2 emissions as reflected in their CMPs. Therefore, carbon emissions related to energy consumption in buildings and universities' own transport fleet are planned in almost all of the CMPs as universities have control over these emissions. Eight universities have measured scope 1 and 2 with 'selected parts of scope 3 emissions' and only one university has measured complete carbon footprint based on scope 1, 2 and 3. The selected parts of scope 3 emissions chosen by eight universities are related to waste and water. Out of these, three universities have calculated emissions associated with staff and student commuting and business travel. Two universities have measured emissions associated with procurement and supply chain. There is no consistent approach as far as scope 3 measurement is concerned. CMPs indicate that majority of the universities have not measured scope 3 and hence, they have not set scope 3 targets. The emissions related to procurement and supply chain, international student travel, business travel and staff and student commuting are overlooked. Many universities are planning to measure and target scope 3 emissions. However, universities have strategies to address various streams of scope 3 carbon emissions.

> Initially scope 1 and scope 2 emissions will be included within the carbon management plan, however moving forward scope 3 emissions will need to be measured and incorporated within the plan. (Loughborough University Carbon Management Plan, p. 18)

HEFCE proposed that universities should commit to scope 3 carbon management with the intention of measuring emissions and setting targets. The majority of the universities (fifteen out of the eighteen) are planning to include scope 3 in their CMPs. Some of the universities have already made estimates of scope 3, but have not developed the reduction targets with accurate carbon footprint and some are about to set the targets. The majority of the universities aim to have significant impact on reducing scope 3 emissions, but there are challenges such as access to data, reliability of data and standard methodology for calculations. Therefore, universities are first focusing on improving their data management practices.

> The absence of a suitable methodology and the difficulty of obtaining meaningful data for huge range of materials procured by the university means that it has not been possible to consider this factor at the present time. The assumptions are used to arrive at the preliminary estimates of the university's scope 3 emissions. Further work is on-going to refine these estimates using more reliable data. (University of Cambridge Carbon Management Plan, p. 17)

Consequently, universities have started to work on scope 3 carbon management with the first step to calculate a carbon emissions baseline.

> Work is being undertaken to gain a greater understanding of the university's scope 3 emissions. At this stage, quantification of a baseline for scope 3 emissions will be the first target. Subsequently, meaningful reduction targets will be established for scope 3 against the baseline. (University of Birmingham Carbon Management Implementation Plan, p. 8)

4.3 *Carbon Reduction Targets*

Each of the eighteen universities has set carbon reduction targets in CMPs. Some universities have very ambitious targets, whereas others have relatively less ambitious. CMPs present the universities' carbon reduction targets along the baseline year. These targets indicate organisational commitment towards carbon management in response to the HE and the national targets. The individual targets of universities contribute towards the overall HE targets, as the HEFCE targets are the main driver for universities. The HEFCE set carbon reduction targets for 2020 and 2050, but universities have set targets for 2020. This may be because 2050 is simply too far away to make a calculated predicted target. Campuses may change dramatically over the next few decades. In addition, some universities have established interim or short-term targets to track the progress of the main carbon reduction target.

4.3.1 Types of Targets

There are two types of targets which universities can adopt: absolute and relative. '*An absolute target is usually expressed in terms of a reduction over time in a specified quantity of greenhouse gas (GHG) emissions to the atmosphere, the unit typically being tonnes of CO_2e. An intensity target is usually expressed as a reduction in the ratio of GHG emissions relative to another business metric*' (WRI and WBCSD 2004, p. 77). For relative targets, the most common indicators are per metre square floor area (Per m^2 area), per Full Time Equivalent (FTE) student and per unit turnover and emissions are expressed as carbon dioxide equivalent (CO_2e). The issue of absolute and relative (intensity-based) targets emerged while analysing the CMPs, however, the distinction between the two types is not very clear.

Universities might mean absolute targets in the plans unless it is mentioned relative against a certain indicator. The majority of the universities (eleven out of the eighteen) have absolute targets. Two universities have set both absolute and relative targets, possibly for different reporting purposes, i.e. internal and external. The University of Derby has 27% absolute carbon reduction target by 2020 from a 2005 baseline.

> The targets to be reported to HEFCE are absolute carbon reductions and do not reflect the dynamic nature of the business and estate. We have therefore applied our own normalised targets against FTE staff and students and in the case of vehicle fleet emissions against mileage. (University of Derby Carbon Management Plan, p. 3)

Three universities have mentioned relative targets in their CMPs. These are research-based universities having higher energy consumption. The University of Birmingham states that relative emissions are reduced despite energy intensive research. This indicates that relative reduction targets allow business growth.

> As a research led university, a significant proportion of the energy consumed is for energy intensive applications. Despite this, there has been a significant reduction in carbon emissions when compared to growth (financially and in terms of staff/student numbers). (University of Birmingham Carbon Management Implementation Plan, p. 5)

De Montfort University's CMP has set the same target as that of the HE sector for scope 1 and 2, which is 43% carbon reduction by 2020 based on 2005/06 baseline (HEFCE 2010a). This is the only university that has developed the target with two interim targets (12% by 2012 and 29% by 2017). The CMP of the Nottingham Trent University states that the targets are minimum figures, but the university aims to exceed.

> Objective and target figures are to be regarded as bare minimums and as such it will be anticipated that these figures will be exceeded. (The Nottingham Trent University Carbon Management Plan, p. 12)

Robinson et al. (2015) suggest that interim carbon reduction targets should be included as one of the HEFCE requirements allowing universities to identify future challenges in meeting the target through early action. The content analysis found that presently, universities have only set targets for scope 1 and 2 carbon emissions and do not have them for scope 3. The University of Lincoln's CMP and the University of Bradford's Ecoversity—One Planet Strategy—have targets of 43% and 50% (scope 1 and 2) by the year 2020, respectively. This indicates universities' commitment towards implementing carbon management.

4.3.2 Carbon Emissions Baseline

Carbon emissions baseline indicates a fixed base year against which carbon management performance is benchmarked. Universities have used the methodology and conversion factors taken from the DEFRA and the Department of Energy and Climate Change (DECC) guidance to calculate carbon emissions. Carbon emissions are usually calculated based on an academic year (1st August–31st July). The majority of the universities (eleven out of the eighteen) have set the baseline of 2005–06 academic year for their calculations and targets, which aligns with the guidelines of the HEFCE. As many universities in the HE sector do not have complete or accurate data for 1990, so HEFCE decided to use 2005–06 as an alternative emissions baseline.

> The institution's carbon emissions baseline has been calculated using data from 2005/6 academic year (1st August 2005 to 31st July 2006). This year has been chosen as it is the earliest year for which reliable and complete data is readily accessible, and is the year that HEFCE is using for the HE sector emissions baseline. The university's financial year also runs concurrently with the academic year from 1st August to 31st July so resource implications and financial savings will be reported concurrently. (The University of Northampton Carbon Management Plan, p. 11)

4.4 Stakeholder Engagement

Stakeholder engagement emerged as one of the key themes in CMPs. Half of the CMPs (nine) mentioned the role of stakeholder engagement in the effective delivery of CMPs. Universities seem to be focusing on stakeholder engagement that allows coordination for maximum impact of strategies, but there is a lot more work that needs to be done. CMPs do not explicitly identify stakeholders relevant to carbon management process and do not present strategies to engage them. The University of Lincoln defined 'stakeholders' in relation to carbon management as below:

> Stakeholders are those parties either within or external to the university who will be affected by the programme to reduce carbon emissions and may influence its success. (University of Lincoln Carbon Management Plan, p. 28)

The universities discussing stakeholder engagement have an agreement that carbon management needs engagement and response at all levels of the organisation. The University of Nottingham emphasises the organisation-wide stakeholder engagement, but it seems that it is not there yet in many universities.

> The plan requires engagement at all levels of the university from individual behavioural changes to institution led initiatives in order that the objectives are understood and that contributions are made to maximise delivery of the targets. We want to foster a 'can do' approach and response across the university. (The University of Nottingham Carbon Management Plan, p. 2)

4.4.1 Staff and Student Engagement

Among the university stakeholders, staff and students is the largest stakeholder group and has received more focus in the CMPs. Universities are attempting to engage staff and students with appropriate engagement strategies. Due to this, the majority of the universities (ten out of the eighteen) have clearly elaborated staff and student awareness and engagement activities in their CMPs.

> The university has an objective to increase energy awareness in staff and students. (University of Leicester Strategy and Implementation Plan, p. 11)

Loughborough University's CMP states that every member of university staff and student needs to get engaged for successful carbon management and achieve the targets set out in the CMP.

> If the university is serious about meeting the challenge of achieving the targets set out within this plan and be seen as a leading low carbon campus within the Higher Education sector, every member of staff and the student body needs to engage in the carbon agenda. (Loughborough University Carbon Management Plan, p. 3)

Queen's University Belfast argues that successful implementation of the CMP requires engagement of staff and students rather than only focusing on technical approaches to carbon management. This indicates that staff and student engagement needs to be coupled with technical interventions for effective implementation of carbon management.

> The successful implementation of the CMP, which builds on the mobilisation of the whole organisation, rather than focusing on a strictly technical approach to carbon saving, requires the commitment of all staff and students. (Queen's University Belfast Carbon Management Plan, p. 4)

The above quotation indicates the key role of engagement. Universities are implementing projects around change management, awareness raising and behaviour change for staff and student engagement. One of the universities' CMPs explains that increasing staff and student awareness and training programmes have proved to be effective across the university. Staff training and education programmes do not seem to be fully developed in the majority of universities. With regard to knowledge and understanding, the majority of staff and students might not understand major parts of CMP due to its complexity and technical terminologies. CMPs are very technical and difficult to understand for ordinary staff and students who do not have experience or education in this subject area. This is one of the major barriers to cultural change in universities. The following quotation reflects the Queen's University Belfast's plans of dealing with human factors.

> A comprehensive awareness raising and educational programme will continue to be developed to help staff and students understand carbon management and its impact on the university. (Queen's University Belfast Carbon Management Plan, p. 5)

The focus of stakeholder engagement is more on staff and students and senior management engagement is overlooked, who are part of senior management team and are involved in strategic decision-making. Mainly lower and middle tier staff is focussed for engagement activities that may not have much influence to bring change in the university. Some of the CMPs provide a 'forward' or statement from a senior executive or VC, but that does not mean they are fully engaged in the carbon management process. CMPs do not clearly discuss how universities plan to engage senior management. Moreover, one university is planning to engage cleaning, security and HR department for carbon management. This indicates the role of every department in the process. Another university recognises the increasing role of student unions in its CMP. Student unions could play their part in engaging student body in an effective way, but there is a lack of active participation.

4.5 Funding and Resources

This theme explores the role of funding and resources to implement CMPs because energy and carbon reduction projects require significant financial investment. Universities have established a number of funding mechanisms for projects in CMPs. Seventeen universities' CMPs include the availability of funding to implement the projects. Each of the seventeen universities uses funding from different sources available. Universities seem to be allocating funding as much as they can to ensure that carbon reduction measures are implemented and targets are met, as set out in CMPs. Loughborough University notes that significant capital investment is required for carbon management.

> Significant capital investment will be required to both assess the feasibility and implement major infrastructure projects that will deliver significant carbon benefits. A full review of the grants and financial support that are available at both local and national level should be undertaken periodically to ensure the university maximises the funding opportunities that are available. (Loughborough University Carbon Management Plan, p. 7)

The above quotation indicates that universities explore multiple funding options to maximise opportunities because one source of funding may not be enough to implement the planned projects. Therefore, universities seem to constantly look for more funding options due to issues around funding. De Montfort University shows its commitment of financial investment:

> The university is committed to investing in new technology to achieve reductions in carbon emissions. This investment will encompass energy efficiency, better design and where possible, renewable technologies. (De Montfort University Carbon Management Plan, p. 13)

4.5.1 Sources of Funding

Universities have various funding sources available to them. The majority of the universities (eleven out of the eighteen) mention the joint HEFCE and Salix Revolving Green Fund (RGF) for the projects. Salix Finance Ltd helps accelerate public sector organisations investment in energy and carbon reduction projects through 'invest to save' mechanism. The principle of the RGF is that the financial savings generated from interest free loans for energy and carbon reduction projects are recycled back to the central fund for reinvestment in other projects. The RGF has two strands: Transformational Fund and Institutional Small Projects Fund. The transformational fund aims to tackle large-scale projects in the range of £1–4 million per institution, which will transform the university's approach to manage energy consumption and reduce carbon emissions (HEFCE 2008b). Four universities have implemented large projects through the transformational fund. The other seven universities used funding through the institutional small projects fund, which is a ring-fenced budget for smaller projects. Four universities are using the 'Long Term Maintenance Programme' budget mainly for energy conservation work in buildings. Three universities have set their budget from the 'Capital Funding Programme'.

Two universities have planned to use 'Renovation and Redevelopment' budgets for redevelopment and carbon reduction. Projects funded from these budgets can serve the dual purpose of energy efficient renovation and redevelopment. Three universities mentioned the possibility of external grants and additional resources. There are different organisations and agencies that can further support carbon reduction projects. Two universities have mentioned in CMPs that there is potential for the suitability of the Energy Performance Contracts (EPCs) in their universities under which the EPC provider offers a financial vehicle to implement energy and carbon reduction projects. None of the universities has implemented EPC model, but universities can adopt this model in future. EPCs aim to address financial, technical and procurement-related issues and help minimise risks. In addition, CMPs do not discuss the costing procedures for the proposed projects and only one university mentioned it. Each project costs in terms of the capital expenditure (Capex) required and the operating expenditure (Opex) and it needs a life-cycle perspective. Currently, most of the universities do not seem to have ring-fenced carbon and energy-related internal budgets. Four universities plan to adopt the strategy of fixed budgets and reinvest savings into future initiatives.

> Establish a carbon management fund pot into which monetary savings from previous carbon management projects can be placed and then used to fund future carbon management projects. (Nottingham Trent University Strategy and Implementation Plan, p. 41)

CMPs do not discuss whether previous investments in projects have met anticipated carbon and cost savings in business cases. The University of Leicester states that there is a tendency to 'cherry pick' projects based on shorter payback period and it is not the right approach. The university states that 'quick win projects' have lesser payback period and are a top priority. The 'long term projects' require higher investment and have longer payback period and universities tend to avoid that. This suggests that implementing carbon reduction projects have a low priority as far as financial investment is concerned.

> The 'Quick Win projects' are those that have lesser payback period. These projects are of top priority to the university as they are quick energy and cost savers and contribute to the reduction of carbon emissions. The university has financial threshold. The 'Long Term projects' require a higher investment and result in longer payback period. (University of Leicester Strategy and Implementation Plan, p. 29)

4.6 Governance

Governance refers to all the processes by which policies and strategic decisions are made (Renz 2007). The governance of a university typically involves the strategic management and decision-making at a senior management level. In this study, governance refers to the processes by which decisions and carbon management strategies are implemented from operational management perspective. It aims to manage a diverse range of elements such as responsibility and communication in the carbon management process. Eleven universities specifically mentioned the term governance in their CMPs. These universities have discussed issues around governance for the successful implementation of CMPs and meeting the carbon reduction targets. At an operational management level, governance for implementing carbon management strategies starts with the monitoring of energy and fuel consumption and then calculates carbon footprint of the university. Governance can also help develop organisation-wide ownership of carbon management. Loughborough University's CMP states the role of governance in the carbon management programme:

> In order to ensure that there is effective and on-going ownership of the carbon management programme, it is important to define a governance or accountability structure for the programme. (Loughborough University Carbon Management Plan, p. 40)

The University of Northampton states that governance of the CMP rests with senior managers in the executive board. The main idea is to oversee and lead carbon management from top and achieve a top-down approach in the university. This could facilitate embedding carbon management in whole organisation. In the University of Northampton, governance and strategic ownership reside with the 'Carbon Management Programme Board' with senior members such as the Pro Vice-Chancellor (Strategic Planning and Resources), Director of Estates, Director of Finance and Dean of the School of Science and Technology. This indicates senior level involvement in governance.

> Governance of the programme as well as the strategic ownership of the university's carbon reduction target, rest with the Programme Board, composed of the University's Pro Vice-Chancellor (Strategic Planning & Resources), Director of Estates, Director of Finance and Dean of the School of Science and Technology. (The University of Northampton Carbon Management Plan, p. 6)

4.6.1 Responsibility

Responsibility is an important element for the governance of carbon management. The majority of the universities (seventeen out of the eighteen) discuss responsibility in CMPs. The content analysis found that responsibility varies in universities and different individuals with different job roles are responsible for carbon management. However, it can be divided into three main categories: middle managers, working/task/steering groups and everyone within a university. The majority of the universities' CMPs (ten) indicate that the overall responsibility lies with a working/ steering/task group in universities and four of the CMPs state that 'everyone' is responsible for carbon management. The universities assign operational responsibility to implement carbon management plan and strategies to middle managers based in estates management department. However, the middle managers argue that 'everyone' within a university should be responsible for it, although it is in their job description. Four CMPs state that everyone in the university, all staff and students, is responsible for carbon management.

> Carbon emissions at De Montfort University need to be seen as the responsibility of the whole university rather than specific and individual departments. (De Montfort University Carbon Management Plan, p. 13)

The working or steering group generally consists of members from different university departments and faculties such as estates, academics, professional services, senior management and others and is mostly chaired by a senior manager or a member of senior management team. These groups are developed to introduce organisation-wide responsibility for carbon management. For example,

> The development and delivery of the Climate Action Plan is supervised by the Sustainability and Environment Advisory Group (Operations). (The University of Edinburgh Climate Action Plan, p. 8)

Seven universities discuss strategic responsibility in CMPs, and it varies from university to university depending upon the organisational structure. Strategic responsibility means strategic oversight by a senior manager, probably a member of the senior management team. Four CMPs mention that Director of Estates has strategic responsibility for carbon management and two of the plans state that it rests with a working or task group of various internal stakeholders. However, only one CMP, the University of Cambridge, states that the Pro Vice-Chancellor (PVC) has strategic responsibility for overseeing carbon management process. This is the only university where PVC is championing carbon management. This is not the case in majority of the universities. The senior management level responsibility could help bring carbon management at strategic level and provide support for resource allocation.

> The Pro Vice-Chancellor for Institutional Affairs has specific responsibility for carbon reduction and sustainability. (University of Cambridge Carbon Management Plan, p. 24)

Members of senior management teams are not fully involved in the process despite their official responsibility, apart from being part of the environmental or carbon management group. Strategic responsibility has not received much attention in the CMPs, and there is no evidence of wider responsibility at a VC/PVC level. This suggests a gap between operational (middle) and strategic (senior) managers in universities.

> Strategic carbon management is the responsibility of the Director of Estates and Commercial Facilities. Day-to-day operational carbon management responsibility lies with the Carbon Reduction Manager (a full-time post created in September 2010). (University of Lincoln Carbon Management Plan, p. 23)

There is a lack of clarity on the divide between operational and strategic responsibility. Five CMPs mentioned operational responsibility for implementing carbon management. The content analysis found that middle managers are responsible for implementation. Three of the CMPs state that middle managers have operational responsibility, whereas two CMPs state that the operational responsibility lies collectively with a carbon management team or group consisting of middle managers in estates. The middle managers include the sustainability manager, environmental manager, energy manager or carbon manager. This is a full time position in majority of the universities and is based in estates or facilities management department. The CMPs reflect that middle managers seem to be more concerned and relevant to carbon emissions and its management by developing targets. This indicates less involvement of senior management in implementation. CMPs are developed by middle managers and approved by senior management team. Leeds Beckett University for example notes that:

> The Sustainability Manager will be responsible for developing and implementing the carbon management strategy and producing interim reports. (Leeds Beckett University Carbon Management Strategy, p. 6)

4.6.2 Communication

Communication is a key for effective governance in universities. Communication is also a key performance indicator in the 'Carbon Management Matrix', which is used to evaluate carbon management performance. Fifteen out of the eighteen universities mentioned communication as a governance issue for implementing carbon management strategies. The content analysis indicates that communication is related to various aspects of carbon management ranging from communicating energy and emissions data, carbon management targets and strategies and novel energy and carbon reduction initiatives to the relevant stakeholders. Most of the universities aim for effective communication practices in their CMPs. There are two types of communication: internal and external. Universities are focused on internal communication for engagement with stakeholders, mainly staff and students.

> Regular and relevant communications to support a transition to a low carbon culture across the university will continue. (University of Lincoln Carbon Management Plan, p. 8)

An effective communication strategy can help raise awareness and change behaviour of different stakeholders across the organisation, which may contribute in meeting carbon reduction targets.

> A key element of ensuring that the carbon reduction target is achieved is through delivering a successful communications strategy to help implement behavioural change across all aspects of the university's activities. (University of Birmingham Carbon Management Implementation Plan, p. 31)

Three universities have separate 'communication strategies' for carbon management. A communication strategy is developed to ensure that all key stakeholder groups are participating and coordinating in delivering the CMP. As far as means of communication are concerned, each of the fifteen universities uses different channels to communicate their carbon management message. The most commonly used methods for communication are websites, magazines/newsletters, university reports, awareness raising/behaviour change campaigns, workshops, environmental champions' network, information screens/digital displays, staff trainings and inductions. However, the use of social media has not received much attention in CMPs. Four universities are carrying out employee inductions to inform new employees about the environmental goals of universities. For example, the Strategy and Implementation Plan of the Nottingham Trent University has integrated energy and environmental management good practices into its induction process for all staff:

> All new staff to have energy and environmental good practice inductions along with traditional fire and health and safety inductions etc. in order to instil energy and environmental awareness. (Nottingham Trent University Strategy and Implementation Plan, p. 41)

4.7 Evaluation and Reporting

Evaluation and reporting of carbon management performance emerged as a main theme. Monitoring the ongoing progress of carbon management and reporting are part of the planning process. This can help explore how carbon management is progressing over a period of time and how it can be improved. Almost all of the universities' (seventeen) CMPs discuss evaluation and reporting in relation to their carbon reduction targets and planned projects. Universities intend to evaluate and report through different channels. There could be a problem due to lack of transparency and traceability, and CMPs do not reflect on it. There is a vast array of regulatory and voluntary reporting mechanisms available to universities. The University of East Anglia's Carbon Reduction Plan states the role of review and evaluation process:

> To ensure continued reduction in emissions is maintained, it is essential that the programme, projects and the carbon reduction plan are regularly reviewed. (University of East Anglia Carbon Reduction Plan, p. 33)

The University of Nottingham plans to develop an annual report to monitor and report the progress and provide an update on the targets.

> An annual report will be produced to monitor and report on our progress and performance achievements against this plan and to provide an update on the CO_2 reduction projects that will ensure that the remaining targets and objectives are met. (The University of Nottingham Carbon Management Plan, p. 2)

The majority of the universities (thirteen out of the eighteen) have adopted similar internal reporting procedures. Middle managers report progress to senior management group or committee. The senior management group is chaired by a senior manager, who is a member of the senior management team. The senior manager can be a Pro/Deputy Vice Chancellor to ensure high-level championing. Then, it will be reported to the university's senior management team and governing council. Cardiff University describes its internal reporting mechanism as:

> The Carbon Management Plan Task and Finish Group are currently developing the CMP, which will then report through the Environmental Management Systems (EMS) Steering Group, which is chaired by the Deputy Vice-Chancellor, which will oversee implementation and further development of the plan. (Cardiff University Carbon Management Plan, p. 3)

The above quotation highlights reporting, which the majority of universities carry out. At an operational estates level, four CMPs mention that universities have adopted a series of Key Performance Indicators (KPIs) to monitor and report the performance. The KPIs include CO_2e emissions, CO_2e emissions per £million turnover or Full Time Equivalent (FTE) staff member and students.

> Carbon reduction is incorporated as a KPI at both institutional and School/Directorate level, with progress in meeting targets included in all School and Directorate Annual Performance Reviews submitted to the University Operating Board in November each year. These are, in turn, reported to the University Management Board, the Planning and Finance Committee and, ultimately, to Senate in December. (Queen's University Belfast Carbon Management Plan, p. 6)

Different universities have different timescales for reporting. Half of the universities report progress annually, both internally and externally. The focus is on internal reporting within the CMPs to gain senior management support. The universities report externally through mandatory and optional policies and schemes in the UK. HEFCE demands that universities measure and report progress against sector-level targets. The University of Northampton defines its objectives with regard to implementation, review and update:

> The university recognises carbon management as a strategic, whole-organisation approach that integrates with our existing strategy and management, enabling the institution to understand the impact of carbon emissions, to identify key risks and opportunities, to formulate a plan to reduce carbon emissions, to effectively implement, review and update the plan into the future, and to communicate success. (The University of Northampton Carbon Management Plan, p. 7)

5 Discussion

This paper presents the analysis of eighteen universities' CMPs to explore carbon management planning in UK universities and identify key elements as part of their journey to a low carbon built environment. The paper speaks to contemporary policy concerns around climate change and carbon management in the context of the HE sector. The novelty of the analysis advances understanding of carbon management area both theoretically and practically by offering insights into how universities are planning carbon management and what are the areas of improvement. The study found that almost all of the universities have CMPs and are starting to understand the strategic role of carbon management. The CMPs studied were quite similar in design, which suggests consistency. CMPs are valuable tools with the potential to place universities on the pathway of implementing carbon management. This finding corresponds with Robinson et al. (2015), who argue that CMPs have helped the sector take carbon emissions seriously. Despite these positive findings, there are concerns. CMPs are overly focused on operational issues rather than strategic. Technical details and projects are dominant while there is little emphasis on non-technical or human factors. This might be because CMPs are designed by environmental and technical managers in estates department. This makes CMPs complex and less effective in communication with stakeholders. CMPs do not seem to be integrated into other university policies and strategies; rather they are a stand-alone document. CMPs face challenges such as future uncertainty and changes. Sometimes, it is difficult to predict the future scenario while developing a CMP, for example increase in student numbers, weather conditions, policy changes and technological advancement. It is difficult to identify projects, as situation may change, for example changes in the HE sector may influence it. Altan (2010) and Lee (2012) state that business growth and expansion of estate are the major challenges for carbon management in universities.

The majority of the universities measured carbon emissions from scope 1, scope 2 and selected parts of scope 3 and have developed plans and strategies. Universities are now in the implementation phase of CMPs. Universities have shown commitment to reduce emissions and have targets to support the HEFCE and national targets. All of the sample universities have set targets, but CMPs do not reflect whether they have achieved actual reductions over a period of time as there is no sector study. However, setting high targets and not making a realistic effort to achieve them should be perceived as 'green wash'. There is no accountability from HEFCE since the change in funding mechanism in the HE sector and universities receive funding direct from students, effectively shifting them into a consumer–provider commercial relationship.

Robinson et al. (2015) argued that having unachievable targets can be harmful as middle managers may lose interest and support of senior management leadership. Ambitious targets are less likely to be achieved and the problem is that realistic but relatively low targets can be criticised and penalised in league tables, such as the People and Planet University League. The majority of the universities have plans to meet absolute targets, which could be difficult due to energy intensive activities and business growth. This corresponds with Lee (2012) who argues that energy intensive research along with growing estates is counterproductive to carbon management. In contrast, universities have relative targets for reporting and allowing business growth. It might be interesting to quantify if universities are successful in meeting absolute targets. The national and HE targets are absolute for scope 1 and 2, although some selected parts of scope 3 are measured by many universities. However, mostly universities are lagging behind in dealing with the measurement, targeting and management of scope 3. It might be because scope 3 is not mandatory and HEFCE has advised to develop targets for scope 1 and 2 (HEFCE 2010a). Universities do not have the data for all scope 3 streams for carbon footprinting, but they understand the important role of indirect scope 3 emissions, especially procurement and travel.

All of the CMPs are signed by a VC or any other senior manager, but their role is limited in the process. Senior leaders not only need to have a vision, but they need to ensure the implementation of the vision (Millar et al. 2012). The analysis suggests that strategic and operational aspects of carbon management are not joined up. Although many universities have planned to adopt a whole-organisation approach, the evidence is still lacking of this in reality (Schultz and Williamson 2005). The whole-organisation approach constitutes students and staff in all faculties and directorates, both academic and professional services engaged in reducing emissions from the direct and indirect sources (scope 1, 2 and 3). In contrast, carbon management has started to emerge at the strategic agenda albeit slowly and CMPs could play a significant role in implementation. Funding and resources is an important part of implementing carbon management strategies. CMPs suggest that there are different funding options available to implement projects and universities are using multiple sources. Some of the projects have received funding, and some are subjected to further funding approval. Some projects do not even get implemented due to lack of funding in universities. Therefore, funding is considered crucial element for the implementation. Universities have competing demands in the current financial climate and may need innovative funding models for carbon management. Stakeholder engagement is a key factor and aims to bring internal and external stakeholders together. Universities are focused on staff and student engagement, and the engagement of senior management is not addressed in CMPs. Many universities have not been able to make CMPs an interactive tool to engage stakeholders due to complexity. CMPs fail to make staff, students and other stakeholders understand what it means to them and how can they contribute to it. CMPs are not visible in universities. Universities seem to have embedded statements into CMPs rather than providing solutions to problems or their causes.

There is a wide range of job titles for individuals responsible for carbon management. However, responsibility varies based on structure of the organisation and seems to be unclear. This matches with the findings of Chen (2011) who state that responsibility of addressing climate change varies largely ranging from the presidents of universities to environmental or sustainability coordinators. The overall responsibility varies and mainly rests with working or task groups consisting of individuals from different university departments. The operational day-to-day responsibility lies with middle managers in estates, whereas strategic responsibility is with senior manager or director of estates who oversight the process. Carbon management is seen as a responsibility of a specific department and/or individual in universities. Predominantly, carbon management is implemented by environmental team in estates department, whereas other departments have no or very little contribution. This suggests that there is a lack of responsibility and management is disjointed and lacking in coherence and coordination. Middle managers do not have authority and involvement in strategic decision-making and they need approval for most of the projects either from the director or senior management team.

Universities review, evaluate and report performance on regular basis through different forums. Updating CMPs and tracking the progress against targets are part of it. This could help managers understand where they are and where they want to go in future to meet the targets. The review and evaluation offer lessons to be learnt for future. There is a lack of consistency and complexity in existing methods for measuring, targeting and reporting within HE and they are all different. Although there are common external reporting mechanisms, but internal reporting varies. The reporting system could be standardised to compare universities. Internal reporting system is emphasised in the CMPs for internal stakeholders. There are legislative and policy drivers for carbon reporting such as Carbon Reduction Commitment Energy Efficiency Scheme (CRC EES). By reporting emissions, universities can improve performance against a benchmark and review CMPs for improvements. The fact that many universities have not updated their CMPs suggests that the plans are static and not all universities are fully committed. This indicates slow progress or managers being busy due to other priorities. This could indicate that universities are drifted away from CMPs since the change in HEFCE policies due to lack of accountability and the plans are out of date (see Table 1). Brite Green (2015) found that many HEFCE funded universities need to revise their CMPs to meet 2020 targets.

6 Conclusion

Carbon management is a crucial issue that needs effective planning and implementation in universities if the ambitious carbon reduction targets are to be met. In the UK, most universities have developed CMPs as part of policy compliance and are now in the implementation stage. However, there is a long way to go if the CMPs are going to enable universities to reach their full potential. UK universities took a lead by setting carbon reduction targets and developing CMPs. HEFCE's work in the area carbon management is innovative and offers lessons to be learnt. Other universities and organisations can learn from the planning experience and knowledge of the UK universities. This study has identified the key elements of the carbon management planning process and universities need to focus in those areas for successful carbon management to meet carbon reduction targets. Despite various advantages of CMPs, there are potential areas of improvements which this study has highlighted. This study will be useful to academics and practitioners aiming to improve carbon management planning in universities and other organisations nationally and internationally. There are also limitations of this study. The authors depended on the analysis of official university documentation and the limitation of using such sources of evidence is acknowledged, as what is said does not always correspond with what managers think or do in universities. Future research can address this issue by gathering primary data from universities to carry out a comparative analysis. With regard to carbon management planning, future research will be necessary to develop a better understanding of how successfully CMPs are implemented against the set carbon reduction targets. Given the increasing emphasis on league tables and student experience in universities, it is clear that any future 'whole organisation' management approach has to embed carbon management into the whole student experience.

References

Altan, H. (2010). Energy efficiency interventions in UK higher education institutions. *Energy Policy, 38*(12), 7722–7731.

Ball, A., Mason, I., Grubnic, S., & Hughes, P. (2009). The carbon neutral public sector: Early developments and an urgent agenda for research. *Public Management Review, 11*(5), 575–600.

Bebbington, J., & Barter, N. (2011). Strategic responses to global climate change: A UK analysis. *Chartered Institute of Management Accountants, 7*(11), 1–10.

Brite Green. (2015). *2020 Carbon Reduction Target: Progress Report of De Montfort University, Brite Green*. London, UK.

Cadez, S., & Czerny, A. (2016). Climate change mitigation strategies in carbon-intensive firms. *Journal of Cleaner Production, 112*, 4132–4143.

Carbon Disclosure Project. (2010). *The carbon management strategic priority: Carbon Disclosure Project (CDP)*, London, UK.

Chen, C. (2011). *Climate change leadership in higher education institutions*. Unpublished Ph.D. thesis, Duke University, USA.

Davies, J. C., & Dunk, R. M. (2015). Flying along the supply chain: Accounting for emissions from student air travel in the higher education sector. *Carbon Management, 6*(5–6), 233–246.

HEFCE. (2008a). *The 2008 grant letter and key performance target measures relating to carbon emissions: Report to the Secretary of State for Innovation*, Universities and Skills, HEFCE, Bristol, England.

HEFCE. (2008b). *Revolving green fund: Consultation on the aims and operation of a revolving green fund*, HEFCE. Bristol, UK.

HEFCE. (2010a). *Carbon reduction target and strategy for higher education in England*, HEFCE, Bristol, England.

HEFCE. (2010b). *Carbon management strategies and plans: A guide to good practice*, HEFCE, Bristol, England.

Her Majesty's Stationery Office. (2008). *Climate Change Act 2008 Her Majesty's Stationery Office (HMSO)*, London, UK.

Higher Education Statistics Agency (HESA). (2014). *HE Statistics for the UK 2012/13*. Available at: https://www.hesa.ac.uk/pr/3226-press-release-208. Last Accessed January 31, 2015.

Hoffman, A. J. (2007). *Carbon strategies: How leading companies are reducing their climate change footprint*. USA: University of Michigan Press.

Klein-Banai, C., & Theis, T. L. (2013). Quantitative analysis of factors affecting greenhouse gas emissions at institutions of higher education. *Journal of Cleaner Production, 48,* 29–38.

Kolk, A., & Pinkse, J. (2004). Market strategies for climate change. *European Management Journal, 22*(3), 304–314.

Lee, S. Y. (2012). Corporate carbon strategies in responding to climate change. *Business Strategy and the Environment, 21*(1), 33–48.

Mazhar, M. U., Bull, R., & Lemon, M. (2012) A review and case study of strategic carbon management in UK Higher Education sector. In *Proceedings of Royal Institution of Chartered Surveyors (RICS) COBRA Conference*, September 2012, Las Vegas, Nevada, USA.

Mazhar, M. U., Bull, R., & Lemon, M. (2017). Critical success factors for embedding carbon management in organizations: Lessons from the UK higher education sector. *Carbon Management, 8*(5–6), 1–14.

Mazhar, M. U., Bull, R., Lemon, M., & Mallaburn, P. (2014). The current state of strategic carbon management within the UK higher education sector: Leading the way forward? In *Proceedings of the 9th International Symposium on Sustainable Leadership* (pp. 148–160), 3–6 June 2014, Salzburg, Austria.

Millar, C., Hind, P., & Magala, S. (2012). Sustainability and the need for change: Organisational change and transformational vision. *Journal of Organizational Change Management, 25*(4), 489–500.

Renz, D. O. (2007). *Non-profit governance and the work of the board.* Kansas City, USA: University of Missouri.

Robinson, O., Kemp, S., & Willaims, I. (2015). Carbon management at universities: A reality check. *Journal of Cleaner Production, 106,* 109–118.

Robinson, O. J., Tewkesbury, A., Kemp, S., & Williams, I. D. (2017). Towards a universal carbon footprint standard: A case study of carbon management at universities. *Journal of Cleaner Production.* https://doi.org/10.1016/j.jclepro.2017.02.147.

Schultz, K., & Williamson, P. (2005). Gaining competitive advantage in a carbon-constrained world: Strategies for European business. *European Management Journal, 23*(4), 383–391.

Stern, N. (2006). *Stern review: The economic of climate change.* UK: Stern Review Report.

The Carbon Trust. (2012). Public sector most committed to carbon reduction. Available at: https://www.carbontrust.com/about-us/press/2012/02/public-sector-most-committed-to-carbon-reduction/. Last Accessed April 10, 2013.

The Carbon Trust. (2013). How the British public sector is leading the world on climate change action. *The Carbon Trust*, London, UK. Available at: http://www.carbontrust.com/news/2013/11/british-public-sector-leading-world-on-climate-change-action. Last Accessed December 14, 2013.

Wehrmeyer, W., Leitner, A., & Woodman, P. (2009). *Lean and green: Leadership for a low-carbon future.* Chartered Management Institute, London, UK.

Weinhofer, G., & Hoffman, V. H. (2010). Mitigating climate change–how do corporate strategies differ? *Business Strategy and the Environment, 19*(2), 77–89.

Wittneben, B., Okereke, C., Banerjee, B., & Levy, D. (2009). Climate change and the emergence of new organizational landscapes. *Organization Studies, 30*(9), 1013–1015.

WRI, & WBCSD. (2004). *The greenhouse gas protocol: A corporate accounting and reporting standard* (revised ed.). USA and Switzerland: The Greenhouse gas Protocol Initiative.

Understanding the Role of Networks in Stimulating Adaptation Actions on the Ground: Examples from Two African Case Studies

Gina Ziervogel, Lorena Pasquini and Jessica Lee

Abstract The networks that support collaboration and knowledge exchange around climate risk and response are emerging as central to climate adaptation. Yet, there is limited empirical knowledge about the conditions by which these networks can go beyond knowledge sharing to achieving on the ground. This paper presents two case studies of networks between universities, the private sector, civil society, and decision-makers at the local government level in different African municipalities: Karonga District (Malawi) and the Eden District (South Africa). The chapter considers conditions that enable collaboration to move from knowledge sharing to adaptation action. It does this by demonstrating how the origins of networks, the characteristics of partner organizations, the needs of each partner, the nature of the relationships between partners and the activities of the network, all combine to affect adaptation responses. For example, the high vulnerability of the Eden District to weather hazards prompted a private insurance company to collaborate with academics and local governments to understand how to reduce flood and fire risk to limit their insurance payouts in the region. The insurers invested in early warning systems and better fire response vehicles, as well as in a more holistic strategy of partnership, which has improved risk responses. The experiences presented in this

G. Ziervogel (✉)
Department of Environmental and Geographical Science and African Climate & Development Initiative, University of Cape Town, Private Bag X3, Rondebosch 7701, South Africa
e-mail: gina@csag.uct.ac.za

L. Pasquini · J. Lee
Global Risk Governance Programme, Faculty of Law, University of Cape Town, Private Bag X3, Rondebosch 7701, South Africa
e-mail: lorena.pasquini@gmail.com

J. Lee
e-mail: jessica22lee@gmail.com

L. Pasquini
African Climate & Development Initiative, University of Cape Town, Private Bag X3, Rondebosch 7701, South Africa

© Springer International Publishing AG, part of Springer Nature 2019
W. Leal Filho and R. Leal-Arcas (eds.), *University Initiatives in Climate Change Mitigation and Adaptation*, https://doi.org/10.1007/978-3-319-89590-1_4

paper highlight why and how networks can integrate knowledge from different partners into decision-making to result in adaptation actions on the ground.

Keywords Adaptation · Collaboration · Networks · Municipalities
Partnerships · Africa

1 Introduction

With increasing global temperatures, the negative impacts of climate change are becoming more apparent, driving an urgent need for climate change adaptation responses (IPCC 2014). Adaptation is a multi-sectoral, multidisciplinary issue requiring collaboration and knowledge exchange from different fields of expertise for responses to be successful (Kinnear et al. 2013; Howes et al. 2015; Moloney and Fünfgeld 2015). "Collaborative governance", which focuses on the formation of interorganizational collaborations (Vangen et al. 2015), has therefore been one of the institutional responses to fostering holistic adaptation responses. The organizations involved in collaborations include government at different levels, academic institutions, non-profit organizations, civil society, and the private sector. Importantly, collaborative governance works across boundaries in multi-sector and multi-actor relationships (O'Leary and Vij 2012), allowing for the circulation and application of "various types of data, information and knowledge to complex situations" (Agranoff 2007: 221).

Alongside the concept of collaborative governance, studies on networks have grown significantly, with networks widely recognized as an important form of multi-organizational governance (e.g. Provan and Kenis 2008; Poocharoen and Ting 2015). Often the claims, hypotheses, and research questions pertaining to these two concepts (collaborative and network governance) overlap or are closely related. In this paper, we view collaborative governance as being "operationalized" through network governance, where networks of organizations are formed among those collaborating on climate change adaptation issues, which is our focus of study. Ferlie et al. (2010) argue that networks are effective in addressing "wicked problems"—defined as situations with no clear solution, involving various individuals and organizations, and where there is often disagreement among stakeholders as to the desired outcomes. These problems—of which climate change adaptation is one—require a broad range of responses, engagement of diverse stakeholders and citizens, and working across boundaries. Drawing on Silvia and McGuire (2010), we see networks as containing multiple actors. These include individuals, agencies, or organizations, with multiple linkages (or relationships) between them.

Despite wide recognition of the important role that collaborative networks play in climate change adaptation (e.g. Silvia and McGuire 2010; Howes and Dedekorkut-Howes 2012; Weiss et al. 2012; Kinnear et al. 2013; Howes et al. 2015), there is limited understanding of the network factors that close the gap between collaborative knowledge exchange across networks and effective adaptation on the ground. This chapter addresses this gap by reflecting on when and where network collaborations go beyond knowledge sharing to achieving actual on-the-ground adaptation actions. Drawing on research on climate change adaptation from two case studies—one in Karonga District (Malawi) and the other in Eden District (South Africa)—we assess how on-the-ground adaptation was facilitated through knowledge-sharing practices. In order to do this, a range of network characteristics was assessed including the characteristics of the actors[1] and the network's structure and processes, which affect the network's values and functions. The findings from the case studies are reflected on to suggest future directions that research and practice might take.

2 Background to the Cases

The Eden District Municipality (EDM) is located on the south coast of the Western Cape, in South Africa, as shown in Fig. 1a, b. In South Africa, district municipalities are made up of a number of local municipalities, seven in the case of the EDM. The district municipality is expected to coordinate development and delivery in the whole district. The EDM area, 23,331 km^2 in extent, is a popular tourist destination, with just over 600,000 people in the area (http://www.localgovernment.co.za). From a socio-economic perspective, it is a highly unequal region, with many people living in informal settlements battling to find employment, many wealthy families living on large estates, and many in between. In addition, it has been impacted by a series of severe droughts and floods over the past decades. From 2009 to 2011, Eden experienced one of the worst droughts in 132 years and had to implement significant water restrictions (Baatjies and Hendrickse 2015). The drought impacted on farmers in the district (Matthews 2015) and on the urban areas. There has also been extensive flood damage. For example, in the town of Sedgefield, there were five severe flood events between 2003 and 2005 that led to significant damage to property and high insurance claims (Reyers et al. 2015). In 2017, severe fires were experienced in the Knysna District, with hundreds of properties in the towns and rural areas being destroyed.

In 2012, a partnership between Santam, a short-term insurance company, and the EDM began. Although there had been limited engagement between the two organizations prior to 2012, a more formal partnership was established that built on a collaboration that Santam had with a consortium of researchers, including the University of Cape Town and the CSIR (the South African Council for Scientific and Industrial Research) (the second author was part of this consortium of researchers).

[1]Actor here refers not to an individual, but to an organization, as this is the level that we examine in this chapter.

(a)

(b)

Fig. 1 **a** Map of South Africa with the Western Cape shaded in pale yellow and the Eden District shaded in red. **b** Map of the seven local municipalities in the Eden District. *Source* http://en.wikipedia.org

Building on this earlier research, a new research project commenced in 2017, led by the University of Cape Town Graduate School of Business, looking at how companies have partnered with municipalities for local development and resilience (the lead author was part of this research project). In the Eden case, key informant semi-structured interviews were undertaken with 18 respondents from the Eden District and Santam during the period March 2017 to May 2017. The interviews explored topics including what the partnership was trying to achieve, actors' motivations and their perceptions of each other's motivations, how the initiative achieved its objectives, the impacts of the partnership, and how issues of fairness and justice were considered. Using this data, actor and network characteristics were assessed.

The second case, the Karonga District, is a lakeshore district located in the northern region of Malawi as shown in Fig. 2. The district covers an area of 3355 km^2 and has an estimated population of 307,216. The settlement of the district is organized by traditional authorities, of which there are 6, and further subdivided into 39 group village heads covering some 336 villages. The administrative headquarters are in Karonga Town, the only town in the district, located some 50 km from the border with Tanzania. Its population of c. 50,000 is projected to reach 72,000 by 2022 (Manda 2014). The Karonga District Council (established under the Local Government Act of 1998) is the governing body for the district. The largely agricultural economy of the district is supplemented by fishery outputs, forestry products, and most recently the emergence of a mining industry (currently, mainly coal mining). The most pressing issues in the district as identified by community members, traditional leaders, and government sectors are the poor quality of education services, the increased occurrence of natural disasters and hazards, and food insecurity. With regard to natural disasters and hazards, the district is particularly vulnerable to earthquakes, floods, strong winds, and drought.

The region registered the largest number of disasters in Malawi between 1946 and 2008 (Manda 2014).[2]

In 2015, the Urban Africa Risk Knowledge (U-ARK; https://www.urbanark.org) project began. The project engages with a consortium of academic and non-academic partners in a large research and capacity-building programme on reducing disaster risk in urban sub-Saharan Africa. The authors of this paper are all part of the U-ARK project. One of the cities selected as a case study was Karonga. A partnership was thus established with the Karonga District Municipality. This brought numerous academics to the region from a few different universities, both local and international. The district municipality was already engaged in multiple partnerships with non-governmental organizations (NGOs) for disaster risk reduction and resilience building, of which one partnership will be examined here. The DISCOVER project ("Developing Innovative Solutions with Communities to Overcome Vulnerability through Enhanced Resilience") is a four-year consortium project of different NGOs, funded jointly by the UK Department for International Development (DFID), Irish Aid, and the Norwegian Embassy. It works in five districts in Malawi (including Karonga), where it seeks to support rural people to adapt to and mitigate the effects of climate change. In Karonga, it is led by the NGO, Self-Help Africa.

The data on which this chapter is based comes from face-to-face, semi-structured interviews conducted with 15 respondents from the Karonga District Municipality and local NGOs during November 2016. Interviews explored a range of topics relating to the experience of the Karonga District with extreme weather events, the knowledge and beliefs of respondents on climate change and adaptation issues, as well as local resources, actions, plans, and collaborations in this regard. Actor and network characteristics were assessed using this data.

3 Eden District

For each case study, we start by establishing the origins of the network which requires examining the needs of each partner: Santam urgently needed to reduce their insurance claims in the Eden District and wanted to draw on research expertise to better understand the nature of the problem, their exposure risk, and the potential approaches for reducing their exposure. During this initial phase, a group of researchers were interested in gathering case study material for understanding the roles of the insurance sector and of local governments in reducing disaster risks, as these were "hot" topics for research agendas at the time. Thus, an initial phase of research conducted by the University of Cape Town and the CSIR fed back information to Santam regarding the EDM's context. Building on this initial phase

[2]The information on which much of this introductory section is based is drawn by the draft Karonga District Socio-Economic Profile 2013–2018, which was made available to the authors from respondents in the region.

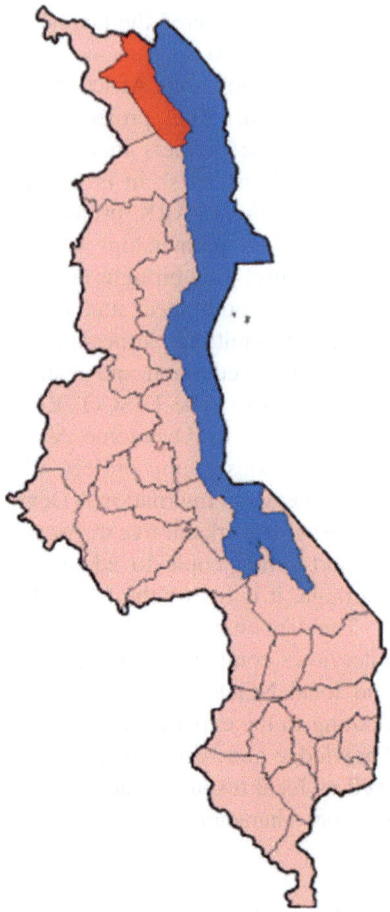

Fig. 2 Map of Malawi with Lake Malawi shaded in blue and the Karonga Lake District shaded in red. *Source* http://en.wikipedia.org

of research, Santam decided to actively support the Disaster Risk Management Department of the EDM. This relationship was perceived as important to Santam because of the potential to decrease their current and future exposure to disaster risk, as well as strengthen access to environmental risk data which could reduce their data access hurdles.

The municipality, while it had not explicitly looked for a partnership, urgently felt the need to reduce their disaster risks and vulnerability, as revealed by both their experience of extreme climate events and the research work conducted by the academic/research partners. Hence, when approached by Santam, the municipality was eager to partner with the insurance company because of their desire to strengthen disaster management capacity, and decrease lives lost and damage to property, in the region.

Through the partnership, Santam supported the municipality in identifying ways to strengthen disaster risk management. The partnership also resulted in concrete actions on the ground. Santam provided both guidance and physical resources to the municipality, thus improving the municipality's flood and fire risk early warning and response systems. The resources that the insurance company provided included fire vehicles, electronic boards (placed along roadsides) to increase risk awareness among the general public, and funding the installation of smoke alarms in informal settlements. The insurance company also helped with training and capacity building of municipal personnel, thus enabling a more coordinated disaster risk response. Santam also benefited from improved access to municipal data, such as flood line data, passed on to their risk modellers. Initial data suggests that fire risk in the area decreased due to the interventions, although this reflection was before the devastating May 2017 fires in the district.

In terms of the relationships between the three groups, the network is not a closed network.[3]Rather, the academic partners supported Santam in the earlier phases of collaboration, and Santam then supported the municipality in later phases. A direct link between the academic partners and the municipality, except in terms of data collection and information exchange, was not established. A further characteristic of the network that needs highlighting is that the partnership between Santam and Eden District has been very focused on the Disaster Management Department of the district municipality as the entry point. Individuals in other departments in the municipality have little knowledge of the partnership, and there is low representation of low-income communities at present.

Although there is no extensive knowledge about the Santam-Disaster Management partnerships within the municipality and with external actors, one of the innovations has been the mode of governance between the private sector and the public sector, and within the public sector itself. As one of the Santam employees said, "[as a company] *we support* [our partners], *we do not prescribe. Whatever solutions are implemented need to comply with the government, the legislated framework*". The respondent is emphasizing Santam's efforts to implement a partnership approach in which they recognize that any implementation actions need to be locally relevant and driven from within local government, rather than "imposing" their private-sector solutions. Further, the partnership has given Santam the awareness that while its primary focus is strengthening its collaboration with the municipality, there is an urgent need to reduce risk more broadly to ensure socio-economic development in the region, particularly for poor communities. Emerging from the initial relationship, a more formal community trust started within Santam to explicitly reduce risks for poor households who might not have insurance, but this is still in its early stages.

Regarding the public sector, the availability of funds from Santam led to the Disaster Management Department convening a cross-departmental group to consider and explore the type of responses that could holistically address fire and flood risk in the region. This meant that budgets were not tied directly to departments, which helped to reduce the "silo" mentality within the municipality. The Santam-Eden

[3]A closed network is one where all the possible links between partners have been realized.

Disaster Risk Management partnership therefore supported cross-silo governance, which is often a limitation in urban adaptation responses (e.g. Pasquini et al. 2013). As one of the Eden District officials from outside of the Disaster Management Department said, "*one of the main obstacles to collaboration is the desire to keep ownership over activities rather than jointly engaging. That is hindering progress. We need education and workshops to address this*". The respondent is highlighting the pressing need for local governments to move beyond working in isolated line functions. Cross-sectoral collaboration within local governments is fundamental to adaptation responses as adaptation is a cross-cutting function.

A key question we are asking of partnerships is how knowledge has been translated into action. In the case of Eden District, there was a strong focus on building a foundation of knowledge and understanding of disaster risk. Academic enquiry helped to establish understanding of both the physical systems (biophysical and infrastructure) and the social and institutional context (Nel et al. 2011a, b). This improved data availability for both Santam and the municipality and helped to provide more understanding on the institutional context of municipal governance with which Santam might engage. Knowledge transfer between Eden District municipality and Santam has continued to be a strong motivation for the partnership.

On reflection, resource transfers between the private sector and the public sector and a shared vision has built relationships and outputs on the ground. The support and funds that were committed were central to shifting how the public sector worked, which in turn fed directly into strengthening adaptation and risk reduction activities. The increase of trust between the partners, through these partnerships, has provided a strong base for current and future engagement, which both parties see as beneficial.

4 Karonga District

As above, establishing the origins of the network requires examining the needs of each partner. The U-ARK project needed case study material for understanding how cycles of risk accumulation could be broken in sub-Saharan cities, and required examples of cities at different scales, located in different contexts. Karonga constituted an ideal case study due to its smaller size and its location in a least developed country. In addition, and more pragmatically, relationships were already present with researchers at the closest local university to the city, Mzuzu University. The Karonga District Municipality, while it had not initiated the partnership, felt itself to be in dire need of assistance to reduce its disaster risks and vulnerability, and was thus open to the collaboration. The district's ethos is one of collaboration with as wide a range of partners as possible, in recognition of its high exposure and sensitivity to a range of risks, and its extremely low adaptive capacity. The district's ethos had already led it to collaborate with different civil society organizations, including the DISCOVER project. For the DISCOVER project, its aim is to bring about tangible and significant increases in the resilience of some of the most vulnerable communities in Malawi. Karonga is viewed as one of the areas in the country with the highest vulnerability.

Through their partnership, the U-ARK project has supported the municipality through an "action-research" approach, seeking to work with the local government structures (as well as other risk managers and local communities at risk) to identify workable actions for addressing risk as part of future urban development and infrastructure investments. For example, participatory mapping has been conducted to develop a cartography of the town to geo-reference patterns of urbanization and risk production. In close connection with local structures, particularly village Disaster Risk Management Committees, U-ARK delivers and communicates project learnings as well as strategic action planning and learning workshops. For the duration of the U-ARK project, a local research assistant has been established in Karonga to support the local government, local stakeholders, as well as visiting U-ARK researchers. Local community members have also been trained and employed as local research counterparts.

In terms of adaptation and resilience-building actions on the ground, the DISCOVER project engages in a broad range of activities. For instance, it aims to enhance the capacity of the local government and communities to prepare and respond to climate-related disasters. The project achieves this aim through working with local government to develop, review, and update disaster risk management plans and hazard vulnerability maps; training communities on disaster response; and establishing community early warning systems for flood and drought-prone areas. The project also focuses on the improvement and diversification of community livelihoods, focusing on crop and livestock production. It provides advice and training on alternative crops and supplies seed loans for crops that may not be available to the community, which beneficiaries later "pay forward" by providing seeds to other community members. Assistance with livestock production follows the same principle, with initial beneficiaries required to "give back" a number of animals from future litters, which are then donated to additional beneficiaries.[4]

In terms of the relationships between the three groups, this network also is not a closed network. Rather, the academic partners are supporting the municipality at the same time as civil society, with a direct link between the academic partners and civil society only having been established in terms of data collection and information exchange. A noticeable network characteristic is the relationship between the district municipalities and both the academic and the civil society partners, who have all been very focused on the village-level Disaster Risk Management Committees as a leverage point for collaborative efforts.

An important point to be drawn from the Karonga case is that despite the diversity of the adaptation responses that are being implemented, their common

[4]The project also trains farmers in animal health care. Other areas of focus are the establishment of village savings and loan groups, and assisting the communities to invest in natural resource-based enterprises, such as bee-keeping, and natural resource management, such as tree planting.

focus is building resilience at the community and household levels. This focus is likely reducible to the "closeness" of the impacts of climate change in the Karonga context. Malawi's economy depends on its agricultural sector, with more than half of the population estimated to live below the poverty line, and over 85% of it to depend on subsistence agriculture (The Heritage Foundation 2017). Livelihoods are thus closely tied to natural resources in Malawi. In this context, the importance of building resilience at the community and household levels becomes apparent, given the severe negative impacts that climate change poses to natural resources (e.g. Thomas and Twyman 2005).

In relation to these conditions of extreme vulnerability, another critical point that emerges from the Karonga case is the acknowledgement by the municipality that partnerships are highly beneficial to them, given their extreme resource constraints. For instance, one councillor, in considering the assistance that NGOs and other stakeholders provide in terms of adaptation and disaster risk reduction, stated "*Yes, especially these people, the Urban ARK,* [they are helpful] [...] *those people are helping. And the DISCOVER* [project] *too because they are teaching people about early warnings on climate change and floods and so forth*". The Karonga case constitutes a glaring example of the changing role of the state, with recent decades witnessing a shift in modes of governing from "government", as hierarchically structured, command-and-control-based forms of governing, to "governance", as governing carried out "in concert" with other sections of society (e.g. Rhodes 1996). Such a shift has been necessitated by the increasing global environmental and sociopolitical changes and challenges that are overwhelming the capacity of different sections of society—including government—to cope alone. Thus, resilience building will require collaboration between different sections of society to meet the scale of the challenges ahead. In Karonga, academic and civil society partners are moving in to "fill the gaps" left by the state's inability to act.

Some of the innovation in the Karonga District Municipality's mode of governance is to openly acknowledge the need for collaboration, and to welcome joint forms of governance. Instead of ignoring or opposing the efforts of academia and civil society, local government attempts to support them where it can, and to include them in its district council meetings. In so doing, strong relationships have been formed between local government and a range of other actors.[5]

Concerning the translation of knowledge into action in the Karonga case, the emphasis has been on (i) transferring knowledge about disaster risks and adaptation actions (particularly coping mechanisms) to the village disaster risk reduction committees and to the local communities, and (ii) directly enhancing household and community-level resilience through supporting small-scale coping and adaptation actions. This emphasis has been driven both by a recognition of households and communities as the risk units of greatest need, and by limited resource availability. Academic projects have, at present, relatively limited ability to implement

[5]Due to limited space in this chapter, we do not discuss all the collaborations that the local municipality has formed with different stakeholders.

on-the-ground actions, as these typically lie outside of the scope of donor funding that is granted to academic institutions and projects. In this context, the emphasis on engaged and applied "action-research" in the U-ARK project, as described in the section on the adaptation activities that are being undertaken in Karonga, is noteworthy. In the case of the DISCOVER project, a critical component of its ability to translate knowledge into actions on the ground was its ability to access international funding. This ability is due to the consortium nature of the project, which brings together international NGOs to work in a local context.

The above point underscores the importance of having connections to international networks. Small and isolated cities and towns lack links to international networks that bring in resources (of various kinds), and it is resource availability that often drives actions on the ground. In this context, the fact that the U-ARK project chose Karonga as a case study means that it is likely Karonga will have greater ability in the future to draw on and attract greater opportunities or funding to implement adaptation action, because it will have greater visibility (through websites, reports, and publications arising from the U-ARK project). We return to this point in the discussion.

5 Discussion

The Eden District and Karonga District cases both highlight how important networks have been in developing (i) an understanding of climate-related risks and (ii) responses that decrease risk. Emerging from the cases are a few key findings. The first is the importance of aligning needs and expectations when networks are established. If a network allows actors to meet their goals more easily than they would outside of the network, then it is more likely to keep partners aligned and thus engaged and committed (e.g. Ansell and Gash 2008; Fleishman 2009). This is akin to viewing collaboration as a "strategic production problem" (Hill and Lynn 2003: 66): why might an autonomous actor choose to allocate effort towards one or both of two outputs, where one output is the provision of an independent product or service, and the other is the provision of a collaborative product for which the actor's productivity is affected by the efforts of other product/service providers. Presumably, autonomous organizations choose to join or form a network that institutionalizes collaboration because they view potential net benefits from doing so.

The Eden case study illustrates the importance of clearly understanding and meeting the desired or needed return on investment of network partners. The partnership originated from the private sector who were driven by their need to ameliorate their increasing losses from insurance payouts associated with extreme weather events. As such, they saw potential benefits from collaborating with university partners to better understand the physical and social risk landscape, and from supporting the municipality to reduce disaster risks. The realization of these potential benefits is presumably maintaining the interest of the insurance company in the collaboration: for net benefits to be realized, an individual organization's share in the benefits of collaboration must justify any additional production and participation costs associated with its involvement (Hill and Lynn 2003: 66).

On the other hand, in Karonga the local NGOs appear to be the critical partners for driving risk reduction and resilience building in the district, and their organizational culture may explain their incentive to collaborate. Culture is a "system of shared values and norms that define appropriate attitudes and behaviours for organizational members" (O'Reilly and Chatman 1996: 160). Hill and Lynn (2003) note that not-for-profit organizations might not be driven by overriding material interests, but that they may rather organize themselves to reduce their resource constraints so that they can be driven and governed according to their socialized values. Thus, a second key finding we highlight is that understanding the different actor characteristics of members in a network is also crucial for network success, as the incentives of network members to participate will depend on their actor characteristics. For example, Santam's goals as a business entity revolve around increasing profits (through reducing insurance payout costs) and likely also enhancing organizational legitimacy through reputational benefits. In both the case studies, the municipalities' incentives to join the network are likely to consist of gaining resources and expertise for delivering "services" to society. In both contexts, though particularly so in Karonga, the municipalities face the unfunded mandate[6] of adaptation and resilience building. These motivations are in line with a worldwide trend that sees government at various levels increasingly drawing on the resources and expertise of public, private, and not-for-profit organizations to more effectively address complex social problems that sit in the "interorganizational domain" (Vangen et al. 2015: 1238). The Karonga case illustrates how areas of extreme vulnerability may be those with the best "antecedents" to the establishment of networks and relationships between stakeholders; to think of it in McGuire and Silvia's (2010) terms, the greater the magnitude and severity of problems faced in the external environment, the greater the level of collaboration between organizations.

Thus, one of the important considerations when forming a network, or when new actors join a network, should be clarity on what the "return on investment" is for different parties (depending on their actor characteristics), and how it can be met. In the case of adaptation, we note that this is likely to be particularly complex where it is hard to attribute adaptation outcomes to specific interventions. On the other hand, it is presumably precisely the recognition of the complex, holistic, and prevalently cross-sectoral nature of adaptation that drives autonomous actors to choose collaborative responses over the pursuit of independent responses. This situation is akin to that of the field of disaster management, where the recognition that disasters often exceed a single entity's ability and/or resources and/or jurisdictional boundaries has led to increasingly collaborative efforts in preparing and responding to disasters (McGuire and Silvia 2010).

[6]An unfunded mandate denotes a responsibility devolved to local government to perform certain actions, with no money provided for fulfilling the requirements.

In this context, the third finding we reflect upon is that universities may be particularly well-placed to be part of networks for reducing climate risks and building resilience, as their required return on investment is likely to be more easily met than for other types of actors, given that it is normally made up of knowledge generation. The theory of teams is concerned with situations in which individual actors have common interests (i.e. they share preferences), but they may have access to different information on which to base their decisions. Thus, their options for actions and strategies may differ (Marschak and Radner 1972). If lack of, or disparate, information precludes agents from pursuing their (common) interests, then a justification exists for the "ideal" role for university partners to synthesize and/or collect a "homogenous" information base, and distribute it among the network partners. This role appears ideal for universities given their general (current) inability to provide financial or other material rewards or resources to other partners.

The role of universities and their research also appears critical for how case study partnerships may develop. In Eden, the private sector engaged with researchers to better define the pathway they should follow for reducing risk within the municipality. Santam built on the understanding from the academic research to plan their engagement with the municipality, and then had the capacity to continue working directly with the municipality, drawing on their private resources. In Karonga, while there existed and exist strong partnerships between the municipality and local NGOs independent of university involvement, the role of universities appears important for the future development of risk reduction and resilience-building activities, not only through their ability to contribute knowledge that can inform current and future resilience-building efforts by partners on the ground, but also by putting Karonga "on the map" of wider donor and international networks and communities (see, e.g., VOA 2017). This appears important in the context of limited donor funding, as will be discussed further below.

A fourth area of reflection lies in thinking about the importance of trust building and the intricacies of the provision of incentives to build trust. One key network value that has arisen in the literature on networks and collaborations is trust (e.g. Bryson et al. 2006; Poocharoen and Ting 2015). The literature points to the need to build trust, especially in collaborations that are not mandated by government but which rely on the mobilization of partners' resources towards the achievement of common goals (e.g. Ansell and Gash 2008), as in the cases examined here. Trust can be built by sharing information and knowledge, demonstrating competency, good intentions, and follow through (Bryson et al. 2006). These trust-building actions should be initiated by the partner most desiring of the relationship. However, the provision of incentives (such as the provision of physical resources by Santam to the municipality), which built up trust between the partners, can be a double-edged sword. Critics of the provision of incentives have highlighted that providing extrinsic incentives may in some way "crowd out" intrinsic motivations that are important to producing a desired action or behaviour, particularly when an extrinsic incentive is provided and then removed (Gneezy et al. 2011). The literature suggests that "the effects of incentives depend on how they are designed, the form in which they are given (especially monetary or nonmonetary), how they interact with intrinsic motivations and social motivations, and what happens after they are withdrawn" (Gneezy et al. 2011: 206). Hence, incentives must be carefully planned.

"Small wins" or immediate outcomes are critical process outcomes essential for building the trust and momentum that can lead to successful collaboration (Ansell and Gash 2008), particularly in cases where potential network partners do not know each other, or if prior antagonism is high and/or long-term commitment to trust building is necessary. In the cases examined, it is important to highlight that "small wins" are likely to be particularly critical to the achievement of resilience-building efforts in the context of extreme poverty—such as in the Karonga District—because climate change adaptation appears to be a peripheral agenda, often placed in competition with other, more pressing socio-economic issues (e.g. Carter et al. 2015). In the Eden District case, it was probably extremely important for the success of the collaboration that Santam provided actual physical resources to the municipality (fire vehicles, electronic boards, smoke alarms), as these resources will have contributed to addressing the municipality's near-term resource scarcity problems, and therefore helped build a foundation of trust between the partners.

A fifth reflection concerns continuity of efforts and long-term sustainability issues. In this context, the imminent funding cycle end for the DISCOVER project in Karonga is worrisome and highlights the importance of two factors. First, the fact that the network in Karonga focused on village-level Disaster Risk Management Committees as leverage points is likely to have positive implications for the long-term sustainability of project outcomes: these Disaster Risk Management Committees are part of decentralized disaster risk management structures under the national disaster risk management policy and thus will continue beyond the end of funding cycles or collaborative ventures. Second, these funding issues underscore how important it is for the local municipality to have connections to international networks, which their connection to their partners has brought them (particularly the Urban ARK project, through bringing local stakeholders to international stakeholder meetings, and exposing the Karonga case through policy briefs, academic journal articles, webinars, etc.). In this context, we note that implementation of adaptation and resilience-building actions on the ground is likely to be greatly assisted by providing or facilitating the access of local governments (the level of government generally most suited to planning and implementing local, context-relevant adaptation strategies) to international networks, as well as local and international sources of knowledge. Municipalities in South Africa perform better on resilience-building initiatives when they are linked into sources of knowledge and external institutions of various kinds (Pasquini et al. 2015). This kind of approach is likely to be particularly relevant to small and relatively rural and/or isolated municipalities. Moreover, these municipalities are normally not targeted by, or able to access, global adaptation and resilience-building networks[7] (which are recognized as driving forces for the development of urban adaptation

[7]For example, C40 Cities Climate Leadership Group (http://c40.org) or the Global Covenant of Mayors for Climate and Energy (http://www.globalcovenantofmayors.org).

agendas, Carter et al. 2015). The Karonga case perfectly exemplifies this situation, prior to the arrival of the Urban ARK project.

There are a couple of additional points we would like to note from our case studies. First, the right players must be part of each specific adaptation network, which will depend on the context. However, at a minimum, all critical stakeholders ought to be included. While the Karonga network arguably revolves around the participation of civil society partners, in the Eden case the lack of involvement from civil society at this point is glaring (although there is a Santam community trust that aims to support poorer communities) and such involvement would be critical to ensuring that the real needs of poorer communities are heard. The second additional point is that local governments are characterized by the problem of institutional "silos" failing to institutionalize collaboration among individuals, departments, sectors, levels of government, and state and non-state actors, thus inhibiting integrated urban development approaches (e.g. Burch 2010; Ziervogel and Parnell 2012). Countering this tendency requires the explicit involvement of municipal actors from a range of departments in collaboration (see also Smedby and Neij 2013). The Eden case shows how an external partner managed to demand that a cross-departmental steering committee help manage the project, which "forced" cross-departmental collaboration to some extent. In keeping with the points made above, it is likely that the foundation of trust built through Santam providing "small wins" to the municipality (through the provision of physical resources to meet the municipality's near-term problems) enabled the company to request this more holistic resilience-building response from the municipality.

Universities appear particularly well-placed to conduct research on vulnerability, disaster risk, and potential adaptation responses. Yet, those working on the ground (including local government officials, private sector, and NGOs) seem better placed to decide on the most appropriate solutions and to implement them. One of the results in examining the case study networks is that these could have been improved by the universities "closing the loops" that were left open. In both sites, the academic partner had collaborative links with one or the other partner, but not with both—their links to one partner were limited to studying them and gaining information, but not collaborating. Examining the structure of the collaboration shows how different partners are or are not interconnected for the purpose of the collaboration. Structural interconnections are key in determining who has genuine influence on the direction of a collaboration, its priorities and its outcomes (Vangen et al. 2015). However, "closing the loops" would require the university partners (and academia in general) to reflect on their role in society and the extent to which they are willing to engage in networks for action on the ground—and the extent to which they deem this appropriate (Van De Ven and Johnson 2006). For instance, to what extent might universities wish to establish themselves as bridging (or boundary) organizations, i.e. organizations designed to facilitate collaboration and knowledge coproduction and exchange between municipalities and other organizations belonging to different communities, scales and policy areas (Vignola et al. 2013)?

6 Conclusion

In conclusion, this chapter has contributed empirical knowledge about the conditions by which networks can go beyond knowledge sharing to achieving on-the-ground adaptation actions. Through understanding the actor characteristics and the network's structure in two municipalities, the chapter has shown why and how networks can integrate knowledge from different partners into decision-making to result in adaptation actions on the ground. A few points emerge as important to consider in future research and practice on understanding and building networks to support climate change adaptation in practice. These include:

1. The importance of aligning needs and expectations when networks are established, and as part of this, understanding the different actor characteristics of the members in a network.
2. The partner who most desires the relationship should be clear on their reason for engaging and can benefit from offering the first incentive, though the provision of incentives requires careful planning. In this context, "small wins" or immediate outcomes are critical for building initial trust and momentum to sustain the collaboration.
3. Recognizing that universities can play an important role in helping understand the problem, context, and possible responses.
4. Recognizing that it is critical for networks to be able to connect local actors to national and international funding, given the need for continuity and long-term sustainability.

All networks are going to have different actors and different configurations. What is clear is that networks can and should be leveraged to plan and implement climate change adaptation. Understanding the nature of the actors and the potential structures that networks can take, can help to provide a better foundation for building successful adaptation in practice. For this reason, more research is needed on the role of networks for strengthening climate change adaptation.

Acknowledgements This work is based upon research supported by the National Research Foundation of South Africa (as part of the Fulcrum Institutions for Socio-Ecological Resilience Project of the Global Change, Society and Sustainability Research Programme), the Urban Africa: Risk Knowledge (Urban ARK) programme (funded by the UK Economic and Social Research Council and Department for International Development under Grant No. ES/L008777/1), and a project supported by the Network for Business Sustainability (South Africa), based at the University of Cape Town Graduate School of Business, titled "Companies partnering with municipalities for local development and resilience: Motives, mechanisms, and outcomes".

References

Agranoff, R. (2007). *Managing within networks: Adding value to public organizations*. Washington DC: Georgetown University Press.

Ansell, C., & Gash, A. (2008). Collaborative governance in theory and practice. *Journal of Public Administration Research and Theory, 18*(4), 543–571. Elsevier Ltd. Available at: http://www.sciencedirect.com/science/article/pii/S0301479713001667.

Baatjies, W., & Hendrickse, R. (2015). Easing the water shortage: The case of George local municipality. *Journal of Emerging Trends in Economics and Management Sciences (JETEMS), 6*(7), 190–198. Available at: http://jetems.scholarlinkresearch.com/articles/Easing%20the%20Water%20Shortage.pdf.

Bryson, J. M., Crosby, B. C., & Stone, M. M. (2006). The design and implementation of cross-sector collaborations: Propositions from the literature. *Public Administration Review, 66*(s1), 44–55. Available at: http://www.jstor.org/stable/4096569?seq=1#page_scan_tab_contents.

Burch, S. (2010). In pursuit of resilient, low carbon communities: An examination of barriers to action in three Canadian cities. *Energy Policy, 38*(12), 7575–7585. Elsevier Available at: http://www.sciencedirect.com/science/article/pii/S0301-4215(09)00475-3.

Carter, J. G., Cavan, G., Connelly, A., Guy, S., Handley, J., & Kazmierczak, A. (2015). Climate change and the city: Building capacity for urban adaptation. *Progress in Planning, 95*, 1–66. Elsevier. Available at: http://www.research.lancs.ac.uk/portal/en/publications/climate-change-and-the-city(d3c610f7-34b4-4b58-a84b-207155a6f0a1).html.

Ferlie, E., Fitzgerald, L., McGivern, G., Dopson, S., & Exworthy, M. (2010). Networks in health care: A comparative study of their management, impact and performance. *NIHR-SDO programme*. Available at: http://www.netscc.ac.uk/hsdr/files/project/SDO_FR_08-1518-102_V01.pdf.

Fleishman, R. (2009). To participate or not to participate? Incentives and obstacles to collaboration. In R. O'Leary & L. B. Bingham (Eds.), *The collaborative public manager* (pp. 31–52). Washington, DC: Georgetown University Press.

Gneezy, U., Meier, S., & Rey-Biel, P. (2011). When and why incentives (don't) work to modify behavior. *The Journal of Economic Perspectives, 25*(4), 191–209. American Economic Association. Available at: http://www.jstor.org/stable/41337236.

Hill, C., & Lynn, L. (2003). Producing human services Why do agencies collaborate? *Public Management Review, 5*(1), 63–81. Sage Publications. Available at: http://journals.sagepub.com/doi/pdf/10.1177/0899764008325247.

Howes, M., & Dedekorkut-Howes, A. (2012). Climate adaptation and the Australian system of government: The Gold Coast example. *Environmental Policy Failure,* 116–130. Springer. Accessed at: https://books.google.co.za/books?id=qkAUDgAAQBAJ&dq=Howes,+M.+and+DedekorkutHowes,+A.,+2012.+Climate+adaptation+and+the+Australian+system+of+government:+The+Gold+Coast+example.+Environmental+Policy+Failure,+pp.116-130.&source=gbs_navlinks_s.

Howes, M., Tangney, P., Reis, K., Grant-Smith, D., Heazle, M., Bosomworth, K., & Burton, P. (2015). Towards networked governance: Improving interagency communication and collaboration for disaster risk management and climate change adaptation in Australia. *Journal of Environmental Planning and Management, 58*(5), 757–776. Taylor and Francis. Available at: http://www.tandfonline.com/doi/full/10.1080/09640568.2014.891974.

Intergovernmental Panel on Climate Change. (2014). *Climate Change 2014–Impacts, Adaptation and Vulnerability: Regional Aspects*. Cambridge University Press. Available at: http://www.ipcc.ch/report/ar5/wg2/.

Kinnear, S., Patison, K., Mann, J., Malone, E., & Ross, V. (2013). *Network governance and climate change adaptation: collaborative responses to the Queensland floods*. Gold Coast, Australia: National Climate Change Adaptation Research Facility. Available at: http://apo.org.au/system/files/34161/apo-nid34161-61476.pdf.

Manda, M. Z. (2014). Where there is no local government: Addressing disaster risk reduction in a small town in Malawi. *Environment and Urbanization, 26*(2), 586–599. Sage Publications. http://journals.sagepub.com/doi/abs/10.1177/0956247814530949.

Marschak, J., & Radner, R. (1972). *Economic theory of teams, New Haven, Cowles Foundation Monographs*. New Haven and London: Yale University Press.

Matthews, S. (2015). Indigenous knowledge systems: What can we learn from Karoo farmers about coping with drought? *Water Wheel, 14*(4), 18–21. The Water Wheel. Accessed at: http://www.wrc.org.za/Lists/Knowledge%20Hub%20Items/Attachments/11276/WWJuly2015_climate.pdf.

McGuire, M., & Silvia, C. (2010). The effect of problem severity, managerial and organizational capacity, and agency structure on intergovernmental collaboration: Evidence from local emergency management. *Public Administration Review, 70*(2), 279–288. Wiley Blackwell. Available at: https://doi.org/10.4135/9781446200506.

Moloney, S., & Fünfgeld, H. (2015). Emergent processes of adaptive capacity building: Local government climate change alliances and networks in Melbourne. *Urban Climate, 14*, 30–40. Elsevier. Available at: https://doi.org/10.1016/j.uclim.2015.06.009.

Nel, J., Archibald, S., Le Maitre, D., Forsyth, G., & Theron, A. (2011a). *Understanding the implications of global change for the insurance industry: The Eden case study*. Stellenbosch, South Africa: CSIR Report. CSIR.

Nel, D., Nel, J., Reyers, B., le Maitre, D., Forsyth, G., & Theron, A., et al. (2011b). *Insurance in a changing risk landscape: Local lessons from the Southern Cape of South Africa*. Geneva: UNEP FI. Accessed at: http://www.unepfi.org/fileadmin/documents/insurance_changing_risk_landscape.pdf.

O'Leary, R., & Vij, N. (2012). Collaborative public management: Where have we been and where are we going? *The American Review of Public Administration, 42*(5), 507–522. Sage Publications. Available at: http://journals.sagepub.com/home/arp.

O'Reilly, C. A., & Chatman, J. A. (1996). Culture as social control: Corporations, cults, and commitment. *Research in Organizational Behavior, 18*, 157–200. JAI Press. Available at: http://faculty.haas.berkeley.edu/chatman/papers/30_cultureassocialcontrol.pdf.

Pasquini, L., Cowling, R. M., & Ziervogel, G. (2013). Facing the heat: barriers to mainstreaming climate change adaptation in local government in the Western Cape Province, South Africa. *Habitat International, 40*, 225–232. Elsevier. Available at: http://ac.els-cdn.com/S0197397513000477/1-s2.0-S0197397513000477-main.pdf?_tid=039a0b42-7787-11e7-a354-00000aacb362&acdnat=1501680922_f851799ff4e046ff48a3a4b2a88fb1de.

Pasquini, L., Ziervogel, G., Cowling, R. M., & Shearing, C. (2015). What enables local governments to mainstream climate change adaptation? Lessons learned from two municipal case studies in the Western Cape, South Africa. *Climate and Development, 7*(1), 60–70. Taylor and Francis. Available at: http://www.tandfonline.com/doi/full/10.1080/17565529.2014.886994.

Poocharoen, O. O., & Ting, B. (2015). Collaboration, co-production, networks: Convergence of theories. *Public Management Review, 17*(4), 587–614. Taylor and Francis. Available at: http://www.tandfonline.com/doi/abs/10.1080/14719037.2013.866479.

Provan, K. G., & Kenis, P. (2008). Modes of network governance: Structure, management, and effectiveness. *Journal of Public Administration Research and Theory, 18*(2), 229–252. Oxford University Press. Available at: https://academic.oup.com/jpart/article/18/2/229/935895/Modes-of-NetworkGovernance-Structure-Management.

Reyers, B., Nel, J. L., O'Farrell, P. J., Sitas, N., & Nel, D. C. (2015). Navigating complexity through knowledge coproduction: Mainstreaming ecosystem services into disaster risk reduction. *Proceedings of the National Academy of Sciences, 112*(24), 7362–7368. PNAS. Available at: http://www.pnas.org/content/112/24/7362.long?tab=author-info.

Rhodes, R. A. W. (1996). The new governance: Governing without government. *Political Studies, 44*(4), 652–667. Blackwell Publishing. Available at: http://onlinelibrary.wiley.com/doi/10.1111/j.1467-9248.1996.tb01747.x/abstract.

Silvia, C., & McGuire, M. (2010). Leading public sector networks: An empirical examination of integrative leadership behaviors. *The Leadership Quarterly, 21*(2), 264–277. Elsevier. Available at: https://doi.org/10.1016/j.leaqua.2010.01.006.

Smedby, N., & Neij, L. (2013). Experiences in urban governance for sustainability: The constructive dialogue in Swedish municipalities. *Journal of Cleaner Production, 50*, 148–158. Elsevier. Available at: https://doi.org/10.1016/j.jclepro.2012.11.044.

The Heritage Foundation. (2017). *The 2017 index of economic freedom: Malawi.* Accessed at: http://www.heritage.org/index/country/malawi.

Thomas, D. S., & Twyman, C. (2005). Equity and justice in climate change adaptation amongst natural-resource-dependent societies. *Global Environmental Change, 15*(2), 115–124. Elsevier. Available at: https://doi.org/10.1016/j.gloenvcha.2004.10.001.

Van de Ven, A. H., & Johnson, P. E. (2006). Knowledge for theory and practice. *Academy of Management Review, 31*(4), 802–821. Academy of management. Available at: http://web.a.ebscohost.com/ehost/detail/detail?vid=0&sid=080557b1-b091-47f2-9e48-e661d11c33b8%40sessionmgr4006&bdata=JnNpdGU9ZWhvc3QtbGl2ZQ%3d%3d#AN=22527385&db=buh.

Vangen, S., Hayes, J. P., Cornforth, C. (2015). Governing cross-sector, inter-organizational collaborations. *Public Management Review, 17*(9), 1237–1260. Taylor and Francis. Available at: http://www.tandfonline.com/doi/full/10.1080/14719037.2014.903658.

Vignola, R., McDaniels, T. L., & Scholz, R. W. (2013). Governance structures for ecosystem-based adaptation: Using policy-network analysis to identify key organizations for bridging information across scales and policy areas. *Environmental Science & Policy, 31*, 71–84. Elsevier. Available at: https://doi.org/10.1016/j.envsci.2013.03.004.

VOA News. (2017). *Report: Top Threat to African Cities is Lack of Good Government.* Accessed at: https://www.voanews.com/a/report-top-threat-to-african-cities-lack-of-good=government/3912657.html.

Weiss, K., Hamann, M., Kinney, M., & Marsh, H. (2012). Knowledge exchange and policy influence in a marine resource governance network. *Global Environmental Change, 22*(1), 178–188. Elsevier. Available at: https://doi.org/10.1016/j.gloenvcha.2011.09.007.

Ziervogel, G., & Parnell, S. (2012). South African coastal cities: Governance responses to climate change adaptation. In A. Cartwright, S. Parnell, G. Oelofse, & S. Ward (Eds.), *Climate change at the city scale: Impacts, mitigation and adaptation in Cape Town* (pp. 223–243). London: Routledge.

Climate Resilience Planning and Organizational Learning on Campuses and Beyond: A Comparative Study of Three Higher Education Institutions

Camille Washington-Ottombre, Sarah E. Brylinsky, Dennis B. Carlberg and Dano Weisbord

Abstract Higher education institutions around the world are confronting challenges associated with climate change, and many are trying to model practices that support climate change mitigation and resilience. This work initiates an inventory of good practices and introduces a theoretical framework for the assessment of resilience planning in order to guide and support institutions of higher education in their efforts to plan for resilience. We analyzed the resilience planning initiatives of three higher education institutions. Smith College, a small liberal arts college, conducted participatory research to help plan for climate change. Large, research-focused Cornell University nurtured collaborations and examined its own capacity for climate action. Boston University played an important role in the efforts of the Boston Green Ribbon Commission to plan for climate resiliency in the city of Boston. We analyzed the climate change mitigation and resilience planning processes of these three institutions using a theoretical framework of organizational change, resilience thinking, and adaptive co-management. Our analysis showed that, although each institution approached the process differently, all three gained significant benefits beyond the scope of climate action. Each experienced a meaningful opportunity to engage with its community and explore the values of

C. Washington-Ottombre (✉)
Environmental Science and Policy Program, Smith College, SR 334, 44 College Lane, Northamtpon, MA 01060, USA
e-mail: cwashingtonottombre@smith.edu

S. E. Brylinsky
Cornell University, Humphreys Service Bldg, Room 129, Ithaca, NY 14853, USA
e-mail: sbrylinsky@cornell.edu

D. B. Carlberg
Boston University, 120 Ashford St., Boston, MA 02215, USA
e-mail: carlberg@bu.edu

D. Weisbord
Smith College, Wright Hall, Northampton, MA 01060, USA
e-mail: dweisbor@smith.edu

© Springer International Publishing AG, part of Springer Nature 2019
W. Leal Filho and R. Leal-Arcas (eds.), *University Initiatives in Climate Change Mitigation and Adaptation*, https://doi.org/10.1007/978-3-319-89590-1_5

stakeholders, and each fostered local social capital and action while building a strong consensus on directions for the future. Aspects of the process supported the institutions' missions of education, research, and outreach and provided opportunities to pause and think deeply about how the institution chooses to engage with the world. All three organizations learned valuable lessons and experienced transformation, but Boston University in particular developed new ways to promote institutional learning on the topic of climate change, a process we describe as triple-loop learning.

Keywords Climate change · Resilience · Organizational learning
Higher education institutions · Campus sustainability

1 Introduction

Higher education institutions (HEIs) around the world are confronting the challenge of climate change and trying to model climate change mitigation and resilience (Lozano et al. 2015). Well-established guidelines and organizations are in place to support and incentivize HEIs' mitigation efforts. In the US, following the American College and University Presidents' Climate Commitment in 2006, international declarations and frameworks have provided structure and guidance for the expansion of mitigation initiatives (Dyer and Dyer 2017; Lozano et al. 2015). HEIs are supported in their efforts by sustainability networks and assessments that provide mitigation metrics and benchmarks (Disterheft et al. 2013; Arroyo 2017; Urbanski and Leal Filho 2015; Washington-Ottombre and Bigalke 2017). Green university rankings drive consumer demand and encourage competition among HEIs to develop ambitious sustainability measures and "decarbonize" (Lauder 2015).

However, even though HEIs are experiencing and will continue to experience the impacts of climate change, a very limited number of guidelines and organizations are in place to support and enhance resilience to climate change (Second Nature 2017). Given diverse and complex impacts of climate change and institution-tailored responses to be formulated by HEIs, it would be counterproductive to formulate "one-size fit all" blueprints for resilience planning, or "panaceas" as demonstrated by Ostrom (2007). To guide and support the resilience planning efforts of HEIs, this work aims at initiating an inventory of climate actions conducted by HEIs and introduces a theoretical framework to assess resilience planning efforts. Resilience can be defined as "the capacity of a system to absorb disturbance and reorganize so as to retain essentially the same function, structure, and feedbacks—to have the same identity" (Walker and Salt 2012). In this case, the "system" is the HEI and the "disturbance" is climate change. HEIs are likely to face an array of threats to their research and educational missions, buildings, operations, campus landscape, transportation system, and procurement as a result of climate change. Therefore, HEIs need to plan for and manage the effects of climate change and invent solutions to a number of strategic, technical, financial, and organizational problems related to climate change.

In this chapter, we explore a theoretical framework of organizational change and resilience planning and management and use it to analyze the planning approaches of three American HEIs with distinct profiles. We conclude by assessing how HEIs can develop as learning organizations through the process of planning for, managing, and mitigating the effects of climate change.

2 Literature Review and Methods

To effectively respond to and mitigate the effects of climate change, HEIs need to transform into learning organizations. An effective approach to climate change for an HEI involves questioning its current operational model. As do other organizations, HEIs need to learn to transition away from fossil fuel and manage the impacts of climate change. Management science has identified three types of organizational learning: *single-*, *double-*, and *triple-loop learning* (Pawlowsky 2001). Single-loop learning is when an organization adjusts its behavior in order to achieve a desired outcome. The resulting change leads to an improvement in how things are done, but the organization does not question its framework of action. For instance, an HEI could engage in single-loop learning when it learns how to achieve a previously unreached recycling objective. Double-loop learning refers to a deeper transformation in which an organization examines the process that led to a problem. This can reframe the organization's view of the problem and affect its decisions about what actions to take. Organizations engaged in double-loop learning are able to transform mental maps, ways to understand and interact with a particular issue, to generate new meanings and actions. For instance, an HEI would demonstrate double-loop learning if it were to shift its focus toward producing less waste rather than solely focusing on recycling. Finally, triple-loop learning means developing the capacity to learn how to learn. It is when organizations learn to invent new processes for generating mental maps. For instance, an organization engaged in triple-loop learning might learn how to track the flow of materials on campus and periodically assess and revise its procurement policy, landscape management, and waste management in order to minimize the production of waste.

To become resilient to climate change, it is crucial that HEIs engage in collective reflection with the goal of achieving triple-loop learning. Climate change is a "wicked" problem that involves high levels of complexity, uncertainty, interdependencies with other sectors and systems, and value-driven analyses and prescriptions (Rittel and Webber 1973). As a "wicked" problem, climate change cannot be "solved," but rather must be continually dealt with and addressed from multiple angles (Grundmann 2016). Therefore, HEIs need to use triple-loop learning in order to address the constantly changing challenges posed by climate change.

HEIs can apply triple-loop learning to address the complex problems associated with climate change by utilizing an *adaptive co-management* approach. Adaptive co-management is an iterative and flexible management approach that focuses on learning by doing and building a synthesis of various scientific and nonscientific knowledge and value systems (Armitage et al. 2009). Within the adaptive co-management framework, collaboration, trust, and power-sharing across various levels of governance are needed to manage complex systems in an adaptive manner. Although there is no single blueprint for implementing an adaptive co-management model to respond to climate change, the following framework describes the process (Fig. 1). First, HEIs need to conduct a *resilience assessment*, in which researchers and community members address two fundamental questions: "resilience to what?" and "resilience of what?" (Walker and Salt 2012). In the context of climate change, the question "resilience to what?" refers to the impacts of climate change. To answer it, HEIs ought to conduct a vulnerability assessment based on local climate projections. Answering the second question "resilience of what?" requires coming to an agreement on the identity of the systems in question. The process of answering these questions can benefit from the deep and inclusive dynamics that occur during the participatory planning exercise described by Forester (1999). Building and reinforcing networks and collaborations are key to achieving resilience in the face of climate change. Networks promote resilience because they support social memory, adaptive capacity and innovations (Folke et al. 2002). After having planned for resilience and built collaborative networks, HEIs can maintain resilience by continuing the iterative questioning process of the adaptive co-management framework.

In this work, we apply the adaptive co-management framework to efforts conducted by three HEIs for developing resilience to climate change. While this paper aims to begin collecting resilience planning experiences, it also intends to

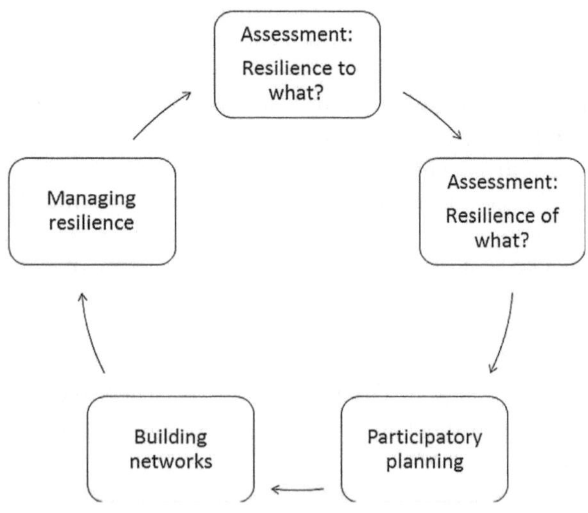

Fig. 1 Adaptive co-management framework for developing resilience to climate change (adapted from Nyberg 1999)

demonstrate that in these cases the adaptive co-management process resulted in organizational learning. Smith College, Cornell University, and Boston University were selected because on the one hand they are all actively engaged in a resilience planning process and on the other they each exhibit distinct types of responses to climate change. Behind these differences are two key factors. First these institutions are projected to experience notably different climate impacts on account of their locations. Second, each of the three HEIs is likely to formulate distinct responses to climate impacts in accordance with their specific institutional profile, mission, and size. For each HEI, we directly participated in resilience planning events and analyzed written planning material, applying the adaptive co-management framework. We also conducted interviews of key campus sustainability actors in each HEI to assess the extent of organizational learning that each experienced had experienced at the time of our study. Given the fact that resilience planning is an ongoing, iterative process, and must be ultimately examined over time, this short-term analysis cannot fully assess long-term organization learning. However, this research provides a close-up snapshot of three HEIs and a foundation for future analysis.

3 Case Study 1: Assessing and Planning for Resilience at Smith College

Smith College (SC) is a small, private, liberal arts college in Western Massachusetts attended by approximately 2500 undergraduate students. This four-year women's college has a long tradition of promoting social change (Chase 1992; Van Dyne 2016). Over the years, SC has developed a strong commitment to sustainability. In 2010, the college drafted a plan to achieve carbon neutrality. From 2010 to 2016, the college's greenhouse gas (GHG) emissions dropped 20% since 1990 levels even though the built environment has continued to grow (Smith College Office of Campus Sustainability 2015). In 2015, the college president and board of trustees formed the Study Group on Climate Change to assess the college's response to climate change. The Study Group on Climate Change conducted community engagement activities, workshops and research, and worked with consultants throughout their mission. Part of their effort to encourage reflection included a student-led participatory planning workshop to assess the priorities of the SC community in regard to climate change.

This workshop which was conducted as a research project for a class was designed to assess the SC community's perceptions of vulnerability and resilience to climate change. According to projections, New England will experience warmer temperatures throughout the year and more erratic precipitation patterns characterized by periods of drought and extreme rainfall or snowfall events (Horton et al. 2014). To assess the college's vulnerability to these changes, students conducted an expert-based spatially explicit vulnerability assessment using ArcGIS mapping software (ESRI 2011). This analysis revealed high vulnerability of residential areas and the campus landscape to increased precipitation and temperature. In order to capture the community's perception of vulnerability, students also conducted a geographic information system-based multi-criteria assessment of vulnerability to climate change on campus. This is a participatory technique that combines geographical data and value judgments (Malczewski 2006). The outcome of this multi-criteria vulnerability assessment differed significantly from that of the expert-based vulnerability assessment in that it highlighted the athletic facilities as being particularly vulnerable to climate change rather than the campus landscape and student housing (Bank et al. 2016).

The student-led participatory planning workshop built on the vulnerability assessment to answer the question "resilience of what?" and pinpoint the focal areas for managing resilience (Bank et al. 2016). Diverse stakeholders such as members of the board of trustees, senior and midlevel staff, faculty, students, and alumnae participated in the workshop. During this two-day workshop, students first conducted a World Café, a participatory large-group dialogue that encompassed the exploration of stakeholders' values, knowledge generation, and information exchange through conversations with diverse stakeholders (Brown and Isaacs 2005). On the second day, they built on this foundation to discuss policy options that SC could develop to respond to climate change. During the workshop, the SC community expressed that SC should be a champion of climate action as it pursues its mitigation and carbon neutrality efforts. Stakeholders unanimously agreed that the college should develop responses that align with its values and missions. Workshop participants felt that future planning efforts should continue to regularly engage the SC community and involve forums for diverse people to debate and exchange ideas on these crucial issues. Finally, the participants emphasized the importance of adopting a trans-sectional approach to climate change that would take into account social justice considerations. The final climate action plan of the Study Group on Climate Change incorporated numerous findings highlighted in the student report of the participatory climate resilience planning workshops (SGCC 2017).

In addition to the concrete impacts of the Study Group on Climate Change report, the participatory workshops contributed to deep cultural transformations within SC. First, students conducted important real-world research using the campus as a living laboratory, which is in alignment with SC's new strategic plan (Smith College 2017). Second, for the first time at SC, this exercise engaged diverse stakeholders in a planning process where values were clarified and made transparent. Those clarifications led to a more efficient planning process with established community support. The engagement and support of diverse stakeholders raised the level of commitment of the SC community to climate resilience and raised the profile of the college as a leader in resilience planning. Finally, the student-led workshops shed light on new possible directions in which to take decisions. Compared to traditional expert-based and bureaucratic decision-making, this exercise allowed SC to include research, education, and the values of diverse stakeholders in the resilience planning process.

4 Case Study 2: Building Networks for Climate Resilience at Cornell University

Cornell University (CU) is a privately endowed research university with a primary campus in Ithaca, a small city in upstate New York. Seven undergraduate units and four graduate units support 14,300 undergraduate and 5200 graduate students. CU is known for agricultural science and is a leader among engineering and architecture schools. As the federal land-grant institution in New York State, the university has a responsibility to make contributions in all fields of knowledge in a manner that prioritizes public engagement to help improve the quality of life in the state, nation, and world. CU was among the first fifty HEIs to commit to carbon neutrality and sustainability through the American College and University Presidents' Climate Commitment in 2007 and was the first Ivy League institution to do so. CU initially sets a goal of achieving carbon neutrality by 2050 and selected a new target date of 2035 in 2014. The university is on track to reach this goal: as of 2017, it has achieved a 35% reduction of GHG emissions from 2001 baseline levels and a 55% reduction from 1990 levels (Cornell University, 2017). In 2014, CU was the first research university to formally incorporate climate adaptation into an existing climate mitigation action plan. Faculty, students, and staff have engaged in learning and research to serve public climate change resilience needs through several initiatives, such as the Tompkins County Energy Roadmap (TCER), a regional blueprint for reducing community-wide greenhouse gas emissions by 80% before 2050, and CU's goal of carbon neutrality by 2035.

Prior to its involvement in the TCER, CU had been a significant stakeholder in the City of Ithaca, the Town of Ithaca, Tompkins County, and the six-county Southern Tier region's sustainability planning activities. CU staff and students are currently engaged with the City of Ithaca's Energy Action Plan, the Climate Smart–Climate Ready conference, the Town of Ithaca Energy Action Plan, the Tompkins County Planning Department, the Tompkins County Hazard Mitigation Plan, the Tompkins County Climate Protection Initiative, and the Cleaner Greener Southern Tier initiative. CU was a founder of the Tompkins County Climate Protection Initiative in 2008, an award-winning climate action and clean energy coalition in Ithaca, NY, which established a consortium of local higher education institutions, businesses, private service centers, and the Town of Ithaca and the City of Ithaca to set common goals and collaborate for climate progress. Although CU has extensive experience leading and taking part in climate change-related collaborations, the TCER planning process revealed new network dynamics and facilitated new collaborations.

During its participation in the TCER, CU engaged in deep reflection about its role relative to other stakeholders and its mission as a federal land-grant institution. In order to effectively engage with community partners, CU had to access its level of capacity and define an appropriate level of engagement. The university used a capacity assessment framework (Table 1) that defined three nexuses of scales of action: level of assessment, level of commitment, and level of capacity (Brylinsky 2017).

Table 1 Capacity assessment framework for campus–community partnerships

Level of assessment			Level of commitment			Level of capacity		
Level of desired outcome expected by partner, with consideration for the data quality and capacity for identifying necessary information to support scientifically accurate findings			Level of collaboration expected of the partner and the type of commitment that would best serve that partner's existing mission, vision, and goals			Partner capacity to contribute to the organizational coordination of ongoing partnership work, outputs, and project management		
Campus	Regional	System	Informal	Semiformal	Formal	Participant	Guide	Leader
Climate action relative to campus infrastructure and people	Climate action that includes the campus and the city/ town	Vulnerability of all shared systems to climate change; bioregions and networks	Sharing information Sharing process	Common goals Common ideas of success	Shared outcomes Public Official	Shows up at meetings, participates in dialogue, attends workshops or trainings	Creates some data based on need; reviews commitments on as needed basis; assists others	Creates formal commitment with time-bound goals; organizes meetings; hosts

CU identified its level of assessment as regional, and level of organizational commitment as formal, building on existing commitments outlined above. The level of capacity of CU was participant; in this instance, it was most appropriate for county legislators to lead the process of research and plan development while CU played a strong role by providing technical experts. As a result of this assessment, faculty and students worked with the Tompkins County Planning Department and the steering committee to develop the TCER. The TCER was unanimously accepted by the County Legislature on Tuesday, April 19, 2016.

Hence, the TCER was almost as much about network and collaboration building for future climate action as it was about greenhouse gas mitigation. CU played a central role in fostering collaboration in the face of conflicting values among stakeholders. For instance, although all stakeholders shared a common vision of transition away from fossil fuels, many supported the use of cheap natural gas as a bridge fuel, whereas others (including CU) believed strongly in the need to transition to fossil-free energy. However, stakeholders were able to overcome their differences and reap benefits from this collaboration beyond this particular planning exercise. CU's experience was in line with the culture of a land-grant institution, in which students, faculty, staff, and senior leadership have a responsibility to provide leadership on climate and sustainability in the service of the public good. Beyond the traditional top-down model of extension services characteristic of land-grant universities, participation in the TCER defined a new model of interaction for CU. The planning exercise promoted CU's leadership role and gave it greater insight into the complexity of local community values and the possibility of collaboration in spite of disagreements. Stakeholders' demand for exemplary behavior from the HEI helped drive the decision to change the carbon neutrality target date from 2050 to 2035. Hence, this planning exercise created new positive feedback loops between CU and regional stakeholders that drove climate action within CU and beyond.

5 Case Study 3: Collaborative Climate Action: Boston University and the Green Ribbon Commission

Boston University (BU) is a large, internationally recognized urban research university with approximately 33,000 undergraduate and graduate students and almost 10,000 faculties and staff distributed across two campuses, 17 schools and colleges and 133 acres in the city of Boston. BU is a residential university in which 75% of undergraduates live on campus. In addition to the university's strong focus on research, BU emphasizes the liberal arts and professional education, with a tradition of practical engagement and service. While BU is reducing its impact on the climate (it has already met its 2020 greenhouse gas reduction goal of 25%), it is also applying its operational and research expertise and resources to prepare its campuses for the effects of climate change.

Research suggests that the area where BU is located, downtown Boston, may become flooded as a result of sea-level rise (Kirshen et al. 2008). In 2014, BU developed a vulnerability assessment for its two campuses and completed a Climate Ready BU report to enhance climate resilience on campus. Using ArcGIS software, students modeled flooding as a result of climate change on campus and in the rest of the city of Boston. Although previous flood modeling had been conducted, this new vulnerability assessment explored specific pathways of water movement and yielded interesting new results. Although it had been commonly believed that the Charles River Campus was protected by the Charles River Dam, located 6.8 feet above mean higher high water, the analysis showed that sea-level rise would cause water to begin to flood the city through the Fort Point Channel, over the seawall 4.2 feet above mean higher high water.

Beyond its campus-based planning efforts, the BU is actively engaged in the Green Ribbon Commission (GRC), a Boston network working toward developing shared solutions to make Boston carbon neutral by 2050 and to prepare the city for the effects of climate change. The GRC is a group of business, institutions, and civic leaders in Boston working to develop shared strategies for fighting climate change in coordination with the city's Climate Action Plan. It is an all-sector, structured leadership network that, while growing out of a municipal initiative, is a group of members and staff affiliated with public, private, and nonprofit organizations in and around Boston. GRC members are at the level of CEO and president, and BU's president is an active member of the Commission representing BU. The GRC is organized into working groups for each of the major sectors in Boston: health care, higher education, and commercial real estate. The higher education sector is represented by BU, Harvard University, Massachusetts Institute of Technology, Northeastern University, and University of Massachusetts Boston.

Within the GRC, the role of HEIs is to provide leadership and information sharing in many areas of expertise such as energy efficiency strategies for laboratories, renewable energy guidelines and purchasing networks, climate modeling, vulnerability assessments and financial and energy modeling. For example, the GRC has worked with the University of Massachusetts Boston to coordinate the Climate Ready Boston report and the Institute for Sustainable Energy at BU for modeling and coordination of Carbon Free Boston, studying the effect of policy on emission reductions needed for Boston to be net carbon neutral by 2050. BU has been a convener for many Higher Education Working Groups and several cross-sector workshops, with the goal of building strong, action-oriented networks to advance sustainability in the city and region. Subject areas have included climate resilience, large-scale renewable energy, and bankable investments in energy efficiency.

The GRC acted as a conveyor of senior leadership at the city level that paved the path for climate resilience in Boston. The GRC's Climate Preparedness Working Group implemented a three-phase planning program to develop Climate Ready Boston. Phase 1 created the Boston Research Advisory Group, which was made up of researchers from HEIs in the city and state to establish a regional climate consensus from which to build a plan. This effort was managed by the University of

Massachusetts Boston, with major contributions from nine HEIs including BU. The contribution of BU included expertise on sea-level rise and coastal storms. Phase 2 developed a vulnerability assessment for sea-level rise, coastal storms, extreme precipitation, and extreme temperatures. Phase 3 is developing resilience strategies for the areas of highest vulnerability including the area of the Fort Point Channel identified in the Climate Ready BU report.

In addition to modeling climate resilience planning at the city level, participants reaped a number of collateral benefits from these collaborations and planning exercises. BU was able to pursue its mitigation efforts and learn from the practice of others, such as in the development of green laboratory practices. As part of the GRC, BU developed sustained research projects and integrated its research, teaching, and outreach missions. The GRC created a new forum that gave BU a voice in the development of the city. The frequent and sustained meetings of senior leaders in academia, government, and business around the topic of climate change that was facilitated by the GRC was unprecedented and facilitated the development of new processes through which the participants learned from each other and fostered sustainability in Boston.

6 Discussion

Analysis of these three case studies shows that the adaptive co-management framework for resilience planning is neither a blueprint nor a rigid framework. Although all three HEIs were productively engaging in climate resilience planning, they were at different stages of the adaptive co-management framework and navigated it differently (Table 2). SC was at the initial stages of resilience, assessment, and participatory planning. CU, on the other hand, was developing collaborative networks while planning for resilience. Finally, BU was building networks and managing resilience at the local and regional scale prior to finalizing its resilience plan. Each of the three HEIs seized different opportunities to work toward climate resilience. They also have responded to directives defined by their own leadership. This observation echoes past literature emphasizing the importance of leadership guidance and orientation in framing and supporting campus sustainability (Helferty and Clarke 2009).

In all three case studies, the HEIs were confronted with the challenge of dealing with diverse stakeholders with distinct values, which can be difficult given the complex nature of climate resilience planning. Acknowledging and addressing those differences were a crucial part of the planning process. SC's initial challenge was to craft a plan that would be accepted and supported by all stakeholders. Being a small campus, it was important to reach a high level of consensus in order to efficiently implement resilience plans. CU was also concerned with diverging values among stakeholders. For instance, some stakeholders believed that natural gas could act as a bridge fuel, whereas others, including CU itself, firmly believed in the need to move away from all fossil fuels. Hence, part of the challenge was to

Table 2 Comparison of three case studies

	Stages of climate resilience planning	Climate resilience as a challenge	Climate resilience as an opportunity	Institution's organizational learning
Smith College	• Assessing resilience • Participatory planning	• Crafting a plan supported by the community • Categorizing and prioritizing climate mitigation and adaptation measures	• Campus-based education and student research • New dialogue among diverse stakeholders • New decision-making processes on campus	• Single-loop: clearer path to carbon neutrality and greenhouse gas mitigation • Double-loop: new mental maps that place climate change, participatory decision-making processes and students at the core of planning process
Cornell University	• Participatory planning • Building networks	• Dealing with conflicting values among stakeholders	• Fulfill its mission as a land-grant university • Sustained research engagement • Alignment of research, teaching, and outreach • Helped accelerate the decision to move the carbon neutrality date from 2050 to 2035	• Single-loop: incentive to become carbon neutral at an earlier date • Double-loop: new perspectives on mission as a land-grant university
Boston University	• Building networks • Managing resilience	• Establishing common benchmarks and metrics	• Sustained research engagement • Alignment of research, teaching and outreach • Learned from the experience of other members	• Single-loop: new ways to achieve mitigation objectives • Triple-loop: new forum and ways to learn and make decisions

push the no–fossil fuel agenda forward while keeping all stakeholders involved, engaged, and feeling legitimate. Similarly, as part of the GRC, BU initially struggled with the other members to define common benchmarks and metrics, which ended up being one of the GRC's significant achievements.

Planning and managing for climate change resilience also created significant opportunities for the HEIs beyond resilience-related outcomes. In many ways, the planning process aligned with the HEIs' general missions and helped them reach several distinct objectives. At SC, the student-run participatory planning exercise allowed the organization to pursue its educational mission and to involve students in meaningful research using the campus as a living laboratory. During the participatory planning exercise, diverse groups of people, from members of the board of trustees, to senior staff, to faculty and students, engaged in deep and productive discussions. This exercise opened up the door for more participatory decision-making processes and for the involvement of students in conducting research on and participating in the governance of SC; this represents a significant cultural shift for the institution. At CU, the participatory planning and network building exercises supported CU's mission as a land-grant university. The planning process provided faculty and students with long-term, real-world research opportunities. SC's and CU's priorities of research, teaching, and outreach were productively aligned with the resilience planning process in a way that enhanced each of their missions. In addition, the planning process and the fact that stakeholders were enthusiastically looking at the HEIs for guidance encouraged the HEIs to have more ambitious carbon neutrality targets. At BU as well, participation in the GRC created long-term research opportunities as well as an alignment between research, teaching, and outreach. It gave the BU a voice in climate resilience planning in the Boston area. In addition, BU learned from other GRC members and improved upon its own sustainability practices.

Finally, all three HEIs experienced organizational learning as a result of their resilience planning processes. SC achieved both single- and double-loop learning. The planning process resulted in a clearer and more achievable path toward greenhouse gas mitigation and carbon neutrality, an example of single-loop learning. The process also led to the implementation of new mental maps related to making decisions in a participatory manner and including students in the decision-making process—a double-loop learning process. Similarly, discussions during the participatory planning process revealed a strong desire within the community to envision campus sustainability in conjunction with social justice issues, a new framework for campus sustainability at that institution. CU also achieved single- and double-loop learning. There, participatory planning exercises provided strong incentives to develop more ambitious carbon neutrality goals, which constitute a single-loop learning experience. The process also led to a double-loop learning experience and created an opportunity to transform traditional land-grant outreach and extension efforts from a top-down process to a circular and iterative top-down/bottom-up experience. Finally, BU experienced a true triple-loop learning experience as part of the GRC, in which new sustained collaborations at the leadership level and new ways to learn and make decisions allowed for more climate-related initiatives.

7 Conclusion

In order to guide and support the resilience planning efforts of HEIs, we began gathering and analyzing resilience planning efforts conducted by three HEIs with the intent to build an inventory of climate actions. In this chapter, we mobilized the adaptive co-management framework as well as insights from the literature on organizational learning and resilience thinking in order to analyze recent resilience planning efforts conducted by these institutions. Three main findings emerge from our study: First, our analysis shows that each HEI was at a different stage within the adaptive co-management framework. Due to this, they were facing distinct challenges and opportunities. However, at the heart of the planning process for all three were efforts to enhance communication and foster collaboration between diverse groups of stakeholders with conflicting values. SC was at the early stages of the adaptive co-management framework and was concentrating on resilience assessment and engagement with its diverse community through participatory planning activities. Having previously formalized its resilience plan, CU focused on its mission as a land-grant university and engaged with local institutions in building robust collaborations for managing resilience at the regional scale despite the presence of disagreements among stakeholders. Finally, BU worked through the GRC, a unique network of diverse institutions in the Boston area, to guide diverse stakeholders in setting commons goals and metrics. With this shared plan, the institution is now able to implement several initiatives that concretely manage resilience to climate change.

Second, regardless of the approach and stage of resilience planning at which each institution found itself, all three HEIs experienced significant benefits beyond the scope of climate change itself. All experienced a meaningful opportunity to engage with their community to make different values transparent. HEIs were able to build a strong consensus relative to directions for the future and foster local social capital and action. Planning for climate action was a time to pause and think deeply about how the HEI engages with the world. Resilience planning also provided an opportunity to pursue the HEI's missions of education, research, and outreach in an integrated manner. Finally, we found that during this process, the three organizations learned valuable lessons. Both SC and CU experienced single and double-loop learning and were able to reach many of their climate action objectives. Only BU, however, truly engaged in triple-loop learning on account of its innovative decision-making forum. In order to fully capture the evolution of organizational learning, future research should periodically analyze the path toward resilience of the three HEIs and assess if the long-term practice of resilience promotes triple-loop learning. Beyond these three cases, more resilience planning initiatives should be analyzed using the adaptive co-management framework in order to construct a library of good practices and to guide institutions in achieving resilience to climate change through organizational learning.

References

Armitage, D. R., et al. (2009). Adaptive co-management for social-ecological complexity. *Frontiers in Ecology and the Environment, 7*(2), 95–102.

Arroyo, P. (2017). A new taxonomy for examining the multi-role of campus sustainability assessments in organizational change. *Journal of Cleaner Production, 140*(Part 3), 1763–1774.

Bank, A., Bobrow, S., Bigalke, S., Davis, A., Edwards, H., Geverlber, & Hewitt, E., et al. (2016). *Report on the study group on climate change participatory workshop*. Retrieved July 25, 2017 from https://www.smith.edu/green/documents/SGCCworkshop_finalreport.pdf.

Brown, J., & Isaacs, D. (2005). *The world cafe*. San Francisco: Berrett-Koehler Publishers.

Brylinsky, S. (2017). *Assessing capacity for and climate change collaborations between the campus and community partners: An organizational needs assessment framework* (manuscript in preparation).

Chase, L. (1992). Imagining utopia: Landscape design at Smith College, 1871–1910. *The New England Quarterly, 65*(4), 560–586.

Cornell University. (2017). *Options report 2017*. Available at: http://www.sustainablecampus.cornell.edu/initiatives/options-for-achieving-a-carbon-neutral-campus-by-2035.

Disterheft, A., et al. (2013). Sustainability science and education for sustainable development in universities: A way for transition. In S. Caeiro et al. (Eds.), *Sustainability assessment tools in higher education institutions mapping trends and good practices around the word* (pp. 3–28). New York: Springer.

Dyer, G., & Dyer, M. (2017). Strategic leadership for sustainability by higher education: The American College & University President's Climate Commitment. *Journal of Cleaner Production, 140*(1), 111–116.

ESRI. (2011). *ArcGIS desktop: Release 10*. Redlands, CA: Environmental Systems Research Institute.

Folke, C., Colding, J., & Berkes, F. (2002). Building resilience for adaptive capacity in social-ecological systems. In F. Berkes, J. Colding, & C. Folke (Eds.), *Navigating social-ecological systems: Building resilience for complexity and change*. Cambridge, UK: Cambridge University Press.

Forester, J. (1999). *The deliberative practitioner, encouraging participatory planning processes*. Cambridge, MA: MIT Press. (313 p).

Grundmann, R. (2016). Climate change as a wicked social problem. *Nature Geoscience, 16*, 562–563.

Helferty, A., & Clarke, A. (2009). Student-led campus climate change initiatives in Canada. *International Journal of Sustainability in Higher Education, 10*(3), 287–300.

Horton, R., et al. (2014). *Climate change impacts in the United States: The Third National Climate Assessment* (J. M. Melillo, et al., Editors). U.S. Global Change Research Program.

Kirshen, P., et al. (2008). Climate change and coastal flooding in Metro Boston: Impacts and adaptation strategies. *Climatic Change, 90*(4), 453–473.

Lauder, A., Sari, R. F., Suwartha, N., & Tjahjono, G. (2015). Critical review of a global campus sustainability ranking: GreenMetric. *Journal of Cleaner Production, 180*(part A), 852–863.

Lozano, R., Ceulemans, K., Alonso-Almeida, M., Huisingh, D., Lozano, F. J., & Waas, T. (2015). A review of commitment and implementation of sustainable development in higher education: Results for a worldwide survey. *Journal of Cleaner Production, 108*(part A), 1–18.

Malczewski, J. (2006). GIS-based multicriteria decision analysis: A survey of the literature. *International Journal of Geographical Information Science, 20*(7), 703–726.

Nyberg, B. (1999). *An introductory guide to adaptive management: For project leaders and participants*. Victoria, British Columbia, Canada: British Columbia Forest Service.

Ostrom, E. (2007). A diagnostic approach for going beyond panaceas. *Proceedings of the National Academy of Sciences, 103*(39), 15181–15187.

Pawlowsky, P. (2001). The treatment of organizational learning in management science. In D. Meinolf et al. (Eds.), *Handbook of organizational learning and knowledge* (pp. 61–88). Oxford, UK: Oxford University Press. (979 p).
Rittel, H. W. J., & Webber, M. M. (1973). Dilemmas in a general theory of planning. *Policy Sciences, 4,* 155–169.
Second Nature. (2017). *Higher education's role in adapting to a changing climate*. Retrieved July 7, 2017 from http://secondnature.org/wp-content/uploads/Higher_Education_Role_Adapting_Changing_Climate.pdf.
SGCC. (2017). *Report of the Smith College study group on climate change: Towards a sustainable future*. Retrieved July 25, 2017 from https://www.smith.edu/climatechange/SGCC-Report-FINAL.pdf.
Smith College. (2017). *Lives of distinction and purpose: A plan for Smith*. Retrieved July 25, 2017 from https://www.smith.edu/planning/docs/StrategicPlan-2017.pdf.
Smith College Office of Campus Sustainability. (2015). *The road to carbon neutrality by 2030*. Retrieved July 5, 2017 from https://www.smith.edu/green/documents/scamp-update-longform.pdf.
Urbanski, M., & Leal Filho, W. (2015). Measuring sustainability at universities by means of the sustainability tracking, assessment, and rating system (STARS): Early findings from STARS data. *Environment, Development and Sustainability, 17,* 209–220.
Van Dyne, S. (2016). "Abracadabra": Intimate inventions by early college women in the United States. *Feminist Studies, 42*(2), 280–310.
Walker, B., & Salt, D. (2012). *Resilience in practice: Capacity building to absorb disturbance and maintain function* (p. 222p). Washington DC: Island Press.
Washington-Ottombre, C., & Bigalke, S. (2017). An aggregated and dynamic analysis of innovations in campus sustainability. *International Journal of Sustainability in Higher Education* (under review).

Camille Washington-Ottombre is an Assistant Professor in Environmental Science and Policy at Smith College (MA, USA). She received her Ph.D. in Forestry and Natural Resource Management from Purdue University (IN, USA). Her research focuses on adaptation to climate change and institutional learning and change. In research funded by the National Science Foundation and the Social Science Research Council, she studied the adaptive capacity of small rural communities in Kenya and Zambia and is now mostly interested in campus sustainability.

Sarah Brylinsky is the Sustainability Communication and Business Integration Manager at Cornell University, where she leads the development of strategic, systems-level sustainability programs for the world's leading land-grant institution. She has launched and supported 100s of campus sustainability programs for the last decade, focusing on climate mitigation, faculty curriculum development, student leadership capacity, and living laboratory strategy, working with campuses in every state and 16 countries around the world. Previously, she has served as Director of Climate Resilience and Educational Programs for Second Nature and supported the American College and University Presidents' Climate Commitment campuses by creating climate action plans and sustainability education programs and helped launch the Campus Sustainability Office of Massachusetts Institute of Technology and the Center for Sustainability Education at Dickinson College. She is the recipient of the AASHE Student Leadership Award, National Institute for Ethical Teaching Curriculum Development award for pedagogy in ecofeminism and recently has invested in national and regional strategies for community climate resiliency development. She holds a B.A. in Communication Management and Design and M.Sc. in Communications Innovation from Ithaca College and is a certified permaculture designer.

Dennis Carlberg AIA, LEED AP BD+C is an architect with over 25 years of experience. In January 2009, he joined Boston University as its first Sustainability Director. HE is also an Adjunct Assistant Professor in the Department of Earth and Environment and Faculty/Advisor for the Earth House, a living-learning community. He comes from a Boston architectural firm where he was a partner, senior designer, and chaired the sustainability committee, which he established in 2000. He began his career at the Solar Energy Research Institute conducting daylighting research to reduce building energy consumption and improve the indoor environment. He co-chairs the Climate Resilience Committee at the Urban Land Institute—Boston, which was formed in 2011 to educate and motivate the Boston area real estate community to actively address climate change, sea-level rise, and climate resilience planning and policy. He also serves on the Climate Ready Boston Infrastructure Advisory Group, the Massachusetts Statewide Resilience Master Plan Technical Advisory Group, the Museum of Science Environmental Sustainability Committee, and the BU Hub Implementation Task Force for general education. He received his Master of Architecture degree from the Massachusetts Institute of Technology where he was awarded the AIA Gold Medal. He received his Bachelor of Arts in Architecture from the University of California, Berkeley.

Dano Weisbord is the Director of Campus Sustainability and Planning at Smith College. He is responsible for long-term planning for carbon neutrality, as well as teaching and learning space, the campus landscape and energy infrastructure. He was the Senior Advisor for Organizational Sustainability for Action Aid Intentional in London UK, and a Senior Project Manager for CLF Ventures in Boston MA. He was a Director of the Global Association of Corporate Sustainability Officers in London, and Lead Author on Sustainable Endowments Institute's *Greening the Bottom Line*. He holds a BFA in Industrial Design from RISD and an MA in Urban and Environmental Policy and Planning from Tufts University.

Language and Climate Change: Towards Language of Sustainability in Promoting Climate Change Mitigation in Malawi

Peter Mayeso Jiyajiya

Abstract The aim of this paper is to analyse the linguistic adequacy of the lexical resources used in climate change discourse in Mulanje district in order to determine how language of sustainability can be achieved to foster citizen participation in climate change mitigation. In Malawi, emphasis has been on creating messages without regard to the effect the language being used may have, thus resulting in misrepresentation and misunderstanding of the concepts and facts. This paper argues that the use of language of sustainability can be a critical factor in attainment of climate change mitigation. This paper draws its conclusions from the first stage of the research project which was aimed at examining the language used in the climate change mitigation project in Mulanje district and its impact on citizen participation in climate change mitigation projects. It was demonstrated that there is disconnection in the contextualisation of the language as it does not reflect the everyday use of the language by the local communities in the project area. This paper suggests that language must be reprocessed to meet the expectations of the local communities and must be contextually appropriate to the context of situation in which the discourse is taking place.

Keywords Climate change · Sustainability · Mitigation measures
Language · Communities

1 Introduction

Climate change has become one of the pressing global challenges that has transcended the scientific interventions and has now become a social and cultural phenomenon. Several initiatives and strategies have been employed by different organisations to curb its devastating effects. However, the challenge in the fight has

P. M. Jiyajiya (✉)
Bingu School of Culture and Heritage, Malawi University of Science and Technology,
P.O. Box 5196, Limbe, Malawi
e-mail: pjiyajiya@must.ac.mw; pmjiya@gmail.com

© Springer International Publishing AG, part of Springer Nature 2019

W. Leal Filho and R. Leal-Arcas (eds.), *University Initiatives in Climate Change Mitigation and Adaptation*, https://doi.org/10.1007/978-3-319-89590-1_6

been to synergise expert influence and non-expert (local communities') knowledge using effective communication strategies. Scholars like Nerlich et al. 2010; Schweizer et al. 2013, have argued that communicators have not creatively communicated to have their messages understood by the target audience. Schweizer et al (2013: 44) point that "…those trying to create a greater sense of urgency have used unsuccessful strategies, i.e. jargon-laden and cautious science-speak…" This observation echoes the challenge being faced in communicating mitigation measures in Mulanje district. The language of communicating climate change has not met the daily linguistic needs of local communities to whom the communication is intended. This has created misrepresentation of information and ideas. What is crucial is the creation of discourses using a language that can "inspire people to seek a better world, a world where their own wellbeing, the wellbeing [of] their community, and other communities is enhanced, and the systems which support life are nurtured rather than destroyed" (Stibbe 2009: 3). Therefore, the purpose of this paper is to analyse the lexical resources used in communicating climate change mitigation in Mulanje district in order to determine how the language of sustainability can be achieved to foster citizen participation. What the paper plans to achieve is the determination of the discourses that can inspire sustainable action among all stakeholders in climate change mitigation exercise. It must be stated that the idea is not to dictate the best possible words or expressions, but to discover discourses that can inspire people to achieve sustainability in climate change mitigation exercise. This can be achieved by close engagement with the people and discover their stories and how they tell them (Stibbe 2016), in the process creating a mutually intelligible discourse that can spur collective action in climate change mitigation exercise.

2 Language and Sustainable Development

Human interaction has always been possible through the use of language, which has been a mediating tool and vehicle for carrying various emotions, feelings and ideas. The use of appropriate language to the context of situation has been a key to sustainable relationships among humans. However, this has not been transferred into development processes in which communicating in the appropriate language to the communities in the project area is pivotal to the successful implementation, acceptance by the local communities and sustenance of the development project. This is probably due to the age-old tendency of disregarding language as an important aspect to social change and any development process. Christie (2005 in Okech 2006) has argued that language is a basic resource needed for negotiating social relationships with others, to construct our sense of our world by shaping values, meanings and understanding. It is language, therefore, that makes any interactive process aimed at achieving sustainable development possible.

Stibbe (2008) has pointed that language plays a vital role in forming social structures and the subsequent impact of those structures on the sustainability of societies. He further adds that "…language persuades people to behave in ways which have differing impacts on the ecosystems which support life" (Stibbe 2008: 1). This underscores the necessity of factoring in the language that would help sustain the climate change mitigation measures, and the language that would penetrate the cultural fabric of the affected communities around the globe in all climate change management projects.

The challenge with climate change activities has been finding solutions that would enable curtailing the devastating effects of climate change. In most cases, finding solutions that fit the local environment and eventually being accepted by the local communities has resulted in a misfit between the needs and interests of experts and the local communities. This is partly due to what Luke (2005) considers the core problem with the current environmental movement: the narrowing of the public sphere and a restricted understanding of the public interest. One contributing factor to this climate change challenge is the means of communicating climate change messages which can contribute "to human understanding and decency" (Docherty et al. 1993 in Penman 2001: 143) rather than just focusing on passing the message. As Nerlich et al. (2010) put it, climate change communicators are urged to move from one-way communication to dialogue and reflexive engagement. But the question is: How can this be achieved when the language being used is highly jargon-laden and more technical?

3 Climate Change Vulnerability, Mitigation and Adaption in Mulanje District

Just like in many countries in the world, climate change impacts have been witnessed in Malawi. This has been evidenced by dry spells, droughts and floods which have adversely affected the sustainable livelihood of rural people. This has complicated the climate change effects since agriculture is the mainstay of many people in Malawi, with more than 90% dependent on it (National Statistical Office 2008). In Mulanje district, the problem of climate change is compounded by the scarcity of land for subsistence farming, since most land is taken by tea plantations. This has led to the exertion of pressure on the natural resources that are found in Mulanje Mountain, which is a protected area. The effects of climate change in the area have impacted on the overly relied rain-fed agriculture which has resulted in low harvests and persistent hunger. Consequently, people have resorted to the mountain for survival and wage labour in the tea estates. This has left most people vulnerable and without options for survival.

However, efforts have been made by a number of organisations such as Mulanje Mountain Conservation Trust (MMCT) to improve the "environmental and socio-economic situation in the area" (Haarstad et al. 2009: 19). A number of resilience and adaptation measures have been employed by the communities in the district. Some of them include irrigation farming, intercropping and relay cropping, and biomass energy production. However, the major challenge has been lack of access to vital and simple information on climate change that would enable them devise sustainable mitigation and adaptation strategies. The increase in climate knowledge and improved abilities to predict climate variability enables people to have relevant information and develop long-term strategies that help reduce the negative impacts of climate variability on their livelihoods. Much as the information about mitigating the changes in climate is being communicated, not much in terms of following the "expert advice" has been achieved. This begs the question about the "effectiveness of current communication efforts" and its impacts on influencing the local communities to "implement change in response to these communications" (Nerlich et al. 2010: 97).

The Mulanje Mountain Forest Reserve (MMFR) has been central to the maintenance of climatic conditions in Mulanje district. It was declared a protected area in 1927 under the British colonial authority and since then, it has been managed by the Department of Forestry (DoF). It was known for its indigenous trees, huge pine plantations and endemic Mulanje cedar tree. However, the clearing of pine plantations by MMCT and DoF to create a buffer to enable uninterrupted growth of Mulanje cedar tree, and the continued over-dependence on the forest resources by some communities, have resulted in drastic and devastating climate variations that have impacted on the production of subsistence crops and subsequently the livelihoods of people in the district. Consequently, there have been conflicts between the environmental organisations and the local communities on how best to mitigate and adapt to the change in the climate due to the depletion of biodiversity in the district.

4 Research Description

This paper is the result of the research project which was conducted in Mulanje district, Malawi. The main aim of the project was to examine the language that the environmental agencies use in the climate change discourse in Mulanje. The focus was to examine and evaluate the aspects and patterns of language use in the communication documents that are used in sensitising and educating local communities in the conservation of the environment. What prompted the inception of the research was the conflict that ensued as a result of the disagreements between MMCT and DoF on the one hand, and the local communities on the other hand. When DoF and MMCT cut down all the pine plantations in Mulanje Mountain, which they called invasive plants, communities protested arguing that pine plantations and other plants in the mountain were beneficial to their climatic conditions and economic and social well-being.

However, despite the conflict, it was realised that both parties were keen to conserve the environment in order to avert climate change effects which they were experiencing. But the question was: Why were they failing to agree on the way forward? Realising that at the centre of human interaction and communication is language, this researcher decided to embark on a research project to identify possible challenges in their engagement and evaluate their discourses. Specifically, the study focused on the following objectives:

- Identify the form in which the conservation information is disseminated.
- Analyse the linguistic adequacy of lexical resources used in environmental conservation discourse.
- Identify linguistic devices in Chichewa and evaluate their impact to dissemination of environmental information.
- Examine the discursive construction of citizen participation in environmental discourse.

The rationale behind this study is to enable the environmental communicators to consider the language they use in framing environmental messages. This would enable them to be aware that some expressions may have a different meaning from what it initially meant. This therefore requires that the message be processed in a form that common people can easily understand. They will also need to understand how technical expertise may well be productively integrated with indigenous knowledge in order to achieve collective action. As Depoe and Delicath (2004: 10) rightly put it "we urge participants in environmental controversies to acknowledge the legitimacy of both technical expertise and local knowledge, and seek a more productive dialogue among multiple discourses in which citizens, experts, and other participants articulate, interrogate, and transform each other's perspectives". This will help policy makers to come up with inclusive policies that will not face resistance from the citizens because citizens will not feel controlled by the few powerful.

This study involved the analysis of documents and interviews that the researcher did with the climate change experts and the local communities. As such, the study used content analysis as the main analytical method, which was supported by critical discourse analysis and multimodality analysis because some documents had other communication modes like pictures. This enabled the researcher to gather more insights into how communication on climate change is being conveyed and whether the discourses are opening up for all stakeholders in the district.

5 Methodology

The study employed qualitative study design. The sampling was purposive because it involved choosing organisation and villages that were either involved in the conservation or benefit from the mountain resources. The sample was drawn from the organisations and agencies that are directly involved in the conservation of the Mulanje Mountain Forest Reserve. These are Mulanje Mountain Conservation Trust (MMCT) and Department of Forestry (DoF). The study also engaged communities from the villages around MMFR. The villages are Nakhonyo, Mbewa and Mandanda, with ten (10) people interviewed in each village, making a total of thirty

(30) respondents. The communities were engaged in order to get their understanding on the communicative aspect of the conservation process and whether meaningful involvement is done by the communities in the climate change mitigation initiatives. The focus on the sampled communities was to test their understanding of the information that is relayed to them by the environmental organisations. The researcher wanted to see how people interpret the expressions about conservation of the environment in Mulanje against the way the agents interpret the same expressions to determine whether there is a gap in understanding of the same concept.

This study used two data collection tools, namely document analysis and interview schedule. The use of these two tools helped to get an in-depth understanding of the language phenomenon under investigation.

6 Limitations of the Study

The major limitation of this study was the ability to reach out to as many people as possible to interview them because of financial constraints since it was a self-funded project. However, the respondents who were interviewed provided adequate and insightful responses that enabled this study to be successful. Another constraint, especially to this paper is that it is limited to one aspect of the study, which is the analysis of the lexical resources. This is because the data obtained from the research were enormous and it cannot be included in this paper which has a limit of number of words, and that this paper needed to explain in detail its purpose.

7 Results and Discussions

The discourse of climate change in Mulanje district has created varied reactions to the measures employed in mitigating the effects of climate variations. The discursive approach in this discourse underscores political correctness of reaching sustainability, in which the experts dictate the way, in the process using expert expressions. This does not augur well with the aim of "discourses which have an impact on sustainability" (Stibbe 2009: 3), which focus on engaging people into taking action to achieve sustainable mitigation measures. Table 1 outlines some of the expressions used in the communication documents that were deemed to be crucial in the discourse of climate change and how the local communities reacted to such terms.

It is clear from the table that most of the respondents did not agree with the meanings assigned to the expressions describing the conservation exercise for ensuring mitigation of climate variations is achieved. If the discourse does not engage all people involved, then it is not sustainable because it reaches to a narrow audience and mostly does not inspire the desirable action. What usually follows is

Table 1 Terms and expressions as conceptualised by the local communities

Terms/expressions	Communities' views
Kukolola zam'khalango (harvesting forest resources)	- The term does not agree with what they are saying. For us, harvesting is for crops while mountain resources we don't harvest because we don't collect everything we have gathered as we do with crops - The way the term has been used differs from what we know. If we are to do the way we know the meaning of the term, they say we are wrong - For us harvesting means getting enough for you without any limitations because you have put your effort, be it forest resources or garden crops - I have no problem with the term because I know they are talking about cutting trees in the mountain
Kuthana ndi zomela zachilendo (eradication/clearing of invasive plants)	- In the mountain there is no plant that they should call invasive because what they are calling invasive plants are what help us and again they strengthen the mountain, and make the mountain beautiful - Any tree found in the mountain is important only that cedar is only found in our mountain that's why we take great care of it but that does not mean that the other trees/plants are not important. That's why we are not agreeing with the Trust (MMCT) people - The issue is not about invasive. These people had an agreement with mining company so that after cutting the entire pine plantation, they paved way for the mining company to start their work. To us the pine they are calling invasive is the one that saves cedar from depletion. What you should know is that cedar is very far and to reach there it takes you 12 h while pine was near. So no one would go far where there is cedar leaving pine near - I agree that some plants are invasive. For example, pine deters the growth of cedar. I've been to the mountain and if you go you will see that where there is pine and cedar, cedar doesn't grow well. Therefore, I understand what this means
Pine ngati mtengo wotetezera cedar (pine as nurse tree)	- We heard that but that does not mean that pine is not important. What we're against is that they had cut all pine trees because it was just planted as a nurse tree to cedar. These are simply their excuses to destroy the mountain - That pine was just planted, I heard that but it is not true that cedar is failing to grow because of pine. Cedar is a spiritual tree, it grows on its own and it doesn't need to be planted and monitor its growth. If you go today and see where MMCT planted cedar 15 years ago, you will see that it is failing to grow yet there is no pine. You can't plant cedar - I know it was just planted as a nurse tree

(continued)

Table 1 (continued)

Terms/expressions	Communities' views
	***Probing question: So do you think it is important to cut down all those trees that were planted simply to conserve Cedar?** - No, it's not like that but pine needs to be cleared especially the one near cedar which is deterring the growth of cedar (three people said this)
Cedar as a national tree	- For us cedar is a precious tree and very important because it is only found in Mulanje therefore it is important to take care of it
Clearing weeds	- For us when we say weeds, we mean unwanted plants in the garden. So if we relate with the mountain, which is not true because every plant in the mountain has its function and is important therefore there is need to take care of them

the eruption of conflicts, which was the case in Mulanje. What is clear from the responses in the table is that the language being used encodes a worldview of those that are the experts in the discipline, and therefore provides little room for engagement with the local communities whose worldview is different from the framers of the communication documents. This is in contrast with what Stibbe (2009) recommends that in order to achieve sustainability, it is not about searching for the "right" word or worldview, but a range of worldviews that will encourage people to act more sustainably and reach a wide range of people. This can be achieved if the local communities affected are closely engaged and have them tell their stories, in turn that would create a discourse with the worldviews that would inspire collective action.

7.1 Linguistic Adequacy of Language

In order to understand the language used by the environmental organisations, the study analysed the discourse in the newsletter, posters and flyers that are used in disseminating messages on conservation. Some information was extracted from the texts mentioned to assess the linguistic adequacy to see whether the language performs its job of delivering the intended messages without vagueness and ambiguity. The lexical resources were analysed to determine their referential adequacy and semantic adequacy.

7.1.1 Referential Adequacy

According to Haugen (1966: 62 cited in Harré et al. 1999), referential adequacy means "the capacity of the language to meet the needs of its users as an instrument of referential meaning". Harré et al (1999) further comment that a language is referentially adequate if it has the lexical resources to discuss a given topic in sufficient detail, "sufficiency" being relative to the task at hand. Chichewa, which is the mostly used language in the conservation projects in Mulanje, as will be observed from the texts to be analysed has shown that in some cases it fails to sufficiently provide the lexical resources that could spur dialogue among stakeholders because it has some terms that are open to numerous interpretations. This in some way impacts on the meaning-making of the task at hand.

Below are some examples of extracts culled from the texts to see how adequate the language is in advancing conservation of Mulanje cedar:

1. *Kololani* nkhuni zanu mosamala (*Harvest* your firewood carefully).
2. *Dulani* mitengo yanu kupititsa patsogolo mphukila (Cut your trees to allow the tree to sprout).

From the above data (1) and (2), the expressions look to be quite easy and straightforward, but the connotations and the referential ability of some words present a vague picture. For instance, in using *kukolola* (harvesting) whatever is in the mountain, people get the impression that one must get enough of the harvest. When asked on what they understand by the term harvesting, 75% of the respondents said that to them it means gathering as much as one would be satisfied with. One respondent actually said "*ifeyo kukolola chilichonse ndekuti ukuyenela upeze zambili zoti zikukwane popanda choletsa chilichonse chifukwa wagwilapo ntchito, kaya ndi zam'nkhalango kaya zakumunda* (For us harvesting means getting enough for you without any limitations because you have put your effort, be it forest resources or garden crops)". This ambiguity stems from the literal translation of the word as used in agricultural realms in which one harvests mature crops from the garden. In addition, the way the second extract has been syntactically constructed presents some semantic problems. The phrase "*Dulani mitengo yanu kupititsa patsogolo mphukila*" does not make much sense because that is not a correct expression in Chichewa. This may result in misinterpretation and misunderstanding. Instead it would have been better if it was phrased as "Dulani mitengo moyenera kuti ithe kuphukira" and even though with this alternative expression, the non-specification of the types of trees to cut down, provides a challenge. Not every tree sprouts, therefore specification of the trees that sprout would guide the local communities on which trees to cut.

In environmental discourse, harvesting is qualified, that is, it may be legal and illegal as illustrated in the extract below:

3. Ngakhale tachita bwino mu zinthu zambiri, tidakakumanabe ndi zovuta zina zomwe zikuononga chilengedwe mu phiri la Mulanje, zina mwa izo ndi monga *kukolola zachilengedwe munkhalango popanda chilolezo* komanso kuzembetsa

nkhuni kupita nazo kumadera ammatauni (In spite of many successes registered, there remains challenges that have great potential for continued loss of the Mulanje Mountain biodiversity such as *illegal forest resource harvesting*, which include *logging and trafficking of firewood* into the urban centres).

4. Ngakhale pali kusintha kwakukulu pothana ndi *kukolola zachilengedwe mosaloledwa, kagwiritsidwe ntchito ka za chilengedwe* kakunka kakukulirakulirabe (Although significant impact had been registered in the reduction of *illegal resource harvesting, resource extraction* from MMFR reached unprecedented proportions).

From the extracts (3) and (4) above, legal harvesting entails having permission to reap from the mountain resources. This somehow contradicts the essence of "harvesting" the mountain resource considering that the authorities encourage co-ownership of the mountain. Relating what the extract stipulates about illegal harvesting and what the respondents said about their conception of harvesting, one gets the impression that there is a likely misunderstanding due to different conceptions of the concept of harvesting. What this suggests is that the respondents question the illegality of harvesting because to them it is deserved harvest. This is why Halliday (1994) considers context as an important communicative aspect in every discourse. Halliday (1994) suggests that when producing discourses, we must consider the context of production of the texts produced. Therefore, the different conceptualisation of the term "harvesting" may result from different contextual underpinnings of the interlocutors in the discourse. However, to some respondents they said they do understand what it means when the message is presented in that way.

What may be challenging again in the other extract (2) given above is that by just saying "dulani mitengo yanu kupititsa patsogolo mphukira" without specifying what type of trees in terms of age, and also how many trees to be cut creates a fuzzy picture of what constitutes legal "harvesting" and illegal "harvesting". But again there are some trees that do not sprout once they are cut down, for example pine. They simply die. The document does not acknowledge that not all trees have "mphukira" in order to advise the local communities on what they should do when they are dealing with such trees. That kind of erasure gives leeway to the villager to cut as much as he/she can, especially for those who believe that harvesting must be gathering enough for oneself. In extract 3 above which is talking about "…*kukolola zachilengedwe munkhalango popanda chilolezo* komanso kuzembetsa nkhuni kupita nazo kumadera ammatauni", there are two co-hyponyms of tree cutting, "… harvesting… logging…" which does not underline the differences in using the two terms. The extract shows that harvesting presupposes logging but in a negative sense.

On the aspect of maintaining and making sure that cedar is growing uninterrupted, the lexical resources that depict other plants as destructive to the growth of cedar are used. As part of the exercise to create a growing space for cedar, the environmental organisations removed all the plants in the areas where cedar was growing. The expressions that have been continuously used in the conservation of Mulanje cedar to explain the actions taken on removing the other plants are given below:

5. Ngakhale kafukufuku waonetsa kuti pakutha pa zaka khumi zoteteza chilengedwe ntchito yaikulu yagwirika *pothana ndi zomera zachilendo monga pine zomwe zimatha kuononga zomera zinzake,* ntchito yaikulu yothana ndi zomerazi idakalipo yochuluka pa Chambe komanso pa Sombani (Although it was recorded that great achievement was made in the *eradication of the invasive Pinus patula* on the Chambe and Sombani basins at the end of ten years of conservation activity; still, there remains more work to be done on yet more difficult *invasive plants*).
6. *Kuchotsa zomera za chilendo* zomwe zimalepheletsa zomera za chilengedwe kuti zimere (*Clearing of invasive alien plant species*).
7. Kulimira mmalo momwe mu mamera cedar (Clearing weeds in areas where cedar is growing).

The term "kuthana (eradication)" as they use it in extract (5) connotes complete removal of any such type of "unwanted" plant (unwanted being any plant other than cedar). However, as it is used in relation to cedar, pine and any other plants, it eludes the essence of promoting biodiversity. When the communities were asked on what they make of such expressions mentioned above, most of them (90%) said that that is the whole reason they have been in conflict with the organisations because to them everything in the mountain is important, especially pine. They said pine grows faster than cedar, and it was being used as a way of conserving cedar because many people used to cut pine more than cedar. Again to the people, pine help with the weather conditions such as rain and preserving the general ecosystem. This shows that the use of these lexical resources gives a different connotation to the people. For example, pine to the people can never be "weed" as it is important to them. Such terms as "invasive" and "alien" alienate people, therefore creating a negative relationship between those controlling the discourse and those it is intended for. The term "invasive" to the environmentalists means anything that is encroaching and/or growing where they believe it is not supposed to grow in relation to the tree species they want to preserve and, therefore, causing uneasiness in the growth of the needed plant. However, to the communities, the word carries a different connotation as it implies other plants, including pine and M'bawa which they value as important species of plants.

In addition to the above terms, the documents have some expression such as "*Kutentha tchire moteteza* kunachitika mu mwezi wa July chaka chino (*Controlled burning* was carried out in July), which functionally means setting fire in the prescribed areas in order to get rid of unwanted plants. "Burning" (kutentha) in itself carries negative connotation regardless of what has been set on fire. The addition of the modifier "controlled" is meant to give a positive connotation of prescription. Mühlhäusler (2001) has argued that sometimes the "controlled or prescribed" fire gets out of control and disturbs the ecosystem. Even if the burning remains controlled, the process of burning and its by-products may be destructive to the adjacent environment.

7.1.2 Social Adequacy

According to Harré et al (1999), social adequacy means that language should be accepted to a maximum number of speakers in the target community, promote social unity and intercommunication and cater for present as well as anticipated future social needs. What this means is that language used in climate change discourse must bring unity of purpose among all the stakeholders and must present similar semantic undertones to all the people as they engage in communication. The challenge with climate change discourse is its nature being a discourse that is technical. The expression of meaning is done using terms that may sometimes misrepresent the information and can easily be misinterpreted, thereby creating disunity. This is due to the different contexts from which the stakeholders are operating. Mostly, the technical language is functional to that discipline and therefore, it is easily understood by those who share the discipline. However, Halliday (1994) advocates the use of context appropriate language so that all stakeholders involved do understand the discourse.

Looking at the above extracts (5), (6) and (7), some expressions such as "harvesting" and "clearing of invasive alien plants" clearly have not been accepted by many speakers considering differentiated meanings that are derived from the expressions. Apparently, the differences in meaning have resulted in conflicts that erupted in the area over what was needed to be done to mitigate climate change variations. What the expressions mean to the communities is different from what they mean to the environmentalists. In that case, the language does not promote social unity. For example, the word "zomera zachilendo (invasive plants)" referring to any plant other than cedar has been a contentious issue between the DoF and MMCT officials on the one hand, and the communities on the other hand. This is due to the fact that when the DoF and MMCT "cleared and/or eradicated" pine as an "invasive (alien) plant" conflicts erupted as people demanded the replanting of pine. They argued that the pine plantations provided employment to the communities as it used to attract tourists and was also a control of the climatic conditions in the area.

These reasons cited by the communities contribute to how they understand the lexical resources used by the environmentalists in their conservation campaigns, hence misinterpretation of the messages. In this case, the environmentalists fail to process their language to suit the socio-economic contexts of the communities in order to clear any misinformation that result from the lexical resources they use in the messages which project a different connotation.

This challenge may also be a result of different mitigation strategies that are advanced by the environmentalists and the local communities. The study established that before the coming of MMCT and their scientific intervention in the conservation process, the local communities had their belief systems that helped in the conservation of Mulanje Mountain biodiversity. For example, the researcher was told that anyone wanting to cut a cedar tree had to strip naked in order to see the tree, and that not everyone was allowed to go into the mountain to avoid angering the spirits. However, over the years, people stopped following these beliefs. As a result, the desire for utilisation of mountain resources resulted in depletion of mountain resources, hence the coming of MMCT. This means different perceptions of the same phenomenon but different lexical resources. In this case, the research found out that some lexical items used by environmentalists are not accepted because they differ in meaning with the local communities. For instance, the documents use the expression "kukolola za m'nkhalango mosaloledwa" (illegal forest harvesting) while the local communities say "kuba za m'nkhalango" (stealing mountain resources) which clearly gives the exact picture of what the activity is.

What the above observations suggest is that language of climate change discourse has reflected the differences in opinion on what needs to be done on the mitigation measures. The use of functional terms by the environmentalists represents their views on what is significant in the climate change mitigation process. This confirms what some scholars (e.g. Balyamujura 1995) pointed, that for environmentalists, the priority is preservation of the biodiversity for future use while for the local communities, the priority is access and utilisation of the biodiversity. In this case, the lexical resources used by the environmentalists reflect that need for preservation of biodiversity. On part of the communities, they believe that having access to the resources and being able to use them in the course of conservation is important. The use of language in this way reflects the worldviews that environmentalists and local communities have that are informed by their priority of needs.

The use of language in this way reflects the control of text by the environmentalists to advance certain ideologies to the people. As van Dijk (1993) puts it, the gatekeepers control what has to be accepted and not accepted. In climate change discourse, control of the messages is in the hands of the environmental organisations who command technical knowledge of the conservation issues. As a result, they work towards controlling the behaviour of the communities towards conservation of the environment. This is done through the use of functional registers that relate to their field. For instance, by saying "*kukolola zachilengedwe mosaloledwa,* kagwiritsidwe ntchito ka za chilengedwe…" and "*Kuchotsa zomera za chilendo* zomwe zimalepheletsa zomera za chilengedwe kuti zimere (*Clearing of invasive alien plant species*)", they are advancing the idea that to them much as harvesting is done on what one deserves, that must be done as a way of reaping although it is unclear as to how much. On the eradication of invasive pine, the ideological underpinning that is revealed is that despite being part of the ecosystem and indeed despite being seemingly significant to the well-being of people, pine can be cleared off because to them that is what would technically help the conservation of cedar.

However, Halliday (1994) has pointed out that the appropriateness of language use is determined by the field (all aspects of physical communication, including setting, topic, purpose and the speaker's intention) and tenor (i.e. people involved in the communication and their relationship, including roles and social positions that participants have). Halliday adds that language is never context-free. According to Halliday's SFG theory, language is not good or bad, it is appropriate to the context of use. In view of the language used by the environmentalists in Mulanje, the language is appropriate to them but it is contextually inappropriate to the local communities. It does not consider the participants and their social roles. Thompson (2004) adds that roles and relationships between participants carry with them social interests. The extracts (5), (6) and (7) above reveal that the language used does not factor in the sociocultural context of the local participants whose understanding of the technical language is limited. This creates misunderstandings and misinterpretation of the messages because the language used has not responded to and is inappropriate to the context of the addressee.

8 Conclusion

The discussion has shown that the linguistic choices made in the communication documents that are being used in Mulanje have resulted in creating multiple meanings due to the use of terms and expressions that are open to varied interpretations. The linguistic choices were meant to serve the ideological objectives of the environmental experts to present their conservation practices as benign and proper way of mitigating climatic variations. This has been sustained by the use of functional registers that are appropriate to their field of practice. This has stifled the engagement of local communities in the discourse that would promote sustainable action from the communities.

The appropriateness of language being used in the climate change mitigation discourse in Mulanje is limited to the producers of the text, in this case the environmental agents. This shows that the communication texts have some lexical resources that are not contextually appropriate to the end-users who are the local communities whose understanding of the conservation practices is limited to their everyday language use. As discussed in the paper, this has ended up creating misinterpretation of the information because the local communities have their own understanding of the lexical resources used in the discourse. This is contrary to what Halliday (1994) postulates that language must be appropriate to the context of use, and therefore mutually intelligible to all the interlocutors involved.

However, it has been shown that the local communities are eager to take part in the mitigation exercise, which provides a leeway for the experts to capitalise on. This could be achieved if both parties engage in a meaningful engagement using mutually intelligible language. In the light of this, it is imperative that in future more trainings and interface meetings need to be organised for the environmental communicators on how they can engage the local people who have their own perspective of the mitigation exercise.

9 Way Forward

What the study has found out as lacking in the climate change discourse is the use of language that local communities can easily identify and relate with on the subject of climate change. Therefore, MMCT and DoF need to create the discourse in the language of everyday communication for the local communities in order to inspire mitigation measures which have interests of their well-being and that of the environment.

Stibbe (2009: 1) contends that "our failure in reaching people is not because there is something wrong with the language we are using, but because our words engage with a narrow audience". This clearly explains the challenge being faced in Mulanje district where the choice of words is creating mistrust and misunderstandings on the issues that all the stakeholders want to get involved in. To overcome this challenge, the communication documents should be written for the communities they are intended for not for the environmentalists. The technical language that MMCT and DoF are using does not speak well with the local communities.

Another thing that needs to be done in Mulanje is for MMCT and DoF to give chance to the local communities to speak out using their everyday language and experiences, rather than being prescriptive. This discursive practice would bring about discourses of sustainability because everyone would contribute to the discourse and eventually take part in all the proceedings. Nerlich et al (2010: 104) point that "scientific communicators are encouraged to adapt their language to suit the tastes, meanings, and concerns of ordinary people". In this case, the language of sustainability would entail making sure that the language enables people to tell their stories regarding climate change. This would bring about discourses that would enhance effective understanding of the messages and hence motivating people into sustainable action.

References

Balyamujura, H. N. (1995). *Economic implications of different land use regimes in Mhala District*. Unpublished Masters thesis, University of Pretoria.

Depoe, S. P., & Delicath, J. W. (2004). Introduction. In S. P. Depoe, J. W. Delicath, & M. A. Elsenbeer (Eds.), *Communication and public participation in environmental decision making* (pp. 1–10). Albany: State University of New York Press.

Haarstad, J., Jumbe, C. B. L., Chinangwa, S., & Mponela, P. (2009). *Environmental and socio-economic baseline study—Malawi*. Oslo: Norad.

Halliday, M. A. K. (1994). *Introduction to functional grammar* (2nd ed.). London: Arnold.

Harré, R., Brockmeier, J., & Mühlhäusler, P. (1999). *Greenspeak: A study of environmental discourse*. Thousand Oaks: SAGE Publications.

Luke, T. W. (2005). The death of environmentalism or the advent of public ecology? *Organisation and Environment, 18*(4), 489–494.

Mühlhäusler, P. (2001). Talking about environmental issues. In A. Fill & P. Mühlhäusler (Eds.), *The ecolinguistics reader: Language, ecology and environment*. London: Continuum.

National Statistical Office. (2008). *2008 population and housing census preliminary report*. Zomba: NSO.

Nerlich, B., Koteyko, N., & Brown, B. (2010). Theory and language of climate change. *Wiley Interdisciplinary Reviews: Climate Change, 1*, 97–110.

Okech, O. (2006). *Language use and mode of communication in community development projects in Nyanza Province, Kenya*. Unpublished Ph.D. Thesis, University of Western Cape.

Penman, R. (2001). Environmental matters and communication challenges. In A. Fill & P. Mühlhäusler (Eds.), *The ecolinguistics reader: Language, ecology and environment* (pp. 143–153). London: Continuum.

Schweizer, S., Davis, S., & Thompson, J. L. (2013). Changing the conversation about climate change: A theoretical framework for place-based climate change engagement. *Environmental Communication, 7*(1), 42–62.

Stibbe, A. (2008). Words and worlds: New directions for sustainability literacy. *Language and Ecology, 2*(3), 1–11.

Stibbe, A. (2009). *The Language of sustainability: Shouting but not being heard. Keynote address at the Language of Sustainability: Shouting but not being heard conference*. London: Institute for Environmental Science.

Stibbe, A. (2016). *Ecolinguistics: Language, ecology and the stories we live by*. London: Routledge, Francis & Taylor Group.

Thompson, G. (2004). *Introducing functional grammar*. London: Francis & Taylor Ltd.

van Dijk, T. (1993). *Elite discourse and racism*. Newbury Park: Sage.

Behavioural Approaches of Rural Women Farmers to Mitigation and Adaptation Measures of Climate Change in Abia State, Nigeria

Ogechi Jubilant Umeh and Ike Nwachukwu

Abstract Farmers' behavioural attitudes towards climate change are complex and poorly understood, making difficult the development of mitigation and adaptation policies that would be accepted and implemented. The purpose of this paper is to assess the behavioural approaches of rural women farmers to adaptation measures of climate change in Abia State, Nigeria. Specifically, this paper ascertains perceived effects of climate change in the study area and adaptation measures practised by the women in the study area and identifies farmers' behavioural responses to adaptation measures to climate change and factors that influence farmers' behavioural approach to adaptation measures to climate change in the study area. Multistage sampling techniques were used in selecting 180 respondents for the study. Descriptive statistics was used for data analysis, while Tobit regression model and Pearson's correlation were used to test the hypotheses. Results show that high sunshine intensity, increased drought, inadequate access to water and crop losses are major perceived changes to climate change, while change of planting time ($\bar{x} = 4.79$), enterprise diversification ($\bar{x} = 4.51$) and crop rotation ($\bar{x} = 4.22$) are major adaptation measures practised by the women in the study area. Seeking for information based on observed changes ($\bar{x} = 5.00$), reverting to indigenous or self-help methods and changing cropping style and time ($\bar{x} = 3.41$) were identified as major behavioural responses of the women to climate change, while Tobit regression result indicated that age (−2.795**), household size (2.705**), level of education (3.389***) and income (2.505**) were major factors that influence the women farmers' behavioural approach to adaptation measures to climate change in the study area. Pearson's correlation analysis shows that there is a significant relationship between the women's behavioural responses to adaptation measures to climate change and adoption of adaptation measures by the women. The study has examined and shown the women's level of conformity, constraints and factors that

O. J. Umeh (✉) · I. Nwachukwu
Department of Rural Sociology and Extension, Michael Okpara University of Agriculture, Umudike, P.M.B 700 Umuahia, Abia State, Nigeria
e-mail: Umehogechi2@yahoo.com

I. Nwachukwu
e-mail: nwachukwuike@ymail.com

© Springer International Publishing AG, part of Springer Nature 2019
W. Leal Filho and R. Leal-Arcas (eds.), *University Initiatives in Climate Change Mitigation and Adaptation*, https://doi.org/10.1007/978-3-319-89590-1_7

influence their adoption of adaptation and mitigation strategies in the study area. It is therefore recommended among others that targeted and timely information be provided to the women through agricultural extension and mass media, the most effective and sustainable indigenous technologies used by the farming communities could be incubated with a view to upscale and out scale them in other to enhance adaptation to climate change and variability by the resource-poor farmers. ICT-driven virtual platform that provides weather advisory services for small-scale farmers should be made available through rural advisory services (RAS). This will help farmers handle climate risk situation promptly.

Keywords Behavioural responses · Rural women · Adaptation
Climate change

1 Introduction

Climate change is perhaps the most serious environmental threat to the fight against hunger, malnutrition, disease and poverty in Africa. This is mainly through its impact on agricultural productivity. Climate change, which is attributable to the natural climate cycle and human activities, has adversely affected agricultural productivity in Africa (Ziervogel et al. 2006). Available evidence shows that climate change is global, likewise its impacts, but the most adverse effects will be felt mainly by developing countries, especially those in Africa, due to their low level of coping capabilities (Anselm and Taofeeq 2010). Nigeria is one of these developing countries (Odjugo 2010).

The Intergovernmental Panel on Climate Change (IPCC 2010) defines climate change as statistically significant variations in weather elements (rainfall, temperature, sunlight, to mention but a few) that persist for an extended period, typically decades or longer. It includes shifts in the frequency and magnitude of sporadic weather events as well as the slow continuous rise in global mean surface temperature. NOAA (2016) reported that the combined average temperature over global land and ocean surfaces for March 2016 was the highest for the months in the 1880–2016 record, at 12.2 °C above the twentieth-century average of 12.7 °C. This surpassed the previous record set in 2015 by 0.32 °C and marks the highest monthly temperature departure among all 1635 months on record, surpassing the previous all-time record set just last month (February 2016) by 0.01 °C (0.02 °F). Overall, the nine highest monthly temperature departures in the record have all occurred in the past nine months. March 2016 also marked the 11th consecutive month a monthly global temperature record has been broken, the longest of such streak in NOAA's 137 years.

As the planet warms, rainfall patterns shift, and extreme events such as droughts, floods and forest fires become more frequent (Zoellick 2009), which result in poor and unpredictable yields, thereby making farmers more vulnerable, particularly in Africa (UNFCCC 2007). Farmers (who constitute the bulk of the poor in Africa) face prospects of tragic crop failures, reduced agricultural productivity, increased hunger, malnutrition and diseases (Zoellick 2009). It is projected that crop yield in Africa may decrease by 10–20% in 2050 or even up to 50% due to climate change (Jones and Thornton 2002), particularly because African agriculture, Nigeria in particular, is predominantly rain-fed and hence fundamentally dependent on the vagaries of weather. As the people of Africa strive to overcome poverty and advance economic growth, this phenomenon threatens to deepen vulnerabilities, erode hard-won gains and seriously undermine prospects for development (Zoellick 2009). There is therefore the need for concerted efforts towards tackling this menace.

Much of climate change agricultural research has tended to concentrate on assessing the sensitivity of various attributes of crop systems (e.g. crop/livestock yields, pest, diseases, weeds)—the biophysical aspects of food production. These studies have identified and recommended many mitigation and adaptation strategies with little or no regard to the aspect of level of conformity, and responses of the affected farmers to utilization of these strategies. These partial assessments most often consider climate change effects in isolation of the actions of the vulnerable, providing little insight into how and what the farmers are actually doing to cope with climate change. To better address the food security concerns that are central to economic and sustainable development agendas, it is desirable to also address these aspects of climate change and agriculture. Nevertheless, UNDP (2011), Warui (2011) and Odogwu (2012) firmly believe that, by harmonizing human development and efforts to manage climate change, it is possible to accelerate socio-economic progress and thus contribute to sustainable human development at the grass-roots level. In order to be effective, adaptation and mitigation technologies need to reach those who are most in need—the poor and vulnerable. This means that targeted efforts must ensure firstly that it is understood that the situation of women may differ from that of men, secondly that technologies are designed in such a way as to be relevant to their circumstances and thirdly to ensure that they are given full access to knowledge, information and technologies related to adaptation.

Studies have shown that women are more potent agents of environmental degradation management. This is because they constitute majority (80%) of the farmers (Mgbada (2013) and Umeh (2015), through continuous cultivation and tendering of the soil, and they are in continuous battle contending with vagaries of weather to achieve a bumper harvest. They play decisive roles in managing and preserving all natural resources including biodiversity, water, land and other natural resources and thus environmental sustainability (UN WomenWatch 2013). It is important to remember, however, that women are not only vulnerable to climate change, but they are also effective actors or agents of change in relation to both mitigation and adaptation. Women often have a strong body of knowledge and expertise that can be used in climate change mitigation, disaster reduction, and adaptation strategies. Furthermore, women's responsibilities in households and communities, as stewards of natural and household resources, position them well to contribute to livelihood strategies adapted to changing environmental realities

(CSW 2008). Therefore, it thus implies that women in Nigeria are not only the worst hit by climate change but are a major factor for due consideration in key decision taking processes concerning climate change issues.

Research has shown that behavioural factors influence the outcome of policy incentives in that they can either complement or constrain the effects of policies. The Theory of Reasoned Action (TORA) originally proposed by Ajzen (2005) provides the conceptual framework for exploring farmers' attitudes and intentions. According to TORA, the intention to adopt a particular behaviour is a function of attitudes towards the behaviour and the subjective norm—the extent to which one is influenced by the views of other people regarding the behaviour. Attitudes are a product of the extent to which one expects the behaviour to result in specified outcomes and the perceived importance attributed to those outcomes.

Thus, it is important to consider farmers' behaviour when seeking to access or improve the effectiveness of intervention programs and policies. As devastative effects of climate change continue to replete our environment and the world over, what is the level of conformity, utilization and behavioural attitude or approach of the women farmers to the mitigation and adaptation measures of climate change? Hence, this work is set to assess and evaluate the behavioural responses of rural women farmers to mitigation and adaptation measures of climate change in Abia State, Nigeria.

2 Objectives

The broad objective of this work is to assess the behavioural approaches of rural women farmers to mitigation and adaptation measures of climate change in Abia State, Nigeria. Specifically, the work will:

i. describe the socio-economic characteristics of rural women farmers engaged in climate mitigation and adaptation measures in Abia State,
ii. ascertain perceived effects of climate change in the study area,
iii. ascertain adaptation measures practised by the women in the study area,
iv. identify farmers behavioural responses to adaptation measures to climate change in the study area and
v. identify factors that influence farmers' behavioural approach to adaptation measures to climate change in the study area.

3 Hypotheses

1. $H_0 1$ = age, household size, level of education, years of farming experience, scale of production, level of income, contact with extension are not significant and do not influence farmers' behavioural approach to climate change adaptation measures in the study area.

2. H_02 = There is no significant relationship between the adaptation measures practised by the women and the women's behavioural responses to adaptation measures to climate change in the study area.

4 Problem Statement

There is a dire need for an emphasis on the proactive rather than the reactive management of climate change risk. This can only be feasible if the potential response and level of conformity to agricultural mitigation and adaptation measures to climate, as well as factors influencing such decisions, are pre-emptively analysed. Most studies on climate change and agriculture in Nigeria and beyond have tended to concentrate on actual and projected impacts as well as to identify farmers' adaptation strategies (Apata et al. (2010) and Ekweanya (2014)). The purpose of this work is to assess the response and level of conformity of rural farmers to agricultural mitigation and adaptation measures to climate, as well as factors influencing such decisions in the south-eastern zone of Nigeria. A better understanding of behavioural barriers can help to enhance higher conformity and adoption of the mitigation and adaptation strategies and policies when those barriers are tackled. This study could also provide a valuable contribution to the deliberation on climate change policies in Nigeria.

Climate change increases the burden of women who already have triple roles because it has led to reduced food production, reduced water availability, reduced income, increase in disease outbreak and other environmental hazards. Study on farm size and variability of climate on arable crop production in Abia State also indicated that temperature has been on the increase, and this explains the increased rate of drought which is a major cause of crop failure in rain-fed-dependent agrarian communities prevalent in the state (Iheke and Olive-Abali 2011). Women bear the burden of food production, fetching water and gathering of firewood. Cropping patterns are changing as a result of climate change. Only very few areas in Nigeria can produce arable crops of short duration such as maize twice during a rainy season due to the unpredictable duration of the season. This affects women more because they are the major food producer in developing countries. In such adverse situation, the women need to take a decisive decision on how to manage the situation.

The need to optimize production and value addition of agricultural products particularly among small-scale producers and entrepreneurs who form the majority of the world's poor is becoming increasingly critical to human survival. WHO (2012) reported that leaders from over 130 countries met in Brazil in June 2012 in the United Nations Rio +20 Conference, to discuss sustainable development, a startling research-based revelation unfolded before them: the number of undernourished women and young children could increase by 20% and will affect one of every five within a decade because of climate change. According to the

World Health Organization and other groups, as countries continue to get hit by massive flooding, droughts, extreme cold snaps, extreme heat, tornadoes and hurricanes, food production will be affected, and this will very likely increase to 20% by 2020, just three years away. And by all indication, Nigeria has joined this statistics because of the rampaging flood.

The WHO (2012) analysis further shows that of the 495 million women and children under five who are already undernourished, 150 million live in Africa (with about 45% leaving in Nigeria), 315 million in Asia and 30 million in Latin America and the Caribbean. It is expected that 465 million more will live in developing countries by 2020, boosting food demand. Specifically, malnutrition in Nigeria is rampant and often severe. The report also reveals that food prices have already been affected by climate change, with a global increase in the first quarter of 2012, which has been due to the extreme cold in Europe which affected wheat crops (WHO 2015). Globally, everyone will be affected by rising food prices and possible food shortages, but the most affected by food shortages will be those in underdeveloped and impoverished countries. Nigeria food inflation is valued at 19.27% in June 2017 (FAO 2017), According to WHO 2015 report, by 2030, climate change could push food prices up by 50–90% more than they would otherwise be expected to rise. Women are the worst hit.

Detrimental effects of climate change can be felt in the short term through natural hazards, such as landslides, floods and hurricanes, and in the long term, through more gradual degradation of the environment. ADBG (2015) reported that African Governments, especially those of the Lake Chad Basin region, are battling to secure the future of the dwindling water body and have expressed the need for adaptation actions as a priority to save the lake and the lives of the over 20 million people living in the area. In Nigeria, more heat plus less rain is already creating drought conditions in parts of northern Nigeria mostly in Kebbi and Niger States. This is troubling when government data show rural households use rainwater for more than half their total water consumption and northern groundwater tables have dropped sharply over the last half-century, owing partly to less rain (NBS 2005). In parts of southern Nigeria such as parts of Delta and Anambra states, flooding caused by sea-level rise is also contaminating freshwater aquifers, rivers and stock-watering points, leaving them with high salinity and more polluted with sediment and sewage (FGN 2005). Specifically, in 1997, the United Nations Development Programme (UNDP) reported that, out of the 5700 erosion sites in Nigeria, 4000 of these were in south-eastern states (Akpokodje et al. 2010) with more than 600 of them active (Igwe 2012). These sites are found mainly in Agulu, Nanka, Alor, Nnewi, Ideani, Oraukwu, Oko, Nkpor, Ekwulobia, Alo, Uke, Ojokoto/Oba, Udi, Enugu, Ukehe, Amucha, Njaba, Umunneochi, Isuikwuato, Ohafia, Abiriba, Arochukwu, Igbere, Ozuitem, Uyo and Calabar (Abegunle et al. 2006; Igwe 2012).

It is glaring that the adverse effects of climate change devastation are already felt in many areas in the south-eastern zone of Nigeria, including agriculture and food security; biodiversity and ecosystems; water resources; human health; human settlements and migration patterns; and energy, transport and industry. In many of these contexts, women are more vulnerable to the effects of climate change than men—primarily as they constitute the majority of the world's poor and are more dependent for their livelihood on natural resources that are threatened by climate change (CSW 2008).

Furthermore, women face social, economic and political barriers that limit their coping capacity.

Women in rural areas are especially vulnerable when they are highly dependent on local natural resources for their livelihood. Those charged with the responsibility to secure water, food and fuel for cooking and heating face the greatest challenges. Secondly, when coupled with unequal access to resources and to decision-making processes, limited mobility places farmers in rural areas especially women in a position where they are disproportionately affected by climate change. It is thus important to identify gender-sensitive strategies to respond to the environmental and humanitarian crises caused by climate change (CSW 2008).

Again in Nigeria, gender remains a key factor in the ability to earn an income, obtain credits from financial institution, access to and control of critical resources, such as land and water, and access to education (Okojie 2002). The influence of gender with respect to being a man or woman arises from the fact that men have greater opportunities to have a better livelihood than women and children conferred on them largely by cultural factors. This makes women and children more vulnerable to the impacts of climate variability. Consequently, to be effective and prepared for the impacts of climate variability and change, Nigeria must take into consideration the peculiarities of men, women and children.

With the above submissions, it is pertinent that there is every need to assess the level of conformity to mitigation and adaptation strategies to climate change by the women who are the vulnerable group worst affected by the climate vagaries. Therefore, bearing in mind that the best approach to save our rural communities and indeed the nation from the looming danger of climate change menace is an integration of mitigation and adaptation strategies, and effectively disseminating same for proper implementation to the grass-roots farmers which comprises mostly women, this study was conceived. The purpose of this study is to investigate the behavioural approaches of rural women farmers to mitigation and adaptation measures of climate change in Abia State, Nigeria.

5 Research Methodology

The research method used in this study is survey method. Multistage sampling procedure was used in selecting 180 respondents for the study. In the first stage, all the three agricultural zones that make up Abia State were selected in other to cover the state effectively. They are Umuahia, Ohafia and Aba agricultural zones. In the second stage, two ADP blocks were randomly selected in each of the zones making it six blocks. Thirdly, two ADP circles each were randomly selected from each of the blocks making it 12 circles, while in the fourth stage, fifteen women farmers were randomly selected from each of the of the circles. This brought the sample size to 180 rural women farmers who are the respondents for the study. These respondents were issued with well-structured questionnaire and interview schedule to obtain answers and information from them on the research topic.

Data for this study were generated from primary sources. The primary sources include well-structured questionnaire, interview schedule and direct tour and observation of ecological disaster sites in the study areas to observe the adaptation/mitigation practices adopted by the rural women in coping with these challenges.

6 Method of Data Analysis

1. To accomplish objectives **one**, descriptive statistics such as frequencies, percentages and mean were employed to analyse the socio-economic characteristics of the women farmers. The socio-economic factors considered are age, educational level, marital status, household size, primary occupation, farming experience, membership of cooperative societies, extension contact, number of extension trainings on environmental management, level of technology and access to credit facilities.
2. 5-Point Likert rating scale was also be used to achieve objectives **two**, **three** and **four**. They are to ascertain perceived effects of climate change in the study area, ascertain adaptation measures practised by the women and identify farmers behavioural responses to adaptation measures to climate change in the study area. The 5-Point Likert rating scale was very high (5), high (4), moderate (3), low (2), very low (1). Variables with mean score of 3.0 (which is the average mean score of the Likert level) and above imply that they are positive and in affirmative to the objective being measured while factors with mean score of less than 3.0 are negative and rarely have an influence on the objective being measured. To determine the mean Likert level = $\sum X/N$. Mean of each item was computed by multiplying the frequency of each response pattern with its appropriate nominal value and dividing the sum by the number of respondents to the items. This can be summarized by the equation below:

$$\text{Mean} = \sum fn/N$$

where

mean	Likert mean score
$\sum$	summation
F	frequency
n	Likert nominal scale
N	number of respondents

$$\text{mean} = 1 + 2 + 3 + 4 + 5/5 = 15/5 = 3$$

7 Hypotheses

Hypothesis one which is identifying factors that influence farmers' behavioural approach to adaptation measures to climate change in the study area was tested using Tobit regression analysis:

$$Y/Y^{*} = (X_1, X_2, X_3, X_4, X_5, X_6, X_7, X_8, \mu)$$

The latent variable $Y = 1$ and $Y^{*} = 0$.

1 influence
0 non–non-influence

where

X_1 age (years)
X_2 household size (number of persons in the household)
X_3 level of education in years (years)
X_4 farming experience (years)
X_5 income (amount in Naira)
X_6 farming status (full time = 1)
X_8 farm size (Ha)
X_9 extension contact (yes = 1)
X_{10} membership of cooperative societies (number of cooperatives)
X_{11} access to credit (access = 1)
X_{12} access to land (access = 1)

Pearson's correlation analysis was used to test hypothesis two which entails relationship between the adaptation measures practised by the women and the women's behavioural responses to adaptation measures to climate change in the study area.

$$\text{PPMCC}(r) = \frac{n\sum XY - (\sum X)(\sum Y)}{\sqrt{n(\sum X)^2 - (\sum X^2)(n\sum Y^2) - (n\sum Y)^2}}$$

where

r correlation coefficient
Y dependent variable
X independent variable
n sample size

t-test of significance on the correlation coefficient was equally carried out with model given as:

$$t = \frac{rx\sqrt{n-2}}{1 - r^2}.$$

where

t t-test of significance
r correlation coefficient
n sample size

8 Limitations of the Study

Some limiting factors were encountered in the course of the research. Some of the ecologically devastated areas could not be easily assessed due to poor road network. Some of the women have migrated from their homes after losing their crops to flood or drought and therefore could not be reached for collection of data. Fund was also a major constraint. Larger area of study and more women could have been reached with adequate funding. Therefore with adequate funding, this study is recommended for out-scaling to cover more states.

9 Results and Discussion

9.1 Socio-economic Characteristics of the Women Farmers

Socio-economic profile of the women was examined and presented in Table 1. The results show that the mean age of the is 44.5 years, indicating that the women are in their prime productive age able to embark on their agricultural production activities given favourable working conditions. This result is in line with the study of Akinbile and Odebode (2002) which states that the population within the age group of 16–55 years constitutes the active workforce in Nigeria. Majority (86.67%) of them are also married with household size of 6.5. There is appreciable level of literary among the women as 55.5% have attained secondary school and mean farming experience of 15.5 years. Level of literacy could have a significant influence on their sourcing and access to information as well as input for their farming operations. These findings agreed with Gordon and Craig (2011) as they opined that education increases level of skill and foster access to information on improved agricultural practices. The number of years of farming experience has shown to equip farmers to make decisions on sourcing of farm inputs and appropriate technologies to employ in their farms Nwaobiala (2014). The results also indicate that the women have mean farm size of 1.5 ha mean monthly income of *N* 37,550. This indicates the women are small smallholder low-income earning farmers and therefore they need to protect the available farm land against vagaries of weather and other adverse working conditions in order to maintain their farming production and sustain themselves and family. 61–11% of the women have contact with

Table 1 Percentage distribution of the socio-economic characteristics of the women farmers engaged in climate adaptation measures in Abia State

Socio-economic variables	Percentage	Mean
Age	59.44	44.5
Marital status	86.67 (married)	
Educational level	55.55 (SSCE/WAEC/GCE)	
Household size	68.89	6.5
Income (N)	53.89	37,550
Farming status	65.56 (full time)	
Farming experience (years)	52.22	15.5
Farm size (HA)	55.67	1.5
Contact with extension	61.11 (fortnightly)	
Membership of cooperative societies	54.44	3.5
Access to government credit	78.33 (none)	
Access to land	60.56 (lease)	

Field survey 2016

extension agents fortnightly, while most (54.44%) of them belong to 3–4 cooperative societies. Access to extension and membership to cooperatives enhance their access and exchange of information on production innovations, weather forecasts, adaptation/mitigation strategies and other important updates concerning their farming operations. This is in agreement with Nwaobiala and Ubor (2016) that Nigerian farmers are small-scaled and such interventions such as extension and input services are necessary, as it can help improve the status of the farmers, guaranteeing timely access to information, fertilizer, seeds and other essential agricultural inputs. Table 1 also shows that most (78.33%) of the women do not have access to government credit facilities, while 60.56% do not own their farm land as they on lease implying that they are self-sponsored farmers. This indicates that they may not have enough funds to upscale their production or adopt new innovations but may resort to indigenous practices and other labour- and money-saving strategies for sustainable farming operations.

10 Perceived Effects of Climate Change in the Study Area

Perceived effects of climate change by the women in the study area were accessed and presented in Table 2. Result shows that high sunshine intensity ($\bar{x} = 4.88$), increased rate of erosion ($\bar{x} = 4.67$), increased drought ($\bar{x} = 4.66$) and crop losses ($\bar{x} = 4.52$) and inadequate access to water ($\bar{x} = 4.47$) are major perceived changes to climate change in the study area. This result agrees with CSW (2013) that inadequate access to water and poor water quality does not only affect women, their responsibilities as primary givers and the health of their families, but also impact agricultural production and the care of livestock and increases the overall amount of labour that is expended to collect, store, protect and distribute water.

Table 2 Mean distribution of perceived effects of climate change in the study area

Variables	Total	SD	Mean
High sunshine intensity	879	1.78334	4.88
Poor soil fertility	800	1.72012	4.44
Crop losses	814	1.73127	4.52
Excessive storm/lightening	538	0.56311	2.99
Increased rate of erosion	841	1.75001	4.67
Decrease in agric. output	778	1.68815	4.32
Increased drought	804	1.74667	4.46
Change in Harmattan period	841	1.75001	4.67
Poor health condition of farmers	681	1.20342	3.78
Inadequate access to water	807	1.74687	4.47

Field survey 2016

11 Adaptation Measures Practised by the Women in the Study Area

Adaptation measures are required to enable farmers to mitigate the harmful effects of unavoidable climate change by maximizing field opportunities and supporting sustainable agricultural and human development. The study indicated from Table 3 that change of planting time ($\bar{x} = 4.79$), enterprise diversification ($\bar{x} = 4.51$), crop rotation ($\bar{x} = 4.22$), planting of resistant varieties ($\bar{x} = 4.10$) and seeking for expert advice ($\bar{x} = 4.03$) are major adaptation measures practised by the women in the study area. Women are not only vulnerable to climate change, but they are also effective actors or agents of change in relation to both mitigation and adaptation. The findings are in line with CSW (2013) that women often have a strong body of knowledge and expertise that can be used in climate change mitigation, disaster reduction and adaptation strategies. Furthermore, women's responsibilities in households and communities, as stewards of natural and household resources, position them well to contribute to livelihood strategies adapted to changing environmental realities.

12 Farmers' Behavioural Responses to Adaptation Measures to Climate Change

Studies of behavioural response pattern to adaptation measures to climate change address the interaction between individuals and their environments. The results can be used as guides in developing effective extension teaching methods and dissemination of related information. Data in Table 4 show that the women farmers in the study area respond differently to the adoption and practice of adaptation measures. Major responses exhibited by the women are: seeking for information based

Table 3 Mean distribution of adaptation measures practised by the women in the study area

Variables	Total	SD	Mean
Adoption of agricultural innovation	579	1.44126	3.22
Good cropping system	771	1.78821	4.28
Enterprise diversification	812	1.99761	4.51
Early planting of crops	845	1.99882	4.69
Mixed cropping	802	1.97288	4.45
Use of organic manure	553	1.39984	3.10
Crop rotation	759	1.87766	4.22
Land rotation	500	1.00344	2.78
Use of cover crops	638	1.22209	3.54
Minimum soil tillage	480	1.00777	2.67
Mulching	753	1.77670	4.18
Change of planting time	863	1.99907	4.79
Planting of resistant varieties	737	1.76555	4.10
Adoption of irrigation techniques	495	1.06791	2.75
Expert advice	725	1.78889	4.03
Communal effort	701	1.58819	3.89
Government assistance	462	1.06791	2.57

Field survey 2016

on observed changes ($\bar{x} = 4.79$), change of cropping time ($\bar{x} = 4.03$), reverting to indigenous or self-help methods ($\bar{x} = 3.67$), consulting fellow farmers for advice ($\bar{x} = 3.33$) and playing ignorant of the situation at hand ($\bar{x} = 3.67$). The result indicates that most of the women try to obtain more information and education to be well equipped to make an informed decision on the adaptation measures to apply in such situation. Some in their bid to seek for information, consult fellow farmers either to be educated before taking up particularly strategy or to compare their situation and take the same decision as the neighbours. However, some of the women play ignorant of the existence of these adaptation measures, thereby falling

Table 4 Mean distribution of farmers' behavioural responses to adaptation measures to climate change in the study area

Variables	Total	SD	Mean
Change of cropping style and time	614	1.41128	3.41
Change of cropping time	725	1.70012	4.03
Seek for information based on observed changes	974	1.98822	5.41
Revert to indigenous or self-help methods	660	1.42215	3.67
Consulting fellow farmers for advice	564	1.41141	3.33
Value addition and storage of remaining crops	473	0.89921	2.63
Playing ignorant of the situation at hand	519	0.92134	2.88
Abandonment of farming enterprise due to loss of crop	493	0.91007	2.72

Field survey 2016

deep into the dangers and challenges of climate change menace. Timely dissemination of information and education of farmers on mitigation and adaptation strategies, weather forecast and other weather-related information is very important and a necessary prerequisite for farmers' informed decision-making. Agricultural extension services' delivery outfit, agricultural advisory services agencies and other related NGOs should as a matter of urgency provide required education and information on climate change issues at the grass-roots level.

13 Factors that Influence Farmers' Behavioural Approach to Adaptation Measures to Climate Change in the Study Area

Influences that act on individual decision-makers on their responses to adaptation strategies were measured, and the results are presented in Table 5. Results show that eight out of eleven factors have influences at different significant levels on the women's responses to adaptation strategies in the study area.

Age (−2.795)**: the coefficient of age has significance but negative effect on the women's response. The vulnerability of women to disasters is increased for a number of reasons. Post-disaster, women are usually at higher risk of being placed in unsafe, overcrowded shelters, due to lack of assets, such as savings, property or land (WHO 2012). Cultural constraints on women's movements may hinder their timely escape, access to shelter, alternative farm land or source of income. Exacerbating this effect is the issue of age which incapacitates the women making them less mobile and more vulnerable to weather vagaries' challenges and incapable of taking pragmatic action towards mitigation or adaptation strategy.

Household size (2.705):** In the absence of well-functioning labour markets, smaller households face higher labour bottlenecks at critical points in their family farming enterprise (Ezeh et al. 2012). The coefficient of household size shows significant positive influence on the women's response to adoption of adaptation measures. This has implications on labour supply and may justify the need to augment family labour with hired labour in disaster management situation. Therefore, household size which also represents labour supply needs to be duly considered.

Level of education (3.389*)**: the coefficient of level of education has a strong influence at 1% significant level on the women's choice of adaptation measures. This is in agreement with Imonikhe (2010) that education enhances farmer's ability to make accurate and meaningful management decisions and also that a farmer's level of education is an important factor in determining his/her ability to understand policies or programmes that affect farming, to accept and adopt agricultural innovations. Furthermore, education and training help to unlock the natural talents and inherent enterprising qualities of farmers and enhance their abilities to understand and evaluate new production techniques.

Table 5 Tobit regression result of the determinants of behavioural responses to climate change adaptation measures in the study area

Parameter	Estimate	Std. error	*t*-value
Age	−0.012	0.007	−2.795**
Household size	0.040	0.023	2.705**
Level of education	0.023	0.058	3.389***
Farming experience	0.001	0.006	0.221
Income	0.000	0.000	2.505**
Farming status	0.007	0.043	0.172
Farm size (Ha)	0.024	0.038	5.646***
Contact with extension	0.009	0.007	1.265
Membership of cooperative societies	0.000	0.000	5.027***
Access to credit	0.137	0.048	2.952***
Access to land	0.000	0.000	3.626***
Pseudo R^2	56.879		
Log-likelihood	221.198		
Pearson's goodness-of-fit test	323.271		

Income (2.505)**: the result indicates that the women's ability to adapt to the demands of climate change is influenced by the extent of their control over economic resources and access to economic and financial resources. In agreement with this finding, Nwibo and Okorie (2013) noted that income level of an individual plays a great role in shaping the type of production enterprise to venture into and subsequent decisions taken in course of the enterprise. Significant resources are needed to cover the cost of the goods, services and technologies required in the implementation of climate change measures in developing and developed countries.

Farm Size (Ha) (5.646*)**: the coefficient of farm size shows high influence at 1% significant level. Socio-economic analysis of the women indicated that the mean farm size of the respondents was 1.64 ha. This is a clear indication that the women farmers in the area are mostly small-scale farmers. This result which is quite a popular finding among previous studies confirms that rural households with small landholding and farm output which is the case among majority of women in the study area depend on targeted interventions from government and related agencies to escape the vagaries and adverse effect of climate change.

Membership of cooperative societies (5.027*)**: coefficient of membership of cooperative societies shows high influence on the decision process of the women. The involvement of more women in cooperative activities may be attributed to their being more gregarious in nature as well as their ability to organize themselves much more easily than males. In agreement with this finding, Badiru et al. (2016) observed that cooperative members gain new skills and training, access to agricultural information, credit, higher quality raw materials, thereby enabling members to engage in multiple activities through which women can make a informed decision and get out of poverty.

Table 6 Pearson's correlation result of the relationship between adaptation measures and behavioural responses of the respondents to climate change

Correlations			
		Adaptation measures	Behavioural responses
Adaptation measures	Pearson correlation	1	0.630
	Sig. (two-tailed)		0.094
	N	160	160
Behavioural responses	Pearson's correlation	0.630	1
	Sig. (two-tailed)	0.094	
	N	160	160

Two-tail correlation significant at 10%

Access to credit (2.952*):** inadequate capital is a major problem confronting small-scale farmers in Nigeria. Meanwhile, access to credit is regarded as one of the key elements in raising productivity (Anyiro and Oriaku 2011). There is generally observed low government support for women farm business. They cannot qualify for loans because they lack the collateral or because their farm business is small. Lack of assistance still keeps the women at the subsistence level of operation. At this level, the rate and prompt adoption of adaptation measures against climate change menace is often times greatly reduced.

14 Relationship Between the Adaptation Measures Practised by the Women and the Women's Behavioural Responses to Adaptation Measures to Climate Change in the Study Area

The two-tailed Pearson's correlation analysis for the study is positive (0.094) at 10% level of significance. This implies that the women's behavioural inclination for adaptation measures to climate change actually influences their adoption and practice of these adaptation measures. Thus, it is important to consider the women farmers' behaviour when seeking to access or improve the effectiveness of policies. This will enhance the level of conformity and responses of the affected farmers to utilization of these strategies (Table 6).

15 Conclusion

The study has examined the level of conformity, constraints and factors that influence behavioural responses of rural women farmers to climate change mitigation and adaptation strategies in Abia State, Nigeria. Results revealed that change of planting time ($\bar{x} = 4.79$), enterprise diversification ($\bar{x} = 4.51$) and crop rotation

($\bar{x} = 4.22$) are major adaptation measures practised by the women in the study area. Moreover, seeking for information based on observed changes ($\bar{x} = 5.00$), reverting to indigenous or self-help methods and change of cropping style and time ($\bar{x} = 3.41$) were identified as major behavioural responses of the women to adoption of climate change adaptation measures. Factors such as age (−2.795**), household size (2.705**), level of education (3.389***) and income (2.505**) were major factors that influence the women farmers' behavioural approach to adaptation measures to climate change in the study area. A better understanding of barriers and behavioural responses of the women when they are faced with climate change menace as well as barriers to effective execution of these practices can help to enhance higher conformity and adoption of the mitigation and adaptation strategies and policies when those barriers are tackled. Eighty per cent of the grass-roots farmers consist of the women. Women's involvement in climate change actions in the rural areas should be encouraged, and this will result in effective diffusion and adoption of environmentally friendly agricultural production global best practices.

From the result of the study, it is therefore recommended that targeted and timely information is provided to the women through agricultural extension and mass media approaches. These information and technologies should be designed in such a way as to be relevant to their circumstances. Again, the most effective and sustainable indigenous technologies used by the farming communities could be incubated with a view to upscale and out scale them in other to enhance adaptation to climate change vagaries by the resource-poor farmers. ICT-driven virtual platform that provides weather advisory services for small-scale farmers should be made available through rural advisory services (RAS). This will help farmers handle climate risk situation promptly, such as prolonged droughts or floods and other related climate change matters. This platform will enable farmers to get alerts about impending weather conditions and recommendations on what they should plant and mitigation/adaptation strategies to apply according to their location.

References

Abegunle, A.A., Adeyinka, S. A., Olawuni, P. O., & Oluodo, O. A. (2006). An assessment of the socio-economic impacts of soil erosion in Southern Nigeria. *Shaping the Change, XXIII FIG Congress*, Munich, Germany, pp. 1–15.

African Development Bank Group (ADBG). (2015). Lake Chad, a living example of the devastation climate change is wreaking on Africa. 03/12/2015 © 2016 African Development Bank Group, p. 2.

Ajzen, I. (2005). Attitudes, personality and behaviour. *Open Journal of Farm Management, 13*(4), 5–7.

Akinbile, L. A., & Odebode, S. O. (2002). Determinants of farmers use of sustainable conservation practices in Osun State, Nigeria. *Challenges of organic farming and sustainable land use in the tropics and subtropics*. http://www.tropentag.de/2002/abstracts/links/akinbile-nfaxzey7.pdf.

Akpokodje, E. G., Tse, A. C., & Ekeocha, N. (2010). Gully erosion geo-hazards in Southeastern Nigeria and management implications. *Scientia Africana, 9*(1), 20–36.

Anselm, A. E., & Taofeeq, A. A. (2010). *Challenges of agricultural adaptation to climate change in Nigeria: A synthesis from the literature*. Retrieved May 3, 2016 from http://factsreports.revues.org/index678.html.

Anyiro, C. O., & Oriaku, B. N. (2011). Access to and investment of formal micro credit by small holder farmers in Abia State, Nigeria. A case study of ABSU Micro Finance Bank, Uturu. *Journal of Agricultural Science, 6*(2), 69–76.

Apata, T. G., Samuel, K. D., & Adeola, A. O. (2010). *Analysis of climate change perception and adaptation among arable food crop farmers in south Western Nigeria*. Paper presented at the conference of International Association of Agricultural Economics, pp. 2–9.

Badiru, I. O., Yusuf, S. K., & Anozie, O. (2016). Adherence to cooperative principles among agricultural cooperatives in Oyo State, Nigeria. *Journal of Agricultural Extension, 20*(1), 8.

Commission on the Status of Women (CSW). (2008) *"Gender perspectives on climate change," Issues paper for interactive expert panel on Emerging issues, trends and new approaches to issues affecting the situation of women or equality between women and men*. Retrieved March 20, 2017 from www.un.org/womenwatch/daw/csw/csw52/issuespapers/Gender, p. 2.

Commission on the Status of Women (CSW). (2013). *"Gender perspectives on climate change," Issues paper for interactive expert panel on Emerging issues, trends and new approaches to issues affecting the situation of women or equality between women and men*. Retrieved March 20, 2017 from www.un.org/womenwatch/daw/csw/csw52/issuespapers/Gender, pp. 1–2.

Ekweanya, A. M. (2014). *Assessment of effects of climate change on arable crop production in Anambra State*. M.Sc. thesis presented to Department of Rural Sociology and Extension, Michael Okpara University of Agriculture Umudike, Nigeria, pp. 4–7.

Ezeh, C. I., Anyiro, C. O., & Chukwu, J. A. (2012). Technical efficiency in poultry broiler production in Umuahia Capital Territory of Abia State, Nigeria. *Greener Journal of Agricultural Sciences, 2*(1), 001–007.

Federal Government of Nigeria (FGN). (2005). *First national communication on climate change*. Retrieved 16/02/2017 from http://www.fgn.ng/nationacommunicationonclimatechange.20, pp. 13–16.

Food and Agricultural Organization (FAO). (2017). *GIEWS—global information and early warning system*. Retrieved 22/06/2017 fromhttp://www.fao.org/giews/countrybrief/country.NGA.

Gordon, A., & Craig, C. (2011). *Rural Non-farm activities and poverty alleviation in sub-Sahara Africa*. Policy series 14, Chatham U. K NRI Retrieved July 19, 2013 for www.article.Sapub.org.

Igwe, C. A. (2012). Gully erosion in south-eastern Nigeria: Role of soil properties and environmental factors. *Intech-Open Access Company*, pp 17–19. https://doi.org/10.5772/51020.

Iheke, O. R., & Olive-Abali, V. I. (2011). Farm size, climate variability and crop production in Abia State, Nigeria. *International Journal of Sustainable Agriculture, 3*(2), 58–64.

Intergovernmental Panel on Climate Change (IPCC). (2010). *Climate change: Impacts, adaptation and vulnerability*. IPCC Working Group II Report, Chapter 19, IPCC.

Jones, P. G., & Thornton, P. K. (2002). *Croppers to livestock keepers: Livelihood transition to 2010 in Africa due to climate change* (pp. 3–4). Geneva, Switzerland: Global Environmental Change, World Health Organization.

Mgbada, J. U (2013). *Agricultural extension; the human development perspective* (2nd ed., p. 3). Computer Edge Publishers. ISBN: 978-8072.

National Bureau of Statistics (NBS). (2005). *Social statistics; federal ministry of water, Abuja, 2005 figures*. Retrieved 29/10/2015 from http://www.nbs.ng/socialstatistics/countrybrief/fmw.

NOAA. (2016). *National Centers for Environmental Information, State of the Climate: Global Analysis for March 2016, published online April 2016*. Retrieved May 7, 2016 from http://www.ncdc.noaa.gov/sotc/global/201603.

Nwaobiala, C. U. (2014). Socio-economic factors influencing farmers' participation on IFAD/FGN/NDDC/community-based natural resource management programme technologies in Abia and Cross River State, Nigeria. *Journal of Agricultural Extension, 18*(1), 49–61.

Nwaobiala, C. U., & Ubor, V. U. (2016). Effectiveness of electronic wallet system of growth enhancement support scheme distribution among arable crop farmers in Imo state, south eastern Nigeria. *Scientific Papers Series Management, Economic Engineering in Agriculture and Rural Development, 16*(1), 2–3.
Nwibo, S. U., & Okorie, A. (2013). Determinants of entrepreneurship among agribusiness investors in South-East, Nigeria. *European Journal of Business and Management, 5*(10), 115–123.
Odjugo, P. A. O. (2010). General overview of climate change impacts in Nigeria. *Journal of Human Ecology, 29*(1), 47–55.
Odogwu, G. (2012). *Flooding menace: Implications for food security*. October 25, 2012: PUNCH Breaking News, information and opinion in Nigeria. Retrieved September 29, 2015 on Africa: Five-Year Review of the Implementation of the World Summit on food security.
Okojie, E. E. C. (2002). Gender and education as determinants of household poverty in Nigeria. *World Institute of Development Economics Research*. Discussion Paper No 2002/37, pp. 5–6.
Umeh, O. J. (2015). Effectiveness of gender mainstreaming on grassroots agricultural and rural development. In I. Nwachukwu (Ed.), *Contemporary issues in extension systems and development* (pp. 91–95). Enugu: University Press.
UN WomenWatch. (2013). *The UN internet gateway on gender equality and empowerment of women*. www.un.org/womenwatch.
UNDP Asia-Pacific Human Development Report. (2011). *Overview of linkages between gender and climate change*. http://www.undp.org/gender.climatechange, pp. 2–4.
United Nations Framework Convention on Climate Change (UNFCCC). (2007). *Climatic change impact, vulnerabilities and adaptation in developing countries UNFCCC Secretariat*, Martin-Luther-King-Straat 8 53175 Bonn, Germany. www.unfccc.int.
Warui, H. (2011). *Climate change adaptation: Taking the message from the office to the farmer for action*. Paper presented at the Climate Change: Local Perspective Global Concern Workshop held from 22nd to 23rd June 2011 at Egerton University, Egerton, Kenya.
WHO. (2012). *Gender inequities in environmental health*. 25th Session of the European Environment and Health Committee (EUR/5067874/151).
World Health Organisation (WHO). (2015). *WHO warns climate change to threaten human health in Nigeria*. Email: warigonc@who.int.
Ziervogel, G., Nyong, A., Osman, B., Conde, C., Cortes, S., & Dowing, T. (2006). *Climate variability and change: Implications for household food security*. Assessments of Impacts and Adaptations to Climate Change (AIACC) Working Paper No. 20, January 2006. The AIACC Project Office, International START Secretariat, Washington DC, USA, p. 6.
Zoellick, R. B. (2009). A climate smart future. *The Nation Newspapers* (p. 18). Lagos, Nigeria: Vintage Press Limited.

Ogechi Jubilant Umeh has a first-class degree in Agricultural Extension and Management in 1998, master's degree in Agricultural Extension and Administration in 2014 and currently a Ph.D. research student also in Agricultural Extension and Administration at Michael Okpara University of Agriculture, Umudike, Abia State, Nigeria. She is currently a lecturer at the same university, working in the Department of Rural Sociology and Extension. Her current research interests include agricultural extension, rural advisory services and gender issues in agriculture.

Professor Ike Nwachukwu is a Graduate of Social Sciences (1983); obtained master's degree in Agricultural Communication (1987) and a doctorate in Agricultural Extension and Rural Sociology (1999); and had series of certificate courses in communication training skill, effective public relations, scientific writing and production of extension materials. He became a Professor of Agricultural Extension and Communication in 2008 and currently a Lecturer and Director at Michael Okpara University of Agriculture Umudike. He has done over 200 publications among books, chapters, journals and scientific papers.

Climate Change Impacts and Research in the Caribbean: Constraints, Opportunities and the Role of Tertiary Institutions

Lisa Benjamin and Adelle Thomas

Abstract Climate change is anticipated to have, and is already having, negative impacts on small island developing states (SIDS). Biophysical impacts include sea level rise, increased coastal erosion, ocean acidification and increased frequency of extreme events, leading to incidents of loss and damage in these states. The impacts of loss and damage are leading to negative consequences for the economies of these states, including shrinking public budgets, decreased availability of finances, negative impacts on industries such as fishing and tourism, as well as negative impacts on livelihoods, leading to what the authors argue is an unvirtuous cycle of eroding development and resilience. Given the severity of these impacts, it would be assumed that increased research on climate change, and its impacts, would have been ramped up in these states. However, Caribbean states suffer from capacity constraints, and research, including systematic observation, particularly of slow onset events, has consequentially been limited. The lack of localised data on climate change further hampers research initiatives in tertiary institutions. This paper seeks to explore the relationship between capacity constraints in Caribbean SIDS and research gaps in climate change. This paper provides a brief analysis of current activities of tertiary institutions in the region in relation to climate change and seeks to provide remarks on how research efforts in the region can be further coordinated and improved by bridging existing capacity constraints.

The authors are co-founders of the Climate Change Initiative; however, the views of the authors in this paper are personal to the authors and do not necessarily reflect the views of The University of The Bahamas.

L. Benjamin (✉)
LL.B. Programme, The University of the Bahamas, P.O. Box N-4912, Nassau, Bahamas
e-mail: lisa.benjamin@ub.edu.bs; lrb1973@yahoo.co.uk

A. Thomas
Climate Analytics, Nassau, Bahamas
e-mail: adelle.thomas@gmail.com

© Springer International Publishing AG, part of Springer Nature 2019

W. Leal Filho and R. Leal-Arcas (eds.), *University Initiatives in Climate Change Mitigation and Adaptation*, https://doi.org/10.1007/978-3-319-89590-1_8

Keywords Climate change · Caribbean SIDS · Capacity constraints
Climate research · Tertiary institutions

1 Introduction

Climate change is anticipated to have, and is already having, negative impacts on small island developing states (SIDS). Biophysical impacts include sea level rise, increased coastal erosion, ocean acidification and increased frequency of extreme events, leading to incidents of loss and damage in these states. The impacts of loss and damage are leading to negative consequences for the economies of these states, including shrinking public budgets, decreased availability of finances, negative impacts on industries such as fishing and tourism, as well as negative impacts on livelihoods, leading to an unvirtuous cycle of eroding development and resilience. Given the severity of these impacts, it would be assumed that increased research on climate change, and its impacts, would have been ramped up in these states. However, Caribbean states suffer from capacity constraints, including human and financial constraints, and research, including systematic observation, particularly of slow onset events, has consequentially been limited. The lack of localised data on climate change further hampers research initiatives in tertiary institutions. The lack of localised data also hampers the ability of policy makers to develop detailed and nationally relevant plans and policies. The impacts of climate change are also exacerbating existing financial constraints, with funding to adapt to, and manage, loss and damage due to climate change primarily coming from public, national budgets. Combined, these constraints and impacts create an unvirtuous cycle of climate, development and resilience constraints, which compound the existing vulnerabilities of these states. However, these gaps also create opportunities for regional tertiary institutions. There are a number of existing tertiary institutions which are carrying out research on environmental change, sustainable development, as well as climate change; however, these efforts are often not coordinated between institutions.

This paper seeks to explore the relationship between capacity constraints in Caribbean SIDS and research gaps in climate change. This paper provides a brief analysis of current activities of tertiary institutions in relation to climate change and seeks to provide remarks on how research efforts in the region can be further coordinated and improved by bridging existing capacity constraints. Section two of this paper explores the existing and anticipated negative impacts from climate change in SIDS and highlights the vulnerability of these states. Section three explores some of the general and specific capacity constraints faced by Caribbean states, including both the economic impacts of climate change and data constraints. Section four provides a brief overview of the work being carried on by existing tertiary institutions in the region. Section five concludes with a discussion of some ongoing research gaps, as well as an overview of possible opportunities for research institutions in the region.

2 Vulnerability of Caribbean SIDS to Climate Change

SIDS in the Caribbean are far from a homogenous group as they vary in physical size, geographic characteristics, population, culture, levels of development and sovereignty. The small islands of the Lesser Antilles are dwarfed by the relatively larger islands of Hispaniola and Cuba. Some of the islands such as Jamaica and Saint Lucia are mountainous while others like The Bahamas barely exceed sea level. Cuba, with a population of 11.2 million far exceeds the 46,396 inhabitants of Saint Kitts and Nevis (UN 2017). Haiti, the first independent nation in the western hemisphere, stands in stark contrast to the islands of Montserrat, Saint Barthelemy and Anguilla that are territories of European countries. Belize and Guyana are not islands but are located in Central and South America. English, Spanish, Haitian Creole, French, Dutch and Papiamento along with a wide variety of creoles, patois and dialects can be heard across the region. The various colonizers of the region and histories of migration have resulted in a mosaic of ethnicities, governance structures, religions and cultural characteristics that is as varied as the islands themselves.

However, despite the many differences between Caribbean SIDS, there are a number of characteristics that can be found in most and that result in extreme vulnerability to climate change (Pelling and Uitto 2001). The relatively small physical size of SIDS in comparison to other countries results in intensive land use and the spatial concentration of infrastructure and populations. For many SIDS, remoteness and limited internal manufacturing results in a dependence on importing goods from external suppliers for critical items such as food, medicine and other goods, often at high cost. SIDS often have large coastal areas, with some nations such as The Bahamas being characterized as completely coastal, due to the small size of the islands. Demographic characteristics including limited human resources and the concentration of populations in coastal areas are also common. Small economies and dependence of industries such as tourism, agriculture and fisheries on natural resources are also characteristics of SIDS. These features of SIDS lead to greater exposure to climate change impacts and limited adaptive capacity which in turn result in high vulnerability to climate change.

The impacts of climate change for Caribbean SIDS are dire, and SIDS have already begun experiencing these impacts (Nurse et al. 2014). Sea level rise and coastal erosion are already beginning to be experienced in the region, threatening populations and infrastructure located in coastal zones. Coral reefs, an important resource for fisheries as well as for coastal protection, have also been in decline due to higher sea surface temperatures and ocean acidity. Mangroves, another important resource for coastal areas, have also been negatively affected due to rising sea levels. Terrestrial and marine biodiversity has been reduced due to changes in habitats spurred by rising air and water temperatures. Limited space in coastal zones has resulted in placement of infrastructure and populations in physically exposed areas, increasing their vulnerability to extreme events and slow onset events. Coral bleaching and beach erosion have resulted in negative perceptions by tourists of some islands as well as decreased accommodation pricing. There have also been impacts on human health with high incidences of leptospirosis being found in the region, a pattern that has been linked to climate variability. Changes in patterns of precipitation have resulted in concern about freshwater availability and impacts on health, sanitation and hygiene.

While Caribbean SIDS are already experiencing these impacts related to climate change, projections of future impacts are even more alarming. Perhaps most disturbing is the impacts of potential sea level rise for low-elevation islands. While estimates of sea level rise fluctuate based on levels of global warming, there is potential for tipping points to be surpassed which may result in melting of ice sheets and catastrophic levels of sea level rise (IPCC 2014). Even with more conservative estimates of sea level rise, the resultant impacts on SIDS that are already experiencing challenges will be dire. Sea level rise, combined with coastal erosion, threatens the very existence of some islands and is highly likely to result in increased incidents of loss and damage, impacts of climate change that cannot be avoided through either mitigation or adaptation (Benjamin and Thomas 2016). With warmer global temperatures, the impacts that SIDS are currently experiencing will be intensified and the limited adaptive capacity of these entities will likely result in significant physical, economic, social and cultural changes.

3 Capacity Constraints of Caribbean SIDS

3.1 General Capacity Constraints

As set out above, Caribbean states are diverse; however, many of these states share a set of characteristics which make them uniquely vulnerable to climate change, but are also uniquely constrained in dealing with the impacts of climate change. These characteristics include geographic vulnerabilities such as:

- Low-lying areas vulnerable to sea level rise and storm surges;
- Geographic positions strongly affected by tropical storms and hurricanes;
- Current high temperatures;
- Scarce land resources;
- Considerable dependence on fresh groundwater resources;
- Large coastal areas (Thomas 2016);

These geographic vulnerabilities are compounded by socio-economic vulnerabilities, which include:

- High concentrations of population and infrastructure in coastal regions;
- Dependence on a narrow range of imports;
- Dependence on natural resources;
- High import costs making them susceptible to fluctuations in international trade and commodity prices;

- Limited opportunities for economic diversification;
- High transportation and communications costs;
- Weak institutional structures;
- Limited human capacity (Benjamin 2010).

Many economies of these states are highly dependent on their natural resources, such as beaches, coral reefs, and mangroves for tourism and fishing industries. The impacts of climate change outlined in Section two are having, and anticipated to have, further detrimental impacts on these resources, further weakening the base of many of these countries' major industries. The existing vulnerabilities of these states, combined with the economic and social impacts of climate change, are further exacerbating their vulnerabilities and thereby increase capacity constraints.

3.2 Impacts of Climate Change Further Erodes Resilience

Caribbean states are struggling financially. A recent IDB report has questioned whether these states are suffering from a 'Caribbean sclerosis' (IDB 2014). The report notes that diseconomies of scale can lead to higher average costs for the public sector, and as a result under-provision of public goods (IDB 2014). This leads to fewer policy formation initiatives and reduced quality of public goods, including educational services (IDB 2014). These findings have been confirmed by the Global Competitiveness Index which ranks values for the Caribbean lower than the rest of the world in basic indicators such as institutions, infrastructure as well as higher education (World Economic Forum 2013–2014). The debt-to-GDP ratio for Caribbean states is also higher than the rest of the world (IDB 2014). It is not surprising, therefore, that given these financial constraints, environmental protection is not a top priority for these states. Ramlogan and Persadie (2004) note that the hardships of life in developing countries mean that environmental protection is largely placed on the back burner, in favour of more pressing needs such as poverty reduction, health care and infrastructural development.

The IDB report also notes that the region suffers from more natural disasters per square metre than the rest of the world (5.1 events per square metre as opposed to 1 event per square metre as the global average) (IDB 2014). The costs of these natural disasters are also higher for smaller economies to absorb (IDB 2014). The impacts of sudden-onset events such as hurricanes have impacts on public safety, livelihoods and public health as well as on infrastructure (UNEP and Sabin Centre 2015). The Caribbean region is experiencing increased incidents of loss and damage, and these incidents are having social as well as economic impacts on these states (Thomas and Benjamin 2017). These impacts put further strain on already stretched and weak public institutions. As Lyster (2015) notes, climate change often leads to uncompensated damages in the developing world as Governments have to absorb large welfare losses from public budgets. She also notes that after disasters, Government face further macroeconomic strains, including depleted tax bases, declining reserves and credit ratings making borrowing more difficult, leading these states to divert capital from other social programmes to post-disaster efforts (Lyster 2015). These costs remove financing from other important development objectives, such as poverty reduction, health care, education and, ironically, climate change adaptation (Benjamin and Haynes 2017). The impacts from climate change are therefore decreasing the ability of these states to adapt and develop resilience to climate change impacts, and leading to an unvirtuous cycle of eroding development and resilience. Existing capacity constraints restrain the ability to recover from extreme events. Extreme events then divert money from existing public programmes, including programmes for sustainable development and climate change adaptation, leading to increased structural, including capacity, constraints.

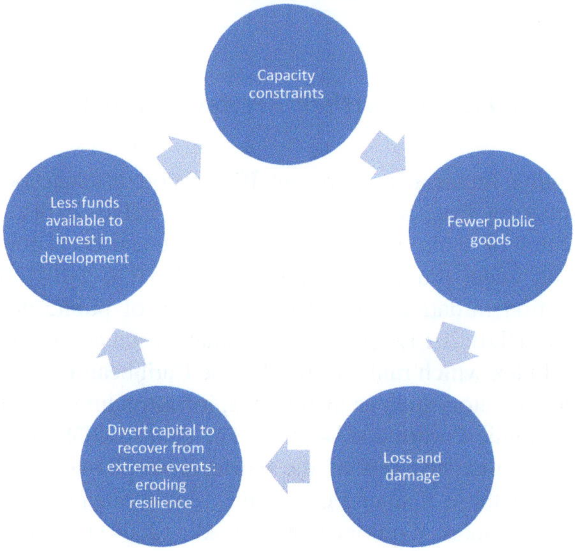

3.3 *Lack of Localised Data*

The human, technical and financial constraints outlined above have led to limited data being collected over the long term within SIDS (Thomas and Benjamin 2017). This lack of data has constrained policy makers, who need a steady stream of data in order to determine the impacts of climate change, the limits of adaptation, and to develop policies and strategies to deal with climate change, including loss and damage (Roberts and Pelling 2016). Therefore, the lack of localised data on climate change in these states has hampered the policy-development process. In a study of INDCs of SIDS, supplemented by interviews, Thomas and Benjamin (2017) found that while these states often had meteorological data on the intensity of extreme events, they also found that in most SIDS there was no ongoing data collection on sea level rise or ocean temperatures and acidity. The lack of data collection for slow onset events was often linked to capacity constraints (Thomas and Benjamin 2017). Interviewees in the study also acknowledged the lack of historical or baseline data in their countries, and that where data did exist, the datasets were often incomplete, or distributed among various government agencies (Thomas and Benjamin 2017). Data collection efforts were also uneven, often concentrated in urban areas, and in archipelagic countries, partial data collection for some islands was even more acute (Thomas and Benjamin 2017). They also found gaps in financial assessments of loss and damage due to climate change, as well as a lack of policies targeted towards loss and damages in these states (Thomas and Benjamin 2017).

Existing capacity constraints in Caribbean states are therefore hampering efforts to collect data on the impacts of climate change, which is consequentially hampering efforts at policy formation or policy updating in these states. Inability to create data-driven, detailed, nationally appropriate plans for climate change adaptation and loss and damage, leaves these states even further exposed to the impacts of climate change, and decreases their resilience, leading to multiple levels of compounded vulnerability. These gaps in data collection and policy formation provide opportunities for tertiary research institutions in the region to fill these capacity gaps and produce localised data. These institutions could work closely with policy makers to develop nationally relevant climate plans and policies for future climate-related action. Section four provides a brief overview of tertiary institutions within the region which are currently working on issues such as environmental change, sustainable development and climate change.

4 Overview of Tertiary Institutions in the Region

To assess the current activities of tertiary institutions in the Caribbean in relation to climate change research, an extensive search of peer-reviewed literature was conducted. First, a list of all tertiary institutions in the Caribbean was compiled. Next, the search term 'climate change' was used to search for literature that has been published since 2000 and where the author(s) listed their affiliation as a tertiary institution in the region. This allowed for identification of individual researchers that are active in climate change research. After identifying researchers, further details on their associated institutions, including research groups within these institutions, were collected using official information available on the Webpages of these organizations. While this methodology facilitated identification of tertiary institutions within the region that are affiliated with active researchers, the methodology could have been improved by the inclusion of interviews or surveys of the researchers. This would have allowed for a more in-depth analysis of the particular constraints facing researchers at different institutions and allow for comparisons between organizations, which may provide further avenues for research in this area.

4.1 University of the Bahamas and the Climate Change Initiative

The Climate Change Initiative (CCI) was originally co-founded by two academics at the University of The Bahamas and was designed to be a group of interdisciplinary faculty largely focused on the human dimensions of environmental change and policy implications. Although the University signed the Presidents' Climate Leadership Commitment (ACUPCC) in 2011, frequent changes in administration since that time led to a decline in support for climate change research and activities. Lack of support from senior management for sustainability initiatives is not unusual at tertiary institutions (Ávila et al. 2017). Due to lack of support for the CCI by the administration at the institution, the Initiative has now expanded its remit beyond the University and is now is a group of experts from around the world that conduct research on climate change in the Caribbean. This group has conducted research on a broad variety of subjects, but largely focuses on climate change, including (i) the role of AOSIS in the UNFCCC negotiations, (ii) responsibilities of carbon major companies, (iii) economic impacts, risks and vulnerabilities in coastal regions, (iv) adaptation opportunities, constraints and limits and (v) human rights approaches to climate change adaptation. More recently, the group has been focused on loss and damage and has published articles on (i) management of loss and damage by SIDS on a global scale and (ii) policies and mechanisms to address climate-induced migration and displacement in Caribbean and Pacific SIDS. The Climate Change Initiative has also been involved in other environmental-based research including integrated coastal zone management and disaster risk reduction.

4.2 University of West Indies: Climate Group

The Climate Group, located at the University of West Indies, Mona, Jamaica campus, is housed within the Physics department of the University. The group consists of faculty, research fellows and students that are focused on researching the physical dimensions of climate change. As climate physicists, this group investigates the mechanisms involved in the mean climate and climate extremes in Jamaica and the wider Caribbean region. The group has expertise in (i) regional rainfall modelling, (ii) determination of renewable energy viability, (iii) impacts of climate on crops and (iv) agricultural drought monitoring. The Climate Group maintains close connections with the Government of Jamaica and has been instrumental in providing nationally scaled climate projections.

4.3 University of West Indies: Centre for Resource Management and Environmental Studies (CERMES)

CERMES is housed within the Faculty of Science and Technology at University of West Indies, Cavehill, Barbados campus. This is a fully functional programme that focuses on sustainable development in the Caribbean. The faculty and students are involved in a broad array of research that is related to sustainable development, including socio-economic monitoring for coastal management, adaptation of water supply systems and management of marine protected areas. Faculty at CERMES has published on a number of topics including implications of climate change for water management in the Caribbean, adaptation options for SIDS and impacts of climate change on fisheries.

4.4 University of Belize: Environmental Research Institute

The Environmental Research Institute focuses on research to support effective management, sustainable use and conservation of Belize's natural resources. The Institute has been involved in a wide array of projects including national monitoring of marine protected areas and natural resources, assessment of climate change impacts on water resources, conservation of biodiversity and identification of coral reef systems. Publications from the Institute have focused on coral bleaching and impacts to mangrove forests following extreme events. Calabash Caye Field Station is also part of the Institute and is a research facility that hosts both foreign and Belizean students with the aim of sustainable use and management of marine reserves.

5 Conclusion

As set out above, there are a number of ongoing research initiatives at a number of institutions within the region. However, many institutions are tasked with researching a plethora of environmental issues of which climate change is just one. The region needs more institutions that can be focused on climate change research in particular. This is tied to lack of expertise and limited human resources in these countries; researchers with a background in climate change are tasked with researching other environmental issues while researchers with backgrounds in other fields are also tapped to include climate change in existing research streams. Academics are also tasked with wearing many hats, often acting as UNFCCC negotiators, as well as sitting on international and domestic committees, often unfunded, while working full time in academia. This leaves many academics in the region unsupported and over-stretched, reducing their ability to produce focused and targeted research on climate change. While this is making the most of limited human resources within the region, there is a need to focus on climate change in particular as it is a nuanced and complex issue. Many Universities in the region still focus on teaching loads as a revenue generator and are reluctant to reduce teaching loads to allow faculty to focus on research. In addition, there is a need for connections between academics and policy makers, so that research that is being produced in the region can inform policy formation. These connections can also facilitate the development of a regional research agenda where both policy makers and researchers can identify areas in need of further research. Collaboration between regional institutions is also a pressing need. Further collaboration could lead to the production of regional in addition to national studies on climate change. Such collaboration would allow for best practices to be identified and lessons learnt to be disseminated. Regional hubs on climate change such as the Caribbean Community Climate Change Centre could drive further and better collaboration by establishing academic portals to collect and disseminate regional research on climate change.

The main lessons learnt from the examination above is that capacity constraints, and in particular a lack of human and financial resources, are hampering research on climate change in the region, leading to a 'double inequity'. Not only are these states largely not responsible for the impacts of climate change they are experiencing, they have little data in order to plan and craft policies to cater for these impacts due to capacity constraints. The unvirtuous cycle of eroding development and resilience further compounds their vulnerability. While some tertiary institutions are conducting research on climate change, the broad research agendas of these units, combined with lack of support and data for focused climate change research, is hampering policy formation efforts. This gap provides opportunities for tertiary institutions to scale up and better coordinate efforts, and to put more emphasis on national and regional research on climate change. This would require increased funding for national and regional organisations dedicated to climate change research. Without funding, it is difficult for research to be completed. With little to no funding available, researchers must rely on existing data or easily acquired data. Also with little time for research, research produced may not be innovative or advance the understanding of climate change in the region. The need for capacity building for SIDS, including Caribbean SIDS, in the area of data collection and research, as well as policies and mechanisms to aid in assessment, monitoring and responses, particularly to loss and damage, map almost directly on to the Warsaw International Mechanism's (WIM) thematic areas on loss and damage (Thomas and Benjamin 2017), and these states should spend negotiating capital ensuring sufficient financing is provided to further research on climate change in the region. The focus of the WIM on research and data gathering provides a future opportunity for tertiary institutions to work with their governments to provide opportunities for regional institutions to fill the gap on climate change research and bridge existing capacity constraints.

References

Ávila, L. V., et al. (2017). Barriers to innovation and sustainability at universities around the world. *Journal of Cleaner Production, 164,* 1268–1278.

Benjamin, L. (2010). Climate change and Caribbean small island states: The state of play. *The International Journal of Bahamian Studies, 16,* 78–91.

Benjamin, L., & Haynes, R. (forthcoming 2017). Climate change and human rights in the commonwealth Caribbean: Case studies of the Bahamas and Trinidad & Tobago. In S. Jodoin, A. Johl, & S. Duyck (Eds.), *Routledge handbook of human rights and climate governance*. Routledge.

Benjamin, L., & Thomas, A. (2016). 1.5 °C to stay alive? AOSIS and the long term temperature goal in the Paris Agreement. *IUCNAEL E-Journal, 7,* 122–129.

Inter-American Development Bank (IBD). (2014). In I. Ruprah, K. Melgarejo, & R. Sierra (Eds.), *Is there a Caribbean sclerosis? Stagnating Economic Growth in the Caribbean* (p. 178). IDB Monograph: New York and Washington DC.

Intergovernmental Panel on Climate Change (IPCC). (2014). Climate change 2014: Impacts, adaptation, and vulnerability. In C. B. Field, V. R. Barros, D. J. Dokken, K. J. Mach, M. D. Mastrandrea, T. E. Bilir, M. Chatterjee, K. L. Ebi, Y. O. Estrada, R. C. Genova, B. Girma, E. S. Kissel, A. N. Levy, S. MacCracken, P. R. Mastrandrea, & L. L. White (Eds.), *Part A: Global and sectoral aspects. Contribution of Working Group II to the fifth assessment report of the intergovernmental panel on climate change* (p. 1132). Cambridge, United Kingdom and New York, NY, USA: Cambridge University Press.

Lyster, R. (2015). A fossil fuel-funded climate disaster response fund under the warsaw international mechanism for loss and damage associated with climate change. *Transnational Environmental Law, 4*(1), 125–151.

Nurse, L. A., Mclean, R. F., Agard, J., Briguglio, L. P., Duvat-Magnan, V., Pelesikoti, N., et al. (2014). Small islands. In V. R. Barros, C. B. Field, D. J. Dokken, M. D. Mastrandrea, K. J. Mach, T. E. Bilir, M. Chatterjee, K. L. Ebi, Y. O. Estrada, R. C. Genova, B. Girma, E. S. Kissel, A. N. Levy, S. MacCracken, P. R. Mastrandrea, & L. L. White (Eds.), *Climate change 2014: Impacts, adaptation, and vulnerability. Part B: Regional aspects. Contribution of working group II to the fifth assessment report of the intergovernmental panel on climate change* (pp. 1613–1654). Cambridge: Cambridge University Press.

Pelling, M., & Uitto, J. I. (2001). Small island developing states: Natural disaster vulnerability and global change. *Global Environmental Change Part B: Environmental Hazards, 3*(2), 49–62. https://doi.org/10.1016/S1464-2867(01)00018-3.

Ramlogan, R., & Persadie, N. (2004). *Developing environmental law and policy in Trinidad and Tobago* (p. 338). Trinidad: Lexicon Trinidad Limited.

Roberts, E., & Pelling, M. (2016). Climate change-related loss and damage: translating the global policy agenda for national policy processes. *Climate and Development,* 1–15. https://doi.org/10.1080/17565529.2016.1184608.

Thomas, A. (2016). Overlooked impacts of hurricane sandy in the Caribbean. In K. O'Neill & D. Van Abs (Eds.), *Taking chances: The coast after hurricane Sandy*. New Brunswick, New Jersey: Rutgers University Press.

Thomas, A., & Benjamin, L. (2017). Management of loss and damage in small island developing states: Implications for a 1.5 °C or warmer world. *Regional Environmental Change*. https://doi.org/10.1007/s10113-017-1184-7.

UN. (2017). Population and vital statistics report: Statistical Papers Series A Vol. LXIX. Retrieved July 27, 2017 from https://unstats.un.org/unsd/demographic/products/vitstats/Sets/Series_A_2017.pdf. Last Accessed July 16, 2017.

UNEP and Sabin Centre for Climate Change Law. (2015). *Climate change and human rights*. http://columbiaclimatelaw.com/files/2016/06/Burger-and-Wentz-2015-12-Climate-Change-and-Human-Rights.pdf. Last Accessed July 27, 2017.

World Economic Forum. (2013–2014). *Global competitiveness report*. http://www3.weforum.org/docs/WEF_GlobalCompetitivenessReport_2013-14.pdf. Last Accessed July 16, 2017.

Greening Theatre Landscapes: Developing Sustainable Practice Futures in Theatre Graduates

Linda Hassall and Stephen Rowan

Abstract This paper provides insight into how sustainability literacies can be embedded in theatre performance and production practices in a university context and consequently offers a model of sustainable theatre education that may adopted by others in the university and tertiary sector. The aim of this paper is to discuss sustainable theatre performance and production, practices in the Contemporary and Applied Theatre Department at Griffith University, Australia. Theatre production is resource vociferous and can be acknowledged as inefficient in its wastage and consumption of power. The negative impact of traditional theatre production practices upon the natural environment has seen inroads into policy change and practices within theatre companies at national and international levels. Further, the adaptation of green changes in higher education sectors suggests that there are initiatives being implemented in subjects other than science and environmental studies (Shephard 2008). The adoption of *Greening the Theatre* at Griffith University suggests the theatre department has a commitment to fostering environmentally sustainable societies through teaching and learning and campus operations. This paper specifically discusses one project: *Dust* (Hassall 2015) as being one of the major outcomes to date of the *Greening the Theatre Project* (Rowan 2015). The *Greening the Theatre Project* embeds green philosophies and practices within its university theatre production outputs. The qualitative practice-led research draws on theories of ecocriticism to generate the creative development methods that informed the development of the performance script. The production practices that were implemented in the staging of *Dust* were informed by Rowan's (2015) case study research of sustainability practices in an independent theatre production context. Consequently, the paper discusses how ecocritical themes were investigated as creative process and further how the per-

L. Hassall (✉) · S. Rowan
School of Education and Professional Studies,
Griffith University, Brisbane 4122, Australia
e-mail: l.hassall@griffith.edu.au

S. Rowan
e-mail: s.rowan@griffith.edu.au

© Springer International Publishing AG, part of Springer Nature 2019

W. Leal Filho and R. Leal-Arcas (eds.), *University Initiatives in Climate Change Mitigation and Adaptation*, https://doi.org/10.1007/978-3-319-89590-1_9

formance practice aligns with the adoption of a green philosophy for production, specifically in design considerations, staging realisation and administration.

Keywords Sustainability · Theatre production · University · Australia

1 Introduction: Theatre Practice and Climate Change Literacies

The broader scientific community have continued to debate climate change, its implications and the human impact on environmental degradation for many years (Mayall 2016, pp. 32–33). While this debate continues, artists and teacher artists in university environments are beginning to take the lead in 'promoting awareness of climate science' and in doing so 'are engendering the cultural shift that is required to initiate positive change' (Mayall 2016, p. 32). Throughout history, artists have been at the forefront 'of recording and reflecting the state of society and the natural world within which society exists' (Abrahams 2016, p. vii). In the contemporary context artists, teachers and scholars who are investigating the way physical, psychological, perceptual, producible and imagined ecologies shape art-making and educational forms are, as Munk (1994) and May (2005) suggest, situating themselves in a vast open field of histories to be reshaped, styles to rewrite and contexts to reperceive. As creative writers, dramatists, performance makers and scholars respond to escalating temperatures, rising sea levels and natural disasters on a daily basis, climate change events loom large in the contemporary imagination (Hassall 2017a, p. 1). Indeed from an ecological and ecocritical position, climate change art may be considered as an exploration of the contemporary 'atrocity', the twenty-first century Frankenstein monster of our own remaking, posing both a moral and a psychological paradox for us all (Hassall 2017b).

In Australia like in many other countries, the damage to the environment is not contained to industrial, mining, agricultural or overfarmed locations. The effect of climate change is gnawing at the heart of what was once considered 'The Lucky Country'. The effects of climate change on the natural environment are proving disastrous for our unique native species and for the inimitable landscape itself. Hassall (2017a, p. 4) and Schultz (2006, p. 10) discuss climate change in Australia as predicted by CSIRO data that suggest that Australia will become warmer by 6° or more by the middle of the century, the land will become more arid, sea levels will continue to rise as the Great Barrier Reef continues to deteriorate, entire communities will be destroyed by increasing bush fires, and devastating cyclone events will become more fierce and frequent. Min and Hens (2007) and Bhend and Whetton (2015) believe that recent warming in Australia like in many other countries can be partially attributed to human influence. Global warming and climate change pay testament to these particular recent historical human actions. It is the material 'ecological result of certain master-narratives' that Western culture has told over a short period of time (May 2007, p. 96) and to which artists and performance educators are responding to in Australian university sectors.

2 Australian Universities and Sustainability Education

Since the latter part of the twentieth century, there has been much discussion and support for Australian universities to provide curriculum diversity pertaining to the environment yet research suggests that there are barriers to sustainability education (Thomas and Nicita 2002, Holdsworth et al. 2008). Thomas and Nicita (2002) suggest this may be due to environmental education being incorporated in courses on an 'ad hoc basis' or 'according to the enthusiasm of and interest of individual academics' (p. 477). Holdsworth et al. (2008) discuss the need for a paradigm shift in the Australian university education sector as necessary in addressing environmental problems and state that traditional pedagogical approaches to teaching and education 'are not necessarily the best models to achieve this' (p. 133). Lee et al. (2013) suggest that many universities in Australia, while committed to higher education sustainability development (HESD), do not have well-developed visions, goals and graduate attributes and posit that 'the journey of embedding HESD in university education [in Australia] is a long road to travel' (p. 27). Reid et al. (2006) discuss how artistic and creative disciplines can contribute to sustainability in the higher education sector and state that artistic disciplines 'can act as a catalyst for the development of individual and corporate ecological stewardship' (p. 90). As an art-making creative investigation, the *Greening the Theatre* project and its artistic outcome, *Dust,* align with *Griffith University's Sustainability Plan* 2017–2020. The plan expresses the universities' intention to embed sustainability across all elements of the university. Further, the plan outlines a commitment to sustainability in corporate and operational practices (pp. 8–23) that are being reflected in the theatre industry outside the university sector.

3 Methodologies

This paper outlines two areas of theatre education sustainability practice: the creation and development of a theatre performance and the green frameworks implemented in producing the work. Consequently, independent methodologies were utilised through the separate phases of the project. Qualitative methodologies utilised in the first phase of this research include practice-led methods specific to scripting processes.

3.1 *Dust: Creative Development Methods*

Methods that informed the thematic, creative development of *Dust* included a review of ecocritical and sustainable literature that was explored to inform both the thematic and the aesthetic development of the play. Creative development methods utilised included play readings and dramaturgical play development cycles to enlist feedback for redrafting.

Artists who have a long history of professional art making outside of academic or research disciplines usually conduct practice-led research. The burgeoning field of creative research in the academy suggests that it is possible to more formally facilitate the link between professional creative behaviours and research inquiry. The practice-led research associated with the development of *Dust* was predominantly conducted as theatre making and was generated by Hassall (2015) in association with an ensemble of undergraduate student artistic collaborators. Experimentation and creative development activities and investigation were conducted in rehearsal studios and theatre spaces and constituted a creative shared life space for the researcher and collaborators. The creative environment promoted creative experiences and dramatic expression pertinent to reflexive and reflective performance development methods. The performance or play became *both* the subject of the research and the method of knowledge production. Subsequently, the creation of the performance text was aligned with industry frameworks of practice experiences wherein the primary focus is on artistic development of the artistic product, e.g. the theatrical and the dramatic contexts of the work. The ecocritical construction and/or analysis inherent within *Dust* became the dramatic subject and the aesthetic.

3.2 *Dust as Ecocritical Discourse*

Theatre and performance makers in university and industry environments have the unique opportunity to tell stories from critical and creative perspectives about our undeniable human predicament and about our tenuous ecological circumstance. Environmental and ecocritical themes are becoming recurring topics of theatrical investigation in contemporary Australian drama discourse (Carleton 2016, 2006; Hassall 2013, 2015), and consequently, these contemporary dramatic investigations filter into higher education arts learning environments. The representation of nature and climate in these plays is effectively constructed through the relationship between theoretical investigation and creative exploration. Increased awareness of the challenges associated with promoting awareness on environmental sustainability and climate change has prompted the development of literary and drama studies in an area known as ecocriticism. Ecocritical ideology in some instances can be deployed in order to describe the multifarious dynamic and interdependent relationships between theatre and its sociopolitical context, the performer/actors relationship to audience and space and theatre's production and reception (May 2007). Theatre allows us the opportunity of exposing hypothetical futures together by transforming nature into sociocultural representation. Theatre therefore provides a representational vocabulary by offering 'a sustained exploration of the political cultural effects of [dramatic] visual rhetoric' (DeLuca and Demo 2000. p. 242). In turn, the dramatic rhetoric can contribute to growing discourses that view artistic works as integral to political analysis.

Dust as a dramatic text draws on theories of ecocriticism to enhance the thematic development of the work. Written from the perspective of an apocalyptic future, *Dust* asks the audience to consider the rapid escalation of climate change and associated events. *Dust* recognises nature as a cultural construction. Viewing current climate change events from future perspectives, the play establishes a symbolic performative vocabulary around ecocritical and environmental analysis by making a claim for sustainability awareness and action. The play is an example of what Shephard (2008) describes as proving valuable in promoting sustainability education processes that contribute to the eclectic mix of activities that are addressing affective outcomes in individuals and groups (p. 93). The play is a representation of a future Australia and as such offers a way of seeing and experiencing climate change through a created hyperreality. In doing so, the play establishes a symbolic vocabulary for environmental analysis by making claims for sustainability awareness through providing a 'viewing of nature as landscape that has failed' (DeLuca and Demo 2000, p. 246). The work provokes audiences to act through its exploration of environmental themes and its call for immediate climate action. The play is performed as an intimate conversation between performer and audience, placing the responsibility for change firmly in the audience's hands. Hudson (2012) warns that there is a distinct divide between scientific models of documentary drama whereby aspects of real-world climate challenges are depicted *and* plays such as *Dust* are written by theatre makers who exploit ecological themes for performative theatrical expression. We were conscious of this criticism when we began the rehearsal production process and aimed to develop a sustainable production in execution as well as theme. Essentially, the *Greening the Theatre* project addresses Hudson's criticisms through sustainable themes in performance *and* in the implementation of sustainable production processes. Thematically, the play poses the exacerbated circumstances of a society that are denied a sustainable future due to extreme weather events, climate change, pollution, depleting resources and environmental and social degradation. The realisation of the dramatic world is achieved through sustainable production practices and processes including the aesthetic considerations of scenography (completely recycled and recyclable) and lighting design (all LED lighting) and through administrative considerations including marketing (all paperless).

While theatre processes and methods may seem a unique way of developing climate literacies and sustainable futures in performing arts graduates, Lorenz (1999) and Duxbury (2010) posit that scientists are often given an unenviable role in predicting the escalation of climate change and are often faced with providing conflicting information when advising on the topic. Duxbury (2010) warns that there are overwhelming problems in providing certain communities with vast quantities of opposing evidences and abstract statistics that have 'the potential to alienate the general public and render them helpless in the face of such overwhelming problems' (p. 295). However, art making, specifically performance as discussed in this case, has the unique ability to direct focus to a particular place, a particular moment and a particular people or community in the constructions of imagined worlds (Morton 2007). Introducing climate change themes into performance and production processes and practices suggests critical information specific to the audience can be selected, organised and disseminated to fill knowledge gaps to not only enhance understanding but to also 'remediate misconceptions' (Fortner 2001, p. 21). While *Dust* addresses sustainability themes, this paper suggests it's significant contribution to sustainability discourse is in its adoption and implementation of green philosophies at all production levels.

3.3 Greening the Theatre: Production Methods

Traditionally, ecocriticism grounds its discourse in text and as such the theory has rarely addressed the 'messy, multi-layered and ambivalent issues that arise from performance production' (May 2007, p. 96). The production methods used addresses May's statement by exploring the intersection between *Dusts* politicised environmental themes and the green philosophy that generated the sustainable practices utilised to stage the work. The *Greening the Theatre* project led by Griffith University's Contemporary and Applied Theatre's (CAT) Production Manager, Stephen Rowan addresses theatre production's resource voraciousness at Griffith University. Case study methods informed by Johnson's (2009) methodological framework that outlines three areas of sustainability—ecology, economy and social—inform the production processes implemented at Griffith University. In defining each of these concepts, Johnson (2009) describes ecological sustainability as the 'reduction of waste, energy consumption, and use of toxic materials' and economic sustainability as achieving a 'reduction in costs while maintaining production quality, providing living wages for artist, and ensuring the long-range financial viability of theatre companies'. He further defines the social aspect of sustainability as the 'creation and maintenance of stable and functional intra-company social relations, expanded partnerships with surrounding communities, promotion of artistic diversity, and the production of social capital' (p. iii). These principles coincide with *Griffith University's Sustainability Plan* 2017–2020. Johnson posits that a 'key premise of the sustainability movement is that these three areas [ecology, economy and social], are inextricably interrelated, and must be addressed as parts of the larger whole…' (p. 10). Utilising this framework, Rowan's (2015) research into sustainable theatre production practices was applied to a case study: *Water Wars* (2011–2015). Rowan then implemented the *Greening the Theatre* project in the Griffith University theatre department as a response to findings associated with this study that suggested

> The sustainability ethos of the *Water Wars* production had a definite impact on members of the production team and as such, there has been a sustainability awareness legacy generated by the production. This awareness may see *Water Wars* used as a blueprint for future shows. The production occurred at a time when artists and arts practices in Australia, and in this instance Brisbane, Queensland, were starting to address the ideas of climate change, greenhouse gas emissions and global warming. (p. 77)

This research informed the sustainable production processes adopted and implemented in the staging of *Dust*. Embracing this green legacy, the *Greening the Theatre* project considered what methods and approaches would best serve the exploration and practices of how theatre artists and practitioners responded to the issue of incorporating sustainability values within their theatre practice at a university level.

The production methods discussed address ecological and social sustainability practices as applied to theatre production and consumption in the university. Through the *Greening the Theatre* project, Rowan addresses unsustainable production practice from three main positions: scenic lighting, set construction and marketing. By implementing strong sustainability methods that consider design and production realisation and administration, students were exposed to learning frameworks that fostered sustainable processes and practices in the university learning and teaching environment. These processes align with green methods of theatre production and consumption that are slowly being incorporated in the Australian theatre organisational sector.

4 Movements in Greening Australian Theatre Practice

May (2007) states 'we need ecocriticisms that are dangerous to business as usual' (p. 104). Theatre-making organisations are highly visible members of the community and as such the way the company artists do business 'shine as a potent role model for our audiences and the larger community' (Johnson 2009, p. 22). Theatre production is resource voracious in relation to both lighting and set construction. It is often the case that sets are built for a production and then discarded as rubbish rather than being recycled for re-use of materials. Fried and May identified the situation of waste in theatre as problematic in 1994. They stated that 'the amount of waste produced in scenic construction is prodigious' (p. 28) and recognising the theatre industry's seemingly cavalier disregard of the notion of sustainability they stated that 'anyone with a concern for the future would feel incongruous designing and/or building sometimes enormous sets that have a useful life of a few weeks. After that, they are taken to the dump…' (p. 28). Johnson (2009) extends the discussion by stating that 'as larger and larger segments of the population become increasingly aware of the threat of environmental degradation, more theatre professionals are becoming concerned about the particular negative impacts of their own work' (p. 24). In Australia a green production philosophy is becoming more apparent with professionals such as Michael Cantor, former Artistic Director of

Melbourne's *Malthouse Theatre*, critiqueing the 'business-as-usual' model of theatre production within Australia. He states

> We know the planet is facing an environmental catastrophe of a scale and impact unlike anything it has faced in the period of human civilisation. Yet we continue to develop and produce our theatre, in our energy intensive theatres, in our energy intensive ways, to our fickle and dwindling audiences, as if we didn't know of this impending catastrophe, business as usual. If there is one thing that is clear about the massive reduction in resources that we as a species must come to terms with in the future, it is that it requires changes to every part of our lives, as individuals responsible for our own carbon footprint, and as participants in a 'culture industry'. (*Greening the Arts* 2010, p. 10)

Ecological or sustainable management in theatre production is arrived at, as Fried and May (1994) suggest, via a slow process of self-assessment of personal philosophies and through researching alternative possibilities, e.g. power and materials and reorganisation (p. 9). Steelquist (1994) argues that there may be no better place than the theatre to include so many people in the hope of an ecological sustainable future (p. 23). However, within theatre practice, the fact remains that there is damage done through unsustainable practices including lighting, set realisation and excessive resource wastage (Johnson 2009). Sustainable practices may be addressed in the organisational implementation and processes associated with the production of theatre as a cultural product. Morris (2007) states

> As a business of waste, the theatre industry reflects a consumptive society. Art reflects life. If there is to be real substantive changes in the world, how the theatre and entertainment industry reflect the practice of society will have an impact. (pp. 18–19)

Going green in theatre practice requires a transformational change that suggests a shift in thinking about how theatre as a business behaves sustainably and consequently how others perceive the behaviour of that business. Green design choices can address the carbon footprint associated with theatre, performance making and the production process, and the negative impact of traditional theatre production practices upon the natural environment has seen inroads into policy change and practices within theatre companies at national and international levels. In Australia, there have been significant initiatives developed that attempt to address the issues of sustainability in theatre practice. From a local perspective, the State Government of Queensland launched an investigation into the status of theatre in the state. The ensuing investigative report *Mapping Queensland Theatre* (2009) (aka the *Baylis Report*) identifies the issue of rising costs associated with theatre companies' compliance with sustainability practices and operations. In addition to identifying theatre as a labour-intensive industry, Baylis notes another pressure on theatre companies is the expectation they become sustainable in practice and that in the not too distant future:

> There will also be increasing pressure from government and the community for all organisations to be more environmentally responsible. Eventually this may lead to more efficient forms of operation, but the more immediate effect is likely to be further increases in costs through compliance regulations and taxation on unsustainable practices. (Baylis 2009, p. 11)

On a national level, 2010 saw the establishment of a 'not for profit' organisation called *ClimARTe* that aims at:

> harness the creative power of the Arts to inform, engage and inspire action on climate change by creating an alliance of Arts practitioners and organisations that advocate for immediate, effective and creative action on climate change [by] promoting and facilitating climate and sustainability related Arts events. (http://www.climarte.org)

The organisation states they are 'providing a forum for the exchange of ideas and information on sustainable Arts practice at an individual and organisational level' (http://www.climarte.org). *Live Performance Australia* (*Australian Entertainment Industry Association* or AEIA) was implemented in 2011 and launched the *Greener Live Performances Project* initiative. This initiative focussed on the ecoefficiency reviews and improvement processes of twenty Queensland-based performing arts organisations. The aim of the initiative was to 'develop sustainability case studies, online resources, training and assessment tools for the live performance industry—including performing arts companies, producers, promoters, festivals, venues, touring coordinators and suppliers' (http://www.liveperformance.com.au). *Greening Our Performance*, (GOP 2011), was set up by *Australian Circus and Physical Theatre Association* (ACAPTA). With high-profile members that include *Malthouse Theatre*, *Tipping Point Australia*, *ClimARTe*, *Circus Oz*, *ArtsHouse* and *Arts Centre Melbourne,* this initiative aims to have its members 'support each other in our endeavour to change the way we do things to reduce our resource use' (n.p.) (http://www.malthousetheatre.com.au). Further to its inclusion as part of *Greening Our Performance*, the *Malthouse Theatre* in Melbourne has also launched its own sustainability policy that aimed to reduce the company's power consumption by 60% between 2011 and 2015 (http://www.malthousetheatre.com.au).

However, the biggest sustainability action plan within the Australian theatre industry to date is that of *Sydney Theatre Company*'s *Greening the Wharf* initiative which saw the company installs the second biggest array of photovoltaic solar panels in Australia, together with a rainwater harvesting programme that is capable of supplying 100% of non-potable water needs of all arts organisations on Pier 4/5 of Sydney Harbour (http://www.sydneytheatre.com.au). The company was further committed to finding greener alternatives to their 'biggest power usage'—theatre lighting systems. *Sydney Theatre Company*'s former head of lighting, Henstock (2011), explains the company's initiative to:

> Introduce a green design policy that involves designers… and gets them to think about green alternatives to the traditional technologies… [this will] open up the door for designers to experiment and explore what they can achieve with these sorts of technologies. (https://www.sydneytheatre.com.au/magazine/posts/2011/…/video-greening-the-wharf)

Tipping Point (Australia) released *Greening the arts—thinkpieces for a zero carbon future and a survey of sustainable arts practices* in 2010. The paper contains six thinkpieces, with contributions from Australia and UK. In *The arts and a zero carbon future—the experience of Tipping Point*, Gingold 2010, embraces McKibben's (2005, 2009) sentiments in arguing that the arts and artists should be contributing to raising awareness of climate change and states:

> We desperately need the help of the most creative people around… whether what results is contemplative, activist, participatory, visceral, cerebral or political is much less important than that it exists. Quite simply, there needs to be a large body of work. (Gingold 2010, p. 5)

It is apparent that similar considerations are being considered in Australian universities. However, movements towards sustainability tend to focus on technologies such as lecture capture and online resources (Sherren 2006). The adoption of a *green* methodology towards theatre practice at Griffith University aims to contribute to environmentally sustainable societies through teaching and learning and campus operations. The *Greening the Theatre* project addresses concepts of green consumerism (Carducci 2009) from theatrical and production perspectives, whereby the *ecological processes* utilised in the manufacture of the theatrical product take precedence over the performance outcome. The project aligns with the Griffith University *Sustainability Plan* 2017–2020 in areas of energy conservation, recycling and waste management and transforming processes including paperless marketing strategies. Fostering the development of sustainable futures in theatre practice does take tenacity because as stated above it is resource voracious. The project's ecocritical framework addresses existing political structures by identifying areas of change in temporal and/or economic structures and in doing so engages graduates with the idea that it is possible to challenge unsustainable consumerist structures (Garrard 2010) in the field.

5 Results—Sustainable Theatre at Griffith University

The CAT department at Griffith University has an extensive production output across all years of the undergraduate Programme. Rowan's (2015) research supports Johnson's (2009) call for theatre professionals to address particular negative impacts of their work on the environment. Consequently, Rowan believes that sustainable theatre futures are a practice concept that can be implemented in the learning and teaching environment. *Dust* became the production model, whereby students engaging in the graduate performance production outcome were made aware of the ecological footprint that they may leave behind once entering the industry. Consequently in collaboration with CAT staff, Rowan and the theatre production academic team became instrumental in adopting sustainable approaches to theatre practices in the university by including:

- Introduction of LED lanterns,
- Utilising all recyclable resources in set design and construction,
- Addressing and restricting disposable resource practices,
- Implementing paperless marketing strategies.

5.1 *LED Lighting*

The first lens through which ecological tensions in the CAT department were examined was that of electricity consumption through use of theatre lights and three-phase power consumption.

When considering technical support in areas of lighting design through the lens of ecology, the notion of electrical power consumption comes into focus as one of the strongest and most recurring themes to emerge from data (Rowan 2015). In production practices, the consumption of energy is almost exclusively related to the use of theatrical lanterns with lighting design processes having scant regard for power consumption. Walters (2011) notes that a typical lighting design might burn a hundred thousand watts at any one time (Walters in Rowan 2015, p. 99). While the *Dust* production was lit entirely with LED lighting technology, over time all energy-burning lanterns in the CAT department will be replaced with green LED lanterns on Rowan's recommendations. The replacement of inferior lighting technologies with LED lanterns was further informed by research conducted by the former head of lighting at *Sydney Theatre Company* (STC), Henstock (2011), who was instrumental in identifying energy-efficient lighting as implemented in the 'greening' of the STC operations. Consequently, the upgrade to LED lanterns in all areas of production lighting technologies within the department has proven to be a huge energy-saving action and consequently promotes an ecological framework for sustainable power consumption in the area. Energy-saving actions in lighting design and production are shared with students in the learning and teaching environment in relation to devising and designing theatre lighting.

The second area addressed by the *Greening the Theatre* project is in the use of sustainable/recyclable materials utilised in the design and construction of sets and props.

5.2 *Design Considerations, Set Construction and Waste Restriction*

The CAT department works philosophically and practically towards design aesthetics that address ecological challenges. Set construction is also resource voracious and typically achieved through the use of a range of materials including timber, steel and other metals, cloth, screws, nails, glues, brackets, hinges, paints, etc. Typically chosen from virgin stock, these items are fashioned and constructed into the specific requirements for a particular production. Used for the season of a particular production set elements are usually dismantled for disposal as landfill, are incinerated or similarly disposed of after use in only one production season. Students are supported and assisted in design practices, whereby tutors support the utilisation of all recyclable materials in minimalistic set aesthetics within the contemporary performance production courses. Consequently, waste disposal and management strategies are minimised through the use of materials that are recycled and/or can be recycled at the end of the season. *Dust* provides an example of how this was achieved in areas of:

- **Sceneographic design**—set design was constructed of found [dead] sticks and branches collected from the Australian bush land surrounding the Griffith University campus. Once the production was over, the students returned the elements to the bush land environment.
- **Construction**—no toxic materials were used in set construction.
- **Costumes**—no new material for costumes of new costumes was purchased. Costumes were sourced only from second-hand clothing stores and returned to costume stock for future repurposing and redesigning.

5.3 Paperless Marketing Strategies

Paper consumption is a factor to be considered in terms of marketing theatre productions within the department, especially in terms of free programme allocation to audience members. Within industry and university contexts, paper consumption includes design, printing and distribution of postcards, flyers and posters as part of production marketing strategies. This method of marketing is typical to production practices. Since 2015 and beginning with the *Dust* project Griffith University, CAT production department has implemented predominantly digital avenues in relation to programming and marketing. This decision was aimed at cutting down on resources used in traditional marketing methods—paper, ink and card.

6 Outcomes

Using a performance research design adapted from Johnson's (2009) social responsibility and sustainability in theatre framework, the Griffith University production as discussed is an innovative case study of successful sustainability methodologies in theatre production and consumption. The green theatre practices and the green ethos as applied to the production as discussed in this paper were showcased as sustainable innovation in theatre within an industry context (Sue Benner Theatre, Metro Arts, Brisbane, 2015) and the *Greening the Theatre* project won the 2016 Arts, Education and Law (AEL) Griffith University Sustainability Team Award in recognition of addressing and implementing sustainable practices in the learning and teaching environment.

Dust was the culmination of implemented green practices and methodologies that grew from sustainable production practices. The performance engaged 600 audience members across 6 performances, further promoting community education surrounding climate literacies. Hassall has been invited to reimagine the production in a high school context to foster climate literacies in secondary school students in November 2017 (Yeronga State High School, Brisbane) and is also in discussion with the artistic director of Australian Theatre for Young People (STC) for a potential production of *Dust* to further promote climate literacies in young people at a national level. Rowan will be consulting on green production applications for both these production outputs.

7 Conclusion

It is evident that climate change and sustainable futures is a topic that needs to be addressed with *all* university graduates with some urgency. For Australian students, the escalating and diverse catastrophic weather conditions in this country and the rate of extinction of species and stressed ecosystems while seemingly difficult to comprehend are in fact a very frightening reality. As the risks of catastrophic climate change accelerate, there is a growing understanding of the role art that can play in making sense of the ecological transformations we encounter (Wiseman 2016). As theatre makers, we are able to respond to these issues by posing provocative questions to our audiences around climate change issues. *Dust* not only thematically explores challenges pertaining to climate change within its dramatic content but boasted a complete 'green' and sustainable production in its theatrical framework including fully recycled scenographic design, use of entirely LED lighting and implementation of paperless marketing strategies. Edwards (2005) suggests that members of a sustainable community understand that the well-being of the individual and the larger community are interdependent. The community in this instance, although part of the broader community, is the community of future theatre makers and theatre audiences.

In promoting understanding and awareness, Duxbury (2010) states 'we begin to address the ways we perceive our surroundings' and open up the possibility that 'all individuals are capable of contributing to the changes required to stem the rapid deterioration of the climate' (p. 295).

Dramatic and performative exploration and production is a unique way to not only explore the topic but can provide a practical kinaesthetic learning experience that embeds sustainable education in both learning and real-life environments (Hassall 2017a). The preparation of future generations of theatre makers for the concerns of social responsibility in a global culture prescribes that scientific investigation and dialogue on global climate change can appear in various formats and include various methods of delivery (Fortner 2001; Burandt and Barth 2010). *Dust* is informed by a politicised, ecocritical perspective that investigates the relationship between the fictive world [performance] and the 'constructs of environment' (Morton 2007, p. 33). *Dust* and its ecocritical framework are an example of how university arts programmes can foster sustainable practice in learning and teaching environments through an alignment between research and arts practice.

The CAT department at Griffith University has responded to the energy crisis through its attention to fostering sustainable futures in its theatre graduates. Rowan's endeavours in this area encourage the fostering of sustainable practice futures in the field outside of the university sector and promote further discussion centred on the sharing of resources and research with emerging theatre makers by the more affluent flagship companies. This paper therefore provides insight into how sustainability literacies can be embedded in theatre performance and production practices in the university and beyond and consequently offers a model of sustainable theatre education that may be adopted by others in university and tertiary arts sectors.

References

Abrahams, G. (2016). In G. Abrahams, B. Johnson, & K. Gellatly (Eds), *Foreword: ART + CLIMATE = CHANGE* (pp. vi–vii). Melbourne: Melbourne University Publishing.

Baylis, J. (2009). *Mapping Queensland theatre, Queensland Government Report, May 2009.* Retrieved from http://www.google.com.au/search?q=Mapping+Queensland+Theatre&rlz=1C1CHNU_enAU475&oq=Mapping+Queensland+Theatre&sugexp=chrome,mod=3&sourceid=chrome&ie=UTF-8.

Bhend, J., & Whetton, P. (2015). Evaluation of simulated recent climate change in Australia. *Australian Meteorological and Oceanographic Journal, 65*(1), 4–18.

Burandt, S., & Barth, M. (2010). Learning settings to face climate change. *Journal of Cleaner Production, 18,* 659–665.

Cantor, M. (2010). The arts and a zero carbon future—The experience of tipping point. *Greening the arts*, Australia, n.p. Retrieved from http://tippingpointaustralia.com/resouces/.

Carducci, V. (2009). Ecocriticism, ecomimesis, and the romantic roots of modern ethical consumption. *Literature Compass, 6*(3), 632–646. https://doi.org/10.1111/j.1741-4113.2009.0068.x.

Carleton, S. (2006). *Constance drink water and the final days of somerset*. Brisbane: Playlab Press.

Carleton, S. (2016). The turquoise elephant directed. In G. Edwards (Ed.), *Griffin Theatre Company, Sydney, Australia*. October: World Premiere Season.

ClimArte, "What we do". (2012). Retrieved from http://climarte.org/about/.

DeLuca, K. M., & Demo, A. T. (2000). Imagining nature: Watkins, yosemite and the birth of environmentalism. *Critical Studies in Media Communication, 17*(3), 241–260. https://doi.org/10.1080/15295030009388395.

Duxbury, L. (2010). A change in the climate: New interpretations and perceptions of climate change through artistic interventions and representations. *American Meteorological Society, 2,* 294–299. https://doi.org/10.1175/2010WCAS1053.1.

Edwards, A. R. (2005). *Sustainability revolution: Portrait of a paradigm shift*. Gabriola Island, BC, CANADA: New Society Publishers.

Fortner, R. W. (2001). Climate change in school: Where does it fit and how ready are we? *Canadian Journal of Environmental Education, 6,* 18–31.

Fried, L., & May, T. M. (1994). *Greening up our houses: A guide to more ecologically sound theatre*. New York: Drama Book Publishers.

Garrard, G. (2010). Ecocriticism. *The Year's Work in Critical and Cultural Theory, Interdisciplinary Studies in Literature and Environment, 19*(1), 5–35.

Gingold, P. (2010). The arts and a zero carbon future—The experience of tipping point. *Greening the arts.* Retrieved from http://tippingpointaustralia.com/resouces/.

Greening our Performance. (2011). Retrieved 4/08/17 from https://climarte.org/resources/greening-our-performance/.

Griffith University Sustainability Plan. (2017–2020). Available at: https://www.griffith.edu.au/__data/assets/pdf_file/0003/945174/EG-approved-Sustainability-Plan-2017-2020.pdf.

Hassall, L. (2013). Salvation. In D. Baker & D. Beattie (Eds.), *TEXT Journal of Literary and Creative Writing-scriptwriting as creative writing research—Special Issue Series Number 19* (pp. 1–20).

Hassall, L. (2015). Dust *produced at the Sue Benner Theatre. Written and Directed by Linda Hassall September 2015*. Brisbane City, Australia: Metro Arts.

Hassall, L. (2017a). Climate literacy and collaborative on-line landscapes: Engaging the climate conversation through drama processes in distance and e-learning environments. In W. Leal, U. M. Azeiteiro, W. Filho, L. Aires (Eds.), *Distance learning and climate literacy and innovations in climate change*. Springer. ID 07-2017-0031. (accepted Book Chapter submitted July 2017 for publication 2018).

Hassall, L (2017b). Contemporary theatrical landscapes: The legacy of romanticism in two examples of Contemporary Australian gothic Drama. In S. Green & N. Krauth (Eds.), *TEXT: TEXT Journal of Literary and Creative Writing-scriptwriting as creative writing research—Special Issue Romanticism and Writing, Summer 2017* (in press for publication in December, 2017).

Henstock, G. (2011). Interview with head of lighting Sydney Theatre Company in Greening the Wharf. Retrieved from https://www.sydneytheatre.com.au/magazine/posts/2011/…/video-greening-the-wharf. Accessed on 3-08-17.

Holdsworth, S., Wyborn, C., Bekessy, S., & Thomas, I. (2008). Professional development for education for sustainability: How advanced are Australian Universities? *International Journal of Sustainability in Higher Education, 9*(2), 131–146. https://doi.org/10.1108/14676370810856288.

Hudson, J. (2012). 'If You Want to Be Green Hold Your Breath': Climate Change in British Theatre. *New Theatre Quarterly, 28*(03), 260–271.

Johnson, C. (2009). *Sustainable theatre: An analysis of theories and practices* (Unpublished Master of Arts dissertation. Arcata, California: Humboldt State University).

Lee, K., Barker, M., & Mouasher, A. (2013). Is it even espoused? An exploratory study of commitment to sustainability as evidenced in vision, mission, and graduate attribute statements in Australian universities. *Journal of Cleaner Production, 48,* 20–28.

Live Performance Australia. (2012). *Reference*. Greener Live Performances. Retrieved from http://liveperformance.com.au/greener_live_performances/overview.

Lorenz, E. N. (1999). *The essence of chaos* (p. 77). Washington: University of Washington Press.

May, T. (2005). Greening theatre studies: Taking eco-criticism from page to stage. *Interdisciplinary Literary Studies, 7*(1), 84–103.

May, T. (2007). Beyond Bambi: Toward a dangerous ecocriticism in theatre studies. *Theatre Topics, 17*(2), 95–110.

Mayall, E. (2016). Debbie Symons and Jasmine Targett: Exhibition of the catchments project. Creative spaces: Carlton connect studio, the University of Melbourne. In G. Abrahams, B. Johnson, & K. Gellatly (Eds.), *ART + CLIMATE = CHANGE* (pp. 32–33). Melbourne: Melbourne University Publishing.

McKibben, B. (2005). *What the warming world needs now is art, sweet art*. Retrieved from http://grist.org/article/mckibben-imagine/.

McKibben, B. (2009). Foreward. In M. W. Robbins (Ed.), *Whole green catalogue*. New York: Rodale Inc.

Min, S. K., & Hens, A. (2007). A Bayesian Assessment of climate change using multimodel ensembles. Part II: Regional and seasonal mean surface temperatures. *Journal of Climate, 20,* 2769–2790.

Morris, D. G. (2007). *Towards a recycled theatre: Industrial ecology theatrical applications for the next industrial revolution* (Doctoral thesis, Washington: Western Washington University).

Morton, T. (2007). *Ecology without nature: Re-thinking environmental aesthetics*. Cambridge MA and London: Harvard UP.
Munk, E. (1994). A beginning and end. *Theater, 25*(1), 5–6.
Reid, J., Carpenter, D., & Meehan, B. (2006). Art for earth's sake: Creative and interdisciplinary collaborations for sustainability in the tertiary sector. In W. L. Filo & D. Carpenter (Eds.), *Sustainability in the Australasian University* (pp. 81–90). Frankfurt and Main: Peter Lang.
Rowan, S. (2015). *Water Wars: Exploring the tensions, challenges and possibilities of applying a sustainability framework to the development and production of a new independent theatre performance* (Unpublished Honours Thesis, School of Education and Professional Studies, Griffith University, Australia, April 2015).
Schultz, J. (2006). Masters of the Universe: How nigh is the end. *Griffith Review: Hot Air. How nigh is the end? Winter, 12,* 6–10.
Shephard, K. (2008). Higher Education for sustainability: Seeking affective learning outcomes. *International Journal of Sustainability in Higher Education, 9*(1), 87–98.
Sherren, K. (2006). Pillars of society: The historical context for sustainability and higher education in Australia. In W. Filho & D. Carpenter (Eds.), *Sustainability in the Australian University Context* (pp. 11–32). Frankfurt: Peter Lang.
Steelquist, B. (1994). In L. Fried & T. M. May (Eds.), *Greening up our houses: A guide to more ecologically sound theatre* (p. 23). New York: Drama Book Publishers.
Sydney Theatre Company, "Sustainability". (2012). Retrieved from http://www.sydneytheatre.com.au/about/sustainability.aspx.
Tipping Point Australia. (2010). *Greening the arts—Thinkpieces for a zero carbon future and a survey of sustainable arts practices*. Retrieved from http://tippingpointaustralia.com/resouces/.
Thomas, I., & Nicita, J. (2002). Sustainability education in Australian Universities. *Environmental Education Research, 8*(4), 475–492. https://doi.org/10.1080/1350462022000026845.
Walters, D. (2011). Lighting designer. In S. Rowan (2015). *Water wars: Exploring the tensions, challenges and possibilities of applying a sustainability framework to the development and production of a new independent theatre performance* (Unpublished Honours Thesis, School of Education and Professional Studies, Griffith University, Australia, April 2015), pp. 41–45.
Water Wars by Ackworth, E. (2011). Directed by Charles, S. *La Boite Roundhouse Theatre, Brisbane, Australian Premiere Season*, August 3–20, 2011.
Wiseman, J. (2016). In G. Abrahams, B. Johnson, & K. Gellatly (Eds.), *ART + CLIMATE = CHANGE* (pp. 6–14). Melbourne: Melbourne University Press.

Linda Hassall is Programme Director of the Contemporary and Applied Theatre Department, Griffith University, Australia. Her practice-led artistic research investigates ecocritical and Gothic landscapes in contemporary performance practices, and her work rearticulates representations of nature in exacerbated cultural, social, physical, psychological circumstances. She is a critically successful playwright and director and has numerous publications including creative works to her credit.

Stephen Rowan has worked in the theatre industry sector for more than 30 years. He is the Senior Technical Officer of the School of Education and Professional Studies, Griffith University and Production Manager of the Contemporary and Applied Theatre Department, Griffith University, Australia. His research interests are in sustainable theatre production practices, and he has recently completed Honours research in the field.

Integrated Rainwater Harvesting Practices for Poverty Reduction Under Climate Change: Micro-Evidence from Ethiopia

Anteneh Girma, Menale Kassie, Siegfried Bauer and Walter Leal Filho

Abstract Rainwater harvesting (RWH) has been practiced and promoted to address the temporal and spatial variability of rainfall, thereby enhancing agriculture production in rainfed systems. The practices could also address the problem of land degradation. However, there is limited practice of approaching RWH from the perspective of managing both the water and land resources. Research on water productivity and impact for poverty reduction mainly focused on irrigated agriculture while it is potential to provide the water needed to produce food for rapidly growing population is the subject of intense debate these days. An important option is to upgrade rainfed agriculture through better land and water management that improves soil moisture conservation and rainwater harvesting that provides supplementary irrigation. In the meantime, studies on impact of agricultural water management focused more on unidimensional poverty while poverty is multidimensional. This study investigates the impact of integrated RWH practices (IRWHPs) on multidimensional poverty in Ethiopia. Results show that the use of IRWHPs has a significant negative impact on the probability that a household is multidimensionally poor. This study suggests that policies that enhance the promotion of IRWHPs would be central for the sustainable intensification of

A. Girma
Sasakawa Africa Association/Sasakawa Global 2000 Ethiopia,
P. O. Box 2634, Addis Ababa, Ethiopia

M. Kassie
International Center of Insect Physiology and Ecology (Icipe), Nairobi, Kenya

S. Bauer
Institute of Project and Regional Planning, Justus Liebig University Giessen,
Giessen, Germany

W. Leal Filho (✉)
Hamburg University of Applied Sciences, Hamburg, Germany
e-mail: walter.leal2@haw-hamburg.de

W. Leal Filho
School of Science and the Environment, Manchester Metropolitan University,
Manchester, UK

© Springer International Publishing AG, part of Springer Nature 2019
W. Leal Filho and R. Leal-Arcas (eds.), *University Initiatives in Climate Change Mitigation and Adaptation*, https://doi.org/10.1007/978-3-319-89590-1_10

smallholder agriculture that simultaneously alleviate poverty and enhance resource sustainability.

Keywords Rainwater harvesting · Impact · Multidimensional poverty Land management

1 Introduction

In sub-Saharan Africa, a variety of climate and non-climatic factors prevent agricultural productivity. These include water availability and soil nutrient depletion among many others. Although the amount and distribution of rainfall are important to crop growth, its seasonal variability during critical crop growth stages is a major production constraint, leading to a gap between actual and potential outputs (Barron and Okwach 2005). Water shortages have profound consequences on smallholder farmer welfare and are a major obstacle to reducing poverty levels in the face of climate change and variability (Hatibu et al. 2006). The water scarcity is often considered as a rainwater storage problem (Eguavoen and McCartney 2013). It is aggravated by large-scale soil erosion and land degradation, which in turn affect the local hydrology, resulting in reduction of evaporation fluxes during the rainy season and reduces inland rainfall (Dile et al. 2016).

Rain water harvesting practices (RWHPs) have been promoted an approach to integrate land and water management, which could contribute to recovery of agriculture production in rainfed systems and the general water resources (Humphreys et al. 2008; Rockström et al. 2010). RWHPs involve a combination of different practices that enhance infiltration and/or reduce runoff, referred as in situ RWHP and those that capture, store, and efficiently use runoff and surface water emerging from farms and watershed, called as ex situ RWHP (Barron 2009; Rockström et al. 2014). Successful application of the combined use of both in- and ex situ practices can contribute to recovery of agriculture production in rainfed systems and the general water resources (Humphreys et al. 2008; Rockström et al. 2010). Integrated rainwater harvesting practices (IRWHP) at the watershed level contributes to a reliable water supply for sustainable food production while enhancing other ecosystem services (Rockström et al. 2004; Barron 2009; Gordon et al. 2010) and supporting sustainable agricultural intensification and resilience against water-related shocks (Dile et al. 2013).

In Ethiopia, many smallholder farmers use different in situ and ex situ RWHPs to mitigate the impact of dry spells and drought. These practices have received government and donor attention following the 1973/74 drought after which several soil and water conservation program was initiated (Shiferaw and Holden 1999; Seyoum 2003). Despite concerted efforts by the government and international community, farmers face constraints in integrating the practices and in exploiting the full value of the land and water resources for their livelihoods (Merrey and Gebreselassie 2011).

Most previous studies related to land and water management (Kassie et al. 2010; Kato et al. 2011; Gebregziabher et al. 2012; Hagos et al. 2012; Abdulai and Huffman 2014) focused on adoption and socioeconomic impacts of single practice while farmers use combination of the practices. Moreover, impact evaluations on land and water management focused on land productivity outcomes, but not on water productivity perspective (Merrey and Gebreselassie 2011). Research on water productivity and impact for poverty reduction mainly focused on irrigated agriculture. However, because the future of irrigated agriculture to provide the water needed to produce food for rapidly growing population is the subject of intense debate these days. An important option is to upgrade rainfed agriculture through better land and water management that improves soil moisture conservation and rainwater harvesting that provides supplementary irrigation (De Fraiture et al. 2007; Molden et al. 2007; Rockström et al. 2010) which is the focus of this paper. The adoption and the economic impacts of these practices are potentially interrelated and could provide better outcomes when they are undertaken jointly (Kassie et al. 2015). Failure to integrate the practices could end up with silted water harvesting ponds (Rämi 2003).

Furthermore, we are unaware of impact studies that evaluate the link between multidimensional poverty and investment in agricultural water management. The review by Castillo et al. (2007) and Namara et al. (2010) shows that previous studies (e.g., Huang et al. 2005; Adetola 2008; Bacha et al. 2011) that assessed the impact of agricultural water management focus on unidimensional poverty while poverty is multidimensional. The use of a multidimensional poverty measure is important because no one indicator (e.g., income) alone can capture the multidimensional nature of poverty. The inadequacy of unidimensional approaches for defining and measuring poverty is widely recognized, and this has led to exploring alternative approaches (Sen 2001; Smith 2004; Carter and Barrett 2006). The 1997 Human Development Report and the 2000/1 World Development Report vividly introduced poverty as a multidimensional phenomenon, and the Millennium Declaration and MDGs have highlighted multiple dimensions of poverty since 2000 (Alkire and Foster 2011). A better understanding of what types of technologies work to reduce poverty in particular contexts and what bottlenecks need to be overcome to optimize impacts would be useful to guide research priority-setting and institutional policy design to better target land and water management practices for sustainable intensification of smallholder agriculture. The paper examines the impact of IRWHP for poverty reduction by constructing a multidimensional poverty index and compares the benefits of using integrated in situ and ex situ practices with use of only in situ practices using endogenous switching regression.

The rest of the paper is organized as follows. The next section presents the data source and estimation procedures. Estimation results will be presented and discussed in Sect. 3. The last section concludes the key findings and the policy implications.

2 Methodology

The data used for this study is based on a farm household survey in Azgo watershed, northeastern Ethiopia, collected during the period June–August 2012. The watershed is purposively selected to measure the impact of integrated land and water management for rainwater harvesting, a positive example of integrating the practices in a country where the majority of the farmers are not using the integrated practices to fully exploit available rainfall. While the purposive sampling limits the generalization of the findings in comparison to the randomly selected watersheds, the latter was constrained by the absence of a substantial number of such watersheds for a random selection. Purposive site selection in evaluation is not uncommon as far as it draws out lessons and principles to the target population.

2.1 *Study Area*

The Azgo watershed is located in Kalu district, south Wollo zone of Amhara region, between 11.9–11.13N and 39.51–39.54E; 1500–2450 m a.s.l. It is a micro-watershed (14,049 km^2), purposively considered to study at landscape scale formed of top, middle, and lower slopes (Fig. 1). The major landscape type in the study area is undulating and rugged. The climate in Azgo is characterized by a bimodal rainfall pattern. The short rainy season occurs from March to May and the long rainy season from July to September. Average annual rainfall ranges from 1082.30 to 2007.50 mm, according to the data registered at the weather stations in the upper and lower watersheds, respectively. Similarly, the temperature range in the area is 15.1–27.5 °C, with the higher temperature in the lower part of the watershed. In the watershed, a rainfed subsistence crop–livestock mixed farming with minimal cash crop is the source of livelihood for majority of the population.

Stratified sampling procedure was employed to select respondents. First, the total population was stratified into users and non-users of IRWHP. Second, based on the researcher resource available and proportionate sampling, 354 farm households operating on 1002 plots were randomly selected for face-to-face interview. The data was collected under close supervision of the first author, using a pretested structured questionnaire by trained and experienced enumerators who are conversant of the local farming and social system. The survey is designed to understand the adoption process and impacts of IRWHP. The survey collected household-, plot-, and village-level data. The plot-level data includes household perceived plot-specific attributes including soil fertility, soil type, and plot slope; plot tenure and land use; plot position in the landscape (upper, middle, and lower); plot spatial distance from main road and homestead; crops grown and types of inputs (water, fertilizer, and seed) used; and types of RWHP used both in situ and ex situ practices. Household-specific data includes food consumption, age, sex, marital status, education level, family size, access to government extension, asset ownership

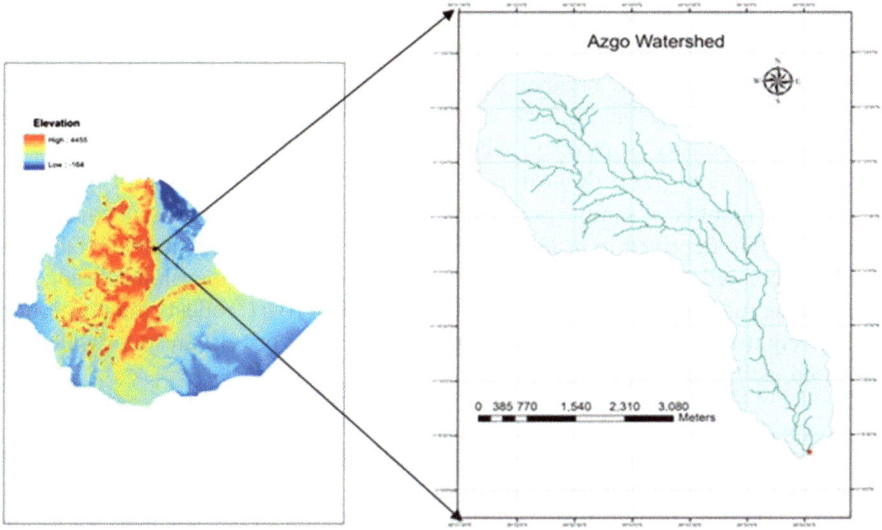

Fig. 1 Location of the study area

(e.g., farm size. livestock), farm income, non-farm income sources, and membership in community-based organization.

Based on these household characteristics, we constructed multidimensional poverty index (MDPI). We follow Alkire and Santos (2014) to develop the MDPI including the dimensions, indicators, and weights. The MDPI has three dimensions: education, health, and living standard and are measured using seven indicators (Table 1). Each dimension is equally weighted; each indicator within a dimension is also equally weighted. Sample households are categorized into poor and non-poor

Table 1 Multidimensional poverty index: indicators and dimension

Dimension	Indicator
Education (0.33)	Formal schooling (0.167)
	Literacy (0.167)
Health/nutrition (0.33)	Per capita food calorie intake (0.33)
Standard of living (0.33)	Livestock asset value in ETB (0.085)
	Housing condition value in ETB (0.085)
	Cooking fuel type (0.085)
	Household asset value in ETB (0.085)

Note (1) Tropical Livestock Unit is defined as an animal of 250 kg live weight (Cattle = 0.7, Goat = 0.1, Sheep = 0.1, Camel = 1.0, Donkey = 0.5) (Jahnke et al. 1988)

(2) Household assets in the food security index denote the value of house, durable furniture, jewelries, transport and communication items, while it does not include the value of the house in the poverty index

(3) The number in parenthesis is weight assigned for each dimension and indicator

based on the MDPI. Similarly we follow Alkire and Santos (2014) to group households into poor and non-poor based on the MDPI.

The MDPI shows the number of multidimensional poor or headcount ratio (suffering deprivations in 33.33% of weighted indicators) and the number of deprivations (intensity of poverty) with which poor households typically contend. The headcount ratio (H) is defined as $H = \frac{q}{n}$, where q and n represent people those are poor and number of sample population, respectively. A household is considered poor if they are deprived in at least 33.33% of the weighted indicators. MDPI can also be expressed as a product of two measures: the multidimensional headcount ratio and the intensity of poverty. The intensity of poverty (I) reflects the proportion of the weighted indicators in which, on average, poor people are deprived. It is expressed as:

$I = \frac{\sum_{1}^{q} C}{q}$, where C is the deprivation score that the poor experience and is computed as the household-level sum of each deprivation of indicator multiplied by its weight. Household with larger household deprivation scores (C) than the MDPI values is considered as multidimensional poor while those with smaller values than the MDPI values are grouped into non-poor.

Identifying the impact of adopting IRWHP is not straightforward, in particular using cross-sectional data wherein users and non-users may be systematically different in terms of their observed and unobserved plot and household characteristics, and which could lead to sample selection bias. Further, they may be systematically selected by development agents based on their propensity to use these practices. To estimate the impact of IRWHP, one needs to take into account the possibility of self-selection. Recent literature on sample selection in the context of discrete outcomes highlights the advantage of using an endogenous switching probit model over alternative econometric methods (Aakvik et al. 2005; Lokshin and Glinskaya 2009). It enables controlling for selection bias and varying the impact of observables and unobservable variables on the outcomes of IRWHP users and non-users (relax the assumption of equality of coefficients among both groups). Following Lokshin and Glinskaya (2009) and Lokshin and Newson (2011), we consider a three-equation model that describes a regime determination rule and two outcome equations where the details are presented in Appendix 1.

3 Result and Discussions

Farm household in the study watershed has undertaken a number of RWHP on their plots (Table 2). The rainwater harvesting practices were implemented both in stand-alone and in combination. The practices categorized into two: in situ and ex situ. The in situ RWHP enhances infiltration and/or reduce runoff, while the ex situ practices capture, store, and enhance efficient use of runoff and surface water emerging from farms and watershed. In this study, the use of IRWHP is represented by a binary variable equal to 1 if the farm household has integrated in- and ex situ RWHP and zero if the households only use in situ practices.

In the study area, both traditional and introduced in situ RWHP exist. These include stone terrace/check dam, soil bund, stone-faced soil bund, tie ridge, trenches, and eye brow. Stone terraces are traditional practices applied in the watershed for more than a century and locally called as '*Kiter* and/or *kab.*' They are made of stone built across a contour in a farmland to control erosion, trap runoff, and sediments. The stone are constructed on 0.5 m width facing the open land. The size and interval of the terraces varies across plots depending on the availability of stone, labor, and topography of the plots. Traditionally, at least four lines of terraces are required in 0.25 ha of land. Soil bunds are the other traditional practice in the watershed and locally called as '*Dhuga.*' They are constructed by digging a degraded channel and embarked the soil to form a ridge so as to safely discharge surplus rainwater from a farmland. They are placed on top of the stone terrace and relay the runoff to artificial waterways, which are bigger in dimension to carry more runoff. The bunds are particularly important during the beginning of the rainy season to divert the runoff to external water harvesting storage units. Stone-faced soil bunds are one of the RWH practices introduced in the watershed through the structure of Ministry of Agriculture or projects operated in the area. They are reinforced soil bunds in one or both sides, constrained mainly in farmlands. The purpose is similar to that of traditional soil bunds but offer stronger resistance against runoff. Tie ridges are small rectangular series of basins formed within the furrow of cultivated fields mainly to increase surface storage and to allow more time for rainfall to infiltrate the soil. It is practice in the lower and middle parts of the watershed with low-to-medium rainfall and well-drained soils on lands with gentle slopes. The construction involves a ridge tier attached to the local plow '*Maresha*'

Table 2 Rainwater harvesting practices implemented in the sample plots (N = 1002)

RWH practices	Number of plots	% of the plot
No structure	79	8.11
In situ practices		
Soil bund	250	25.67
Stone-faced soil bund	281	28.85
Contour trench	42	4.31
Tie ridge	18	1.85
Stone terrace/check dam	660	67.76
Planting trees	14	1.44
Hillside trench	18	1.85
Eye brow	12	1.23
Hillside terrace	24	2.46
Ex situ practices		
Water harvesting pond	93	9.55
Spring diversion	44	4.52

as traditional/manual construction is time- and labor-consuming. Thus, the practice is promoted by the office of agriculture, and the distribution of the ridger involves credit and subsidy. In the watershed, two types of trenches are promoted, respectively, on farmland and hilly areas. Trenches in farm fields are referred as contour trenches and locally called '*Agidim Zig Boy.*' Contour trenches are ditches dung along the slope in such a way that the slope is broken and therefore reduce the velocity of water runoff and enhance infiltration of the surface runoff and soil moisture. It is practiced on plots with higher slope, and the trenches are interrupted at certain interval to hold the rainwater. On hillsides, the trenches are large and deep pits constructed along contour with the same purpose of collecting and storing rainfall water. They also support the downstream lands through protection of flood and water captured by the trenches reaches the underground aquifer. They are constructed with hillside terraces and eyebrow basins and also constitute an integral part of a hilly area closure or control grazing lands. Hillside terraces are similar to stone terraces but constructed in steep degraded slopes to support area closure plantation and protect downstream fields. They are integrated with trenches and eye brown basin to harvest more rainwater to support plantation of multipurpose tree and fodder tree on the hilly areas or grazing lands. Eye brow are larger circular and stone-faced structures for tree planting in degraded and steep hilly areas. They are combined with hillside trenches and trenches in closed hilly area or grazing land management.

The ex situ practices include water harvesting pond and spring development and diversion. Water harvesting pond is commonly used in the watershed. The promotion of ex situ water harvesting system in the watershed is a recent phenomenon compared to in situ practices. The first generation of ex situ, RWH tanks, is made of cements for demonstration purpose as part of the national food security strategy. However, because of the costs of construction and the seepage problem, trapezoidal-shaped RWH ponds are now being promoted and widely practiced in the watershed. The ponds are excavated into a 3 m depth and truncated at its bottom where the upper part is 8 by 8 m^2 and the bottom is 5 by 5 m^2. Farmers construct the ponds in areas where least digging or earth fill is required and hold a large volume of water, on the average 150 m^3 and can irrigate 0.25 ha of land. The structures are constructed around homesteads. The ponds use surface runoff from small catchment within and between homesteads (footpath, small grazing land, hilly areas). Spring development and diversion is a traditional practice in the watershed where farmers develop an earth's surface that would naturally emerge a groundwater.

The unconditional distribution of the poverty outcome by IRWHP use status is presented in Table 3. The result highlights that it seems users are less poor compared to their counterparts of non-users of IRWHP. About 55.46 and 36.36% of the users and non-users are non-poor, respectively. However, proper analysis of the causal impact needs further analysis.

In the understanding of the impacts of IRWHP on outcome variables, we followed two-step procedures as discussed in the methodology section. In the first step, we estimated the determinants of IRWHP using probit model (Eq. 1 in

Table 3 Unconditional distribution of welfare outcomes by IRWHP use status (%)

Welfare outcomes	User (N = 119)	Non-user (N = 231)	Chi-square test
Poor households	44.54	63.64	11.698***
Non-poor households	55.46	36.36	

Note *, **, and *** denote significance level at 10, 5, and 1%

Appendix 1) to mainly generate the selection correction term ($\hat{\lambda}$). The dependent variable is equal to 1 if the farm household has integrated in- and ex situ RWHP and zero if the households only use in situ practices. In the second stage, the outcome Eqs. (4a–4d) estimated using probit model. The dependent variables for outcome equations are binary households' poverty status. A household is multidimensional poor (=1) if a household has larger deprivation scores (C) than the MDPI values and zero otherwise.

The endogenous switching probit estimates are presented in Table 4. The first column presents the selection equation and provides the estimates for the determinants of IRWHP use. The next columns present the impact of explanatory variables on multidimensional poverty status for users and non-users of IRWHP.

The results show that the decision to use IRWHP is influenced by a number of household and farm characteristics and other support services (Table 4). These include age, family size, off-farm activities, extension service, cash crop area, plot slope, and access to main road. Farm household characteristics are expected to influence family labor supply for the construction of the different RWHP, thereby influencing the decision to use these practices. Household family size and composition, age of the household head, and household members' participation in off-farm activities affect the amount of labor available for farm operation including RWHP use. Accordingly, older household heads are significantly less likely to integrate RWHPs. This is probably because aging means losing energy, shorter planning horizons, and more risk-averse for using new technology (Kassie et al. 2013). On the other hand, older age could contribute for a greater accumulation of physical assets like land size. But, large farm size is associated with decreased probability of using IRWHP.

Household engagement in off-farm activities could affect household labor availability for farm operation and influence the decision to use IRWHP since the latter is a relatively labor-intensive activity. Participation in off-farm is significantly less likely to use IRWHP. It seems short-term benefits from off-farm activities seem to attract farmers' labor more than investments in land and water management. This is in line with previous studies (Shiferaw and Holden 1989; Gebremedhin and Swinton 2003).

Reported plot location and characteristics also affect the decision to use IRWHP. Farmers' perception on the slope of the plot is influence the decision to use IRWHP. Plain slope plots do not require much investment in conservation practices, especially for in situ practices since erosion is more serious on steeper plot than on flat plots. Previous empirical work in Ethiopia reported a positive

Table 4 Endogenous switching probit regression estimation for impact of IRWHP use decision on probability of multidimensional food security and poor

Explanatory variables	Selection/adoption equation	Probability of multidimensional poor	
		IRWHP users	Non-users
Ln (household head age)	−0.836*** (0.326)	−4.822** (2.069)	0.026 (0.541)
Sex	0.393 (0.335)	4.431** (1.882)	−0.317 (0.309)
Ln (family size)	0.883*** (0.208)	4.253 (2.211)	0.110 (0.360)
Off-farm income source	−0.946*** (0.203)	−4.816 (0.3440	−0.119 (0.381)
FBO	−0.110 (0.234)	−0.472 (0.268)	−0.081 (0.316)
Ln (farm size)	−0.255* (0.173)	−2.929 (0.724)	−1.454*** (0.381)
Livestock size (TLU)	0.030 (0.096)		
Share of cash crop land	0.851*** (0.307)	−0.229 (0.391)	−0.975* (0.491)
Government extension service	0.611*** (0.190)	3.015* (1.558)	−0.117 (0.291)
Share of high fertile soil land	0.140 (0.186)	1.065** (0.471)	−0.031 (0.244)
Share of poor fertile soil land	0.421 (0.414)	3.929*** (1.489)	0.196 (0.467)
Share of flat slope land	−0.448** (0.205)	−2.056* (1.121)	−0.281 (0.301)
Share of steep land	0.289 (0.317)	1.185*** (0.714)	−0.187 (0.471)
Road distance	0.009 (0.006)	0.055** (0.024)	0.010 (0.009)
Share of own cultivation		−1.575 (1.005)	−0.368 (0.403)
IRM		9.112** (3.952)	0.448 (0.854)
Constant	0.403 (1.356)	−4.433 (3.812)	1.015 (2.146)
Log-likelihood	−178.058	−57.529	−123.418
Wald chi2(18)	77.99***	48.49***	56.00***
N	350	119	231

Note *, **, and *** denotes significance level at 10, 5, and 1%; robust standard errors in parentheses

relationship between slope and likelihood of using physical conservation measures (Shiferaw and Holden 1989; Bekele and Drake 2003). Plot with cash crop (*Khat* in this watershed) is significantly important in enhancing IRWHP use. This is because it has higher return compared to other crops. A study in eastern Hararge highlands of Ethiopia showed that *Khat* provides farmers with the incentive and capacity to make substantial investments (Ketema and Bauer 2012). The positive role of economic incentives and natural resource management is not debatable, but it requires favorable institutional environment and organization setup to link economic incentive to natural resource management (Getnet et al. 2014).

The results further show that there is a significant positive association between government extension and use of IRWHP. This result corroborates with most of the earlier studies, which have shown that access to extension services play a significant role to enhance the technical capability of farmers or facilitate input for introduced technologies, thereby increasing the likelihood of using the practices (Asrat et al. 2004; Anley et al. 2007; Teklewold and Köhlin 2011).

The poverty impact of IRWHP for poverty reduction is presented in Table 5. The result confirms that use of IRWHP significantly increases the probability of multidimensional food security status and reduces the probability of being multidimensional poor. This is in agreement with the descriptive statistics results (Table 3).

Table 5 shows that both users and non-users would benefit from the use of IRWHP in terms of improving household food security. However, non-users would have benefited more if they were using combination of in situ and ex situ (IRWHP) practices. The counterfactual result revealed a positive impact with regard to multidimensional poverty outcome of implementing IWHRP. However, it is only significant for current users of IRWHP. The result reveals that households who have already integrated the practices would have increased the probability of being multidimensional poor by 14% if they had not used IRWHP. This is the ATT, and it is statistically significant. In terms of improving the poverty outcome, the actual

Table 5 Impacts of IRWHP on household welfare: endogenous switching probit results

Outcome variable	Farm household IRWHP use status	Decision stage		Average treatment effect (ATT/ATU)	Impact (%)
		To use	Not to use		
Multidimensional poverty status (MDP) (1 = poor; 0 = non-poor)	IRWHP users	(a) 0.445 (0.027)	(c) 0.581 (0.020)	−0.137*** (0.017)	−14.0
	IRWHP non-users	(d) 0.632 (0.023)	(b) 0.635 (0.015)	0.003 (0.020)	

Note *, **, and *** denote significance level at 10, 5, and 1%; standard errors in parentheses (**a**) and (**b**) respectively denote the actual expected probability of MDPI of IRWHP users and non-users observed in the data; (**c**) and (**d**) denote, respectively, the counterfactual expected probability of MDPI for IRWHP users and non-users; ATT and ATU refer to the effect of treatment (i.e., use of IRWHP) on the treated (i.e., users of IRWHP) and untreated (i.e., non-users of IRWH practices), respectively

Impact presents the percentage changes in the probability of multidimensional poverty due the use of IRWHP (last column of the Table). It is calculated as the percentage difference between expected values of (**a**), (**c**) and (**b**), (**d**) for users and non-users, respectively. For example, the 58.1% of non-users are multidimensional poor (**c**) while they would have reduced the poverty status to 44.5 (**d**), which a 14% less probability of multidimensional poor as a result of IRHP use users would have escaped out of poverty significantly more than the farm households that did not use in the counterfactual case (c). This indicates that there are some important sources of heterogeneity that makes the users 'better non-poor' than non-users irrespective to the use of IRWHPs.

4 Conclusion and Policy Implications

The uses of integrated rainwater harvesting practices have positively contributed for poverty reduction. The results on determinants of use of IRWHP show that household resource and characteristics and access to government extension service have a significant positive effect on the probability of using IRWHP; suggesting considering these factors in the process of promoting IRWHP is important to speed up the adoption process.

The analysis on the link between IRWHP and household poverty highlights that IRHWP increases the probability of multidimensional food security status and reduces the probability of being multidimensional poor. These results imply that rainwater harvesting practices would have wider impact if appropriate scaling strategies are employed for wide spread use of the practices. It is, however, important to note that the biophysical condition (rainfall pattern, topography, and the cropping pattern) and market structure are important factors in the process of identifying suitable locations in promoting IRWHP.

The study empirically shows the need for critical attention in the policy framework to promote and support watershed-based IRWHP for sustainable intensification of smallholder agriculture that simultaneously reduces poverty while enhancing resource sustainability. The significance of the extension variable suggests the need for improving information dissemination to empower farmers in terms of capacity to acquire and use knowledge and technology while developing their leadership to sustainably manage land and water resources for long-term benefits. Improving the planning process and targeting IRWHP are also important to increase benefits and speed up uptake of these practices. Linking farmers to institutions (e.g., input and markets) might also be important to enhance benefits from sell of cash crop and access to labor market that facilitate adoption of IWRHP.

Appendix 1 Estimation of Endogenous Switching Probit Model

We consider three-equation models that describe a regime determination rule and two outcome equations to measure the determinants of use and impact of IRWHP for poverty reduction.

The regime determination rule is the decision whether to use IRWHP or not (R_i). A farmer will decide to use/adopt IRWHP if the expected benefit/utility from adopting is higher than from not using these practices. The decision rule is given as:

$$R_i = \begin{cases} 1 & \text{if } \gamma Z_i + \mu_i > 0 \\ 0 & \text{otherwise} \end{cases} \quad (1)$$

The outcome equations (Y_i), the probability of multidimensional food security is defined:

$$\text{Regime 1}: Y_{1i}^* = X_{1i}\beta_1 + \sigma_{1\mu}\hat{\lambda}_{1i} + \varepsilon_{1i} \quad \text{if } R_i = 1 \text{ (IRWHP users)} \quad (2)$$

$$\text{Regime 2}: Y_{0i}^* = X_{0i}\beta_0 + \sigma_{0\mu} + \varepsilon_{0i} \quad \text{if } R_i = 0 \text{ (IRWHP non - users)} \quad (3)$$

where Y_{1i}^* and Y_{0i}^* are the latent variables for the observed binary outcomes Y_1 and Y_0; Z_i and X_i are vectors of observable factors influencing the decision to use or not to use IRWHP and the probability of multidimensional food security and poverty status. $\lambda_{1i} = \frac{\varphi(Z_i\alpha)}{\Phi(Z_i\alpha)}$ and $\hat{\lambda}_{0i} = \frac{\varphi(Z_i\alpha)}{1-\Phi(Z_i\alpha)}$ are the inverse Mills ratios (IMR) computed from the selection equation (Eq. 1) to correct for selection bias in a two-step estimation procedure, i.e., endogenous switching regression. β and σ are parameters to be estimated, and ε is an independently and identically distributed error term. The error terms (μ_i, ε_{1i} and ε_{0i}) are assumed to be jointly normally distributed with mean-zero and correlation matrix expressed as:

$$\text{cov}(\mu, \varepsilon_1, \varepsilon_0) = \begin{bmatrix} 1 & \rho_{u0} & \rho_{u1} \\ & 1 & \rho_{10} \\ & & 1 \end{bmatrix} \quad (4)$$

where ρ_{u0} and ρ_{u1} are, respectively, the correlations between the error terms of selection (adoption) Eq. 1 and outcome Eqs. 2 and 3; ρ_{10} is the correlation between ε_1 and ε_0. If the off-diagonal elements are nonzero, then the error terms of the adoption and the outcome equations are correlated, indicating evidence of endogenous switching or presence of sample selection bias. As we do not have a credible instrument, we base identification on functional forms. As argued by Lokshin and Glinskaya (2009), the system of Eqs. (1–3) is identified by nonlinearities even if the variables in X and Z overlap completely.

Conditional Expectations and Treatment Effects

We use the predicted probabilities from Eqs. (2) and (3) to compare the average treatment effect of IRWHP on users and non-users of the practice. Following Di Falco et al. (2011) and Kassie et al. (2015), the conditional expectations, treatment, and heterogeneity effects of the probability of multidimensional food secure and/or poor are defined as follows:

$$E(Y_{i1}|R=1;x) = x_{i1}\beta_1 + \sigma_{1\varepsilon}\hat{\lambda}_{i1} \text{ (Actual expected outcomes for users of IRWHP)} \quad (4a)$$

$$E(Y_{i0}|R=0;x) = x_{i0}\beta_0 + \sigma_{0\varepsilon}\hat{\lambda}_{i0} \text{ (Actual expected outcomes for non - users of IRWHP)} \quad (4b)$$

$$E(Y_{i0}|R=1;x) = x_{i1}\beta_0 + \sigma_{0\varepsilon}\hat{\lambda}_{i1} \quad \text{(Counterfactual expected outcomes if users had not used IRWHP)} \quad (4c)$$

$$E(Y_{i1}|R=0;x) = x_{i0}\beta_1 + \sigma_{1\varepsilon}\hat{\lambda}_{i0} \quad \text{(Counterfactual expected outcomes if non - users had used IRWHP)} \quad (4d)$$

Equations (4a) and (4b) represent the actual expected outcomes observed in the sample, while Eqs. (4c) and (4d) are the counterfactual expected outcomes. The counterfactual is the expected outcomes if the characteristics (x_1) of users had the returns of the characteristics (β_0) of non-users and vice versa. The effect of the IRWHP on household welfare is calculated using these conditional expectations. The expected change in the probability of multidimensional food secure and poor due to the use of IRWHP can be then calculated as the difference between Eqs. (4a) and (4c). These estimates are called as average treatment effect (ATT) in the impact assessment literature and are given as:

$$\text{ATT} = (E(Y_{i1}|R=1,x) - E(Y_{i0}|R=1,x) = x_{i1}(\beta_1 - \beta_0) + \lambda_{1i}(\sigma_{1u} - \sigma_{0u}) \quad (5)$$

Similarly, we calculate the effect of IRWHP use decision for the non-user of the practice as the difference between Eqs. (4d) and (4b) and is given as average treatment effects on the untreated (ATU):

$$\text{ATU} = (E(Y_{i1}|R=0,x) - E(Y_{i0}|R=0,x) = x_{i0}(\beta_1 - \beta_0) + \lambda_{0i}(\sigma_{1u} - \sigma_{0u}) \quad (6)$$

References

Aakvik, A., Heckman, J., & Vytlacil, E. (2005). Estimating treatment effects for discrete outcomes when responses to treatment vary: An application to Norwegian vocational rehabilitation programs. *Journal of Econometrics, 125*(1), 15–51.

Abdulai, A., & Huffman, W. (2014). The adoption and impact of soil and water conservation technology: An endogenous switching regression application. *Land Economics, 90*(1), 26–43.

Adetola, A. (2008). Factors influencing irrigation technology adoption and its impact on household poverty in Ghana. *Journal of Agriculture and Rural Development in the Tropics and Subtropics (JARTS), 109*(1), 51–63.

Alkire, S., & Foster, J. (2011). Counting and multidimensional poverty measurement. *Journal of Public Economics, 95*(7), 476–487.

Alkire, S., & Santos, M. (2014). Measuring acute poverty in the developing world: Robustness and scope of the multidimensional poverty index. *World Development, 59,* 251–274.

Anley, Y., Bogale, A., & Haile-Gabriel, A. (2007). Adoption decision and use intensity of soil and water conservation measures by smallholder subsistence farmers in Dedo district, Western Ethiopia. *Land degradation & development, 18*(3), 289–302.

Asrat, P., Belay, K., & Hamito, D. (2004). Determinants of farmers' willingness to pay for soil conservation practices in the southeastern highlands of Ethiopia. *Land Degradation and Development, 15*(4), 423–438.

Bacha, D., Namara, R., Bogale, A., et al. (2011). Impact of small-scale irrigation on household poverty: Empirical evidence from the Ambo district in Ethiopia. *Irrigation and Drainage, 60* (1), 1–10.

Barron, J. (2009). Background: The water component of ecosystem services and in human well-being development targets. In J. Barron et al. (Ed.), *Rainwater harvesting: A lifeline for human well-being* (pp. 4–13).

Barron, J., & Okwach, G. (2005). Run-off water harvesting for dry spell mitigation in maize (Zea mays L.): Results from on-farm research in semi-arid Kenya. *Agricultural Water Management, 74*(1), 1–21.

Bekele, W., & Drake, L. (2003). Soil and water conservation decision behavior of subsistence farmers in the Eastern Highlands of Ethiopia: A case study of the Hunde-Lafto area. *Ecological Economics, 46*(3), 437–451.

Carter, M., & Barrett, C. (2006). The economics of poverty traps and persistent poverty: An asset-based approach. *The Journal of Development Studies, 42*(2), 178–199.

Castillo, G., Namara, R., Munk Ravnborg, H., et al. (2007). Reversing the flow: Agricultural water management pathways for poverty reduction. In D. Molden (Ed.), *Water for food, water for life: A comprehensive assessment of water management in agriculture* (pp. 149–191). London/ Colombo: Earth scan/International Water Management Institute.

de Fraiture, C., Wichelns, D., Rockstrom, J., et al. (2007). Looking ahead to 2050: Scenarios of alternative investment approaches. In D. Molden (Ed.), *Water for food, water for life: A comprehensive assessment of water management in agriculture* (pp. 91–145). London, UK/ Colombo, Sri Lanka: Earth scan/IWMI.

Di Falco, S., et al. (2011). Does adaptation to climate change provide food security? A micro-perspective from Ethiopia.*American Journal of Agricultural Economics, 93*(3), 825–842.

Dile, Y., Karlberg, L., Temesgen, M., et al. (2013). The role of water harvesting to achieve sustainable agricultural intensification and resilience against water related shocks in sub-Saharan Africa. *Agriculture, Ecosystems & Environment, 181,* 69–79.

Dile, Y., Rockström, J., & Karlberg, L. (2016). Suitability of water harvesting in the Upper Blue Nile Basin, Ethiopia: A first step towards a mesoscale hydrological modeling framework. *Advances in Meteorology, 2016.*

Eguavoen, I., & McCartney, M. (2013). Water storage: A contribution to climate change adaptation in Africa. *Rural 21, 47*(1), 38–41.

Gebregziabher, G., Namara, R., & Holden, S. (2012). Technical efficiency of irrigated and rain-fed smallholder agriculture in Tigray, Ethiopia: A comparative stochastic frontier production function analysis. *Quarterly Journal of International Agriculture, 51*(3), 203–226.

Gebremedhin, B., & Swinton, S. (2003). Investment in soil conservation in northern Ethiopia: The role of land tenure security and public programs. *Agricultural Economics, 29*(1), 69–84.

Getnet, K., Pfeifer, C., & MacAlister, C. (2014). Economic incentives and natural resource management among small-scale farmers: Addressing the missing link. *Ecological Economics, 108,* 1–7.

Gordon, L., Finlayson, C., & Falkenmark, M. (2010). Managing water in agriculture for food production and other ecosystem services. *Agricultural Water Management, 97*(4), 512–519.

Hagos, F., Jayasinghe, G., Awulachew, S., et al. (2012). Agricultural water management and poverty in Ethiopia. *Agricultural Economics, 43*(s1), 99–111.

Hatibu, N., Mutabazi, K., Senkondo, E., et al. (2006). Economics of rainwater harvesting for crop enterprises in semi-arid areas of East Africa. *Agricultural Water Management, 80*(1), 74–86.

Huang, Q., Dawe, D., Rozelle, S., et al. (2005). Irrigation, poverty and inequality in rural China. *Australian Journal of Agricultural and Resource Economics, 49*(2), 159–175.

Humphreys, E., Peden, D., Twomlow, S., et al. (2008). *Improving rainwater productivity: Topic 1 synthesis paper.*

Jahnke, R.A., (1988). A simple, reliable, and inexpensive pore-water sampler.*Lymnology and Oceanography, 33*(3), 483–487.

Kassie, M., Jaleta, M., Shiferaw, B., et al. (2013). Plot and household-level determinants of sustainable agricultural practices in Rural Tanzania. *Technological Forecast and Social Change, 80*(3), 525–540.

Kassie, M., Teklewold, H., Jaleta, M., et al. (2015). Understanding the adoption of a portfolio of sustainable intensification practices in eastern and southern Africa. *Land Use Policy, 42,* 400–411.

Kassie, M., Zikhali, P., Pender, J., et al. (2010). The economics of sustainable land management practices in the Ethiopian highlands. *Journal of Agricultural Economics, 61*(3), 605–627.

Kato, E., Ringler, C., Yesuf, M., et al. (2011). Soil and water conservation technologies: A buffer against production risk in the face of climate change? Insights from the Nile basin in Ethiopia. *Agricultural Economics, 42*(5), 593–604.

Ketema, M., & Bauer, S. (2012). Determinants of adoption and labour intensity of stone terraces in eastern highlands of Ethiopia. *Journal of Economics and Sustainable Development, 3*(5), 7–17.

Lokshin, M., & Glinskaya, E. (2009). The effect of male migration on employment patterns of women in Nepal. *The World Bank Economic Review, 2009,* 481–507.

Lokshin, M., & Newson, R. B. (2011). Impact of interventions on discrete outcomes: Maximum likelihood estimation of the binary choice models with binary endogenous regressors. *Stata Journal, 11*(3), 368–385.

Merrey, D., & Gebreselassie, T. (2011). *Promoting improved rainwater and land management in the Blue Nile (Abay) basin of Ethiopia.*

Molden, D., Lisa, S., de Fraiture, C., et al. (2007). Summary for policy makers. In D. Molden (Ed.), *Water for food, water for life: A comprehensive assessment of water management in agriculture* (pp. 1–35). London/Colombo: Earthscan/International Water Management Institute.

Namara, R., Hanjra, M., Castillo, G., et al. (2010). Agricultural water management and poverty linkages. *Agricultural Water Management, 97*(4), 520–527.

Rämi, H. (2003). *Ponds filled with challenges.* UN OCHA-Ethiopia: Water harvesting-experiences in Amhara and Tigray.

Rockström, J., Falkenmark, M., Folke, C., et al. (2014). *Water resilience for human prosperity.* Cambridge University Press.

Rockström, J., Folke, C., Gordon, L., et al. (2004). A watershed approach to upgrade rainfed agriculture in water scarce regions through water system innovations: An integrated research initiative on water for food and rural livelihoods in balance with ecosystem functions. *Physics and Chemistry of the Earth, Parts A/B/C, 29*(15), 1109–1118.

Rockström, J., Karlberg, L., Wani, S., et al. (2010). Managing water in rainfed agriculture—The need for a paradigm shift. *Agricultural Water Management, 97*(4), 543–550.

Sen, A. (2001). *Development as freedom.* Oxford Paperbacks.

Seyoum, M. (2003). Overview of the Ethiopian rainwater harvesting association (ERHA). In McCornick et al. (Ed.), *Integrated water and land management research and capacity building priorities for Ethiopia* (pp. 155–167).

Shiferaw, B., & Holden, S. (1989). Resource degradation and adoption of land conservation technologies in the Ethiopian highlands: A case study in Andit Tid, North Shewa. *Agricultural economics, 18*(3), 233–247.

Shiferaw, B., & Holden, S. (1999). Soil erosion and smallholders' conservation decisions in the highlands of Ethiopia. *World Development, 27*(4), 739–752.

Smith, L. (2004). Assessment of the contribution of irrigation to poverty reduction and sustainable livelihoods. *International Journal of Water Resources Development, 20*(2), 243–257.

Teklewold, H., & Köhlin, G. (2011). Risk preferences as determinants of soil conservation decisions in Ethiopia. *Journal of Soil and Water Conservation, 66*(2), 87–96.

Impact of Renewable Technology on Lignocellulosic Material of Palm Fruit Fibre: Strategy for Climate Change and Adaptation

Onifade Tawakalitu Bola, Wandiga Shem Oyoo, Bello Isah Adewale, Jekayinfa Simeon Olatayo and Harvey J. Patricia

Abstract *Purpose*: This study aims to explore low temperature and pressure to extract the lignocellulosic content of palm fruit fibre and pyrolyse it for energy (bio-oil) and chemical feedstock productions as alternative technology to improper disposal causing environmental pollution as strategies for climate change and adaptation. *Design/Methodology/Approach*: Burning of refuse and agricultural residue releases harmful gases or carbon emission to the climate which causes global warming. Renewable technology can therefore be introduced to utilize agricultural residues to produce energy and chemical feedstock. Vacuum pyrolysis of the lignocellulosic material prepared from palm fruit fibre was performed by considering pyrolysis temperature of 200, 250, 300, 350 and 400 °C and biomass particle size of 0.25, 0.30, 0.42 and 0.55 mm. The bio-oil produced at different process parameters was then characterized by gas chromatography–mass spectrometry (GC-MS) and ultimate analysis. The response was efficiently used for modelling and optimization of the process parameters. The results showed that experimented and predicted data are in reasonable agreement with the values of coefficient of determination, R^2 (0.8098) and Adj. R^2 (0.700). *Findings*: It was found that the optimal conditions for pyrolytic product were temperature 300 °C, biomass particle size 0.55 mm and retention time between 14 and 18 min. Temperature and particle size had effect on the yielded gaseous products, liquid and char materials. And various chemical compounds were produced at different process parameters. *Research Limitation/Implications*: The contribution of renewable

O. T. Bola (✉) · H. J. Patricia
Department of Agricultural Engineering, Ladoke Akintola University of Technology, Ogbomoso, Nigeria
e-mail: tbonifade@lautech.edu.ng

B. I. Adewale
Department of Pure and Applied Chemistry, Ladoke Akintola University of Technology, Ogbomoso, Nigeria

W. S. Oyoo
Institute for Climate Change and Adaptation, University of Nairobi, Nairobi, Kenya

J. S. Olatayo
Faculty of Engineering & Science, University of Greenwich Central Avenue, Kent, UK

© Springer International Publishing AG, part of Springer Nature 2019

W. Leal Filho and R. Leal-Arcas (eds.), *University Initiatives in Climate Change Mitigation and Adaptation*, https://doi.org/10.1007/978-3-319-89590-1_11

technology framework is a process of educating for climate stewardship, at this stage of practical and theoretical. It is good approach of converting residues to useful products as one of the ways of solving climate change problem with the relevant stakeholders that its true potential can be assessed. *Social Implications*: Educating the public for climate stewardship will be more effective if it adopts an approach which seeks a co-production of knowledge. Clean environment with less global warming will be attained. *Originality/Value*: This agricultural residue can be recommended for biofuel, biogas and used as chemical feedstock for industrial purpose. The char can also be used to increase soil organic components. These benefits can therefore be used for policy making on residue disposal and as strategy for climate change and adaptation or mitigation in Nigeria and other African countries.

Keywords Palm fruit fibre · Pyrolysis · Renewable energy · Lignocellulose Climate change

1 Introduction: Climate Change and Energy

The world's climate and weather conditions are changing. Global temperatures are rising and causing more extreme weather disasters, like flooding and heat waves (UKEA 2015). "Adaptation" means changing the way of doing things to prepare for the potential impacts of climate change.

This implies that human will be better protected against consequences like flooding and other weather events (UKEA 2015). Weather variability and the effect on society vary across the country, and the effect of climate change on our economy, infrastructure, society and environment also varies from place to place. These help to make strategic plans for and adapt to climate changes (UKEA 2015). It will also result in better preparation for new opportunities, like the chance to grow different crops and innovations on renewable technologies to convert wastes to energy; health risks during heat waves can be reduced through effective planning and responses by health and social services, and poor rainfall or drought which causes low crop productivity can be improved by new system of irrigation. However, FAO (2001) reported that a long-term change in weather had occurred in the semi-arid and sub-humid zones of West Africa. And it appears that little or nothing can be done to improve variability in weather since most of its causes are natural.

A slight change in temperature at the right moment can cause outbreaks of epidemic diseases or insect pests, which can destroy entire landscapes, forest or farmlands as shown in Fig. 1. Everything in nature is related, changes in one area trigger changes in other areas (Osagie 2002; Adejuwon and Odekunle 2006). For example, the immediate survival of many coastal areas, population, forests and wildlife may now depend on our ability to study, understand and share the small changes that are observed in the environments and ecosystems around man. The current talk about weather variability has been correlated with the activities of man which have in turn generated micro-weather variation, pollution of environment by forest fuels burnt daily from industries and automobiles all of which generate heat, thereby altering the heat balance as seen in Fig. 2 (Osagie 2002; Adejuwon and Odekunle 2006).

Climate change is caused by the rate at which energy is received from the sun, and the rate at which it is lost to space determines the equilibrium temperature and climate of Earth. This energy is distributed around the globe by winds, ocean currents and other mechanisms to affect the climates of different regions (USEPA 2006). Consequently, the use of conventional drying for agricultural materials must be adapted to revolutionize the climate to reduce the rate of energy received from the sun and greenhouse gas emissions as reported by Onifade and Jekayinfa (2015). Conventional (air) drying is mostly used for dehydration operation in food and chemical industry, because drying occurs under controllable conditions and partially depends on climatic conditions and also reduces drying time (Onifade and Jekayinfa 2015).

Climate change can be most generally defined as a change in the statistical properties of the climate system when considered over long periods of time, regardless of cause (GCC 2001). Accordingly, fluctuations over periods shorter than a few decades refer specifically to climate change caused by human activity, as opposed to changes in climate that may have resulted as part of Earth's natural processes (NASA 2011). In this situation, especially in the context of environmental policy, the term *climate change* has become synonymous with anthropogenicglobal warming. Within scientific journals, *global warming* is stated as surface temperature increases, while *climate change* includes global warming and everything else that will be affected by increasing greenhouse gas levels, UNFCCC (1994). Climate change also refers to a change of climate which is directly or indirectly attributed to human activity that affects the composition of the global

Fig. 1 Land pollution *Source* Research study site

Fig. 2 Air pollution

atmosphere and which is in addition to natural climate variability observed over comparable time periods.

The root cause of energy situation is that in the recent years, while there has been no limit on growth in energy demand, the production of energy from the existing sources has come under increasing mounting pressure. It seems probably, therefore, that the shortage of fuel (energy) supplies is an actual fact of life and the fossil fuel supplies in our planet may not last forever (Bello et al. 2009). There is energy crisis in our hand; henceforth, source of energy supply is far below energy consumption. However, there are various renewable energy sources which could reduce the demand on fossils fuels. The renewable technology saves time, and it is usually more environmentally friendly, especially on air emissions (Schroeder 2009). There is great potential in the utilization of renewable biomass for energy and chemicals production to develop human life prosperity to address the crisis of energy security and sustainable economics in the world (Schroeder 2009; Ren et al. 2013).

Nigeria as a country is blessed with abundant agricultural resources; the materials that are left behind are called agricultural wastes or residues (Onwualu 2010). Agriculture is practised almost in every nation of the world, and large volume of crop residues that remain after harvest of crops is produced. Some of the agricultural residues generated are palm fruit fibre, physic nutshell, palm kernel shell, palm tree bark, maize cob, maize husks, maize stalk, cassava stalk, guinea corn stalk, cassava peels, millet stalks, plantain peels, groundnut husks, oil palm empty bunches, palm kernel cake and cowpea shells (Jekayinfa and Scholz 2007). Fortunately, in Nigeria, large quantities of wastes are generated annually, but unfortunately, these residues are not properly utilized as they are left to rot away or decompose (Jekayinfa and Omisakin 2005). A large percentage is dumped at refuse dump or landfill sites or burnt which has environmental implications (USEPA 2006). It is unbelievable that biomass, in general, can displace most of the other fuel or source of energy. This does not distract, however, from the fact that biomass fuel has a very significant part to play in meeting our energy needs. It is a unique part with the realization of the finite limitation of the fossil resources; a sense of urgency has developed in the exploration of alternative energy and feedstock sources.

Palm fruit fibre has great potentials to be used as biomass to develop renewable energy. It is unbelievable that biomass, in general, can displace most of other fuel or source of energy. This does not distract, however, from the fact that biomass fuel has a very significant part to play in meeting our energy needs. It is a unique part with the realization of the finite limitation of the fossil resources; a sense of urgency has developed in the exploration of alternative energy and feedstock sources. The black solid product obtained after pyrolysis process is known as char. This has potential soil-quality benefits which include an increase in organic matter and soil carbon. Higher concentrations of carbon in the soil reduce crusting, rapid soil surface drying and compaction and improve soil structure, water infiltration, water-holding capacity and microbial activity. Decades of crop production have depleted organic matter and soil carbon from many fields in western Nebraska, according to Hergert (UNL 2016). As abundant renewable resources, cellulosic material is being recognized as a possible supplement for long-term requirement (Bello et al. 2009). And biofuel sources such as materials like irrigated grasses would grow well on soil containing char (UNL 2016). The main benefit of converting waste to energy is to improve the economy of the nation, and it also has beneficial environmental implications (Onwualu 2010).

Pyrolysis is defined as the thermo-chemical conversion of biomass to char, bio-oil and gas, in the absence of oxygen and other reactants (Balat et al. 2009; Goyal et al. 2006). It always occurs before combustion and gasification where complete or partial oxidation is allowed to proceed. Pyrolysis is a complex non-equilibrium process where the biomass undergoes multistage decomposition resulting in large changes in specific volume (Tsai et al. 2006). The reaction rate, order and product yields depend on parameters such as temperature, heating rate, pre-treatment, catalytic effects. (Bridgewater 1996; Bridgewater et al. 2002). The reaction mechanism can be approximated by combining the yields from the three lignocellulosic compounds, despite synergetic effects. Therefore, the study of individual components forms the basis of the expected reaction pathways (Van de Velden et al. 2010). Cellulose is the focus of much research because it is the dominant lignocellulosic compound, and therefore, cellulose decomposition is best understood. However, if the gases and oil coming from the process are then burnt, this may also generate emissions. Also, this renewable method is quicker to build and the energy produced may be eligible for more renewable obligation certificates than incineration, increasing the potential from any power generated. Any fuel produced will not make up for the energy spent in manufacturing new products. Re-use and recycling are still better like incineration and landfill; energy savings from waste prevention and recycling are likely to be greater than the energy produced. The agricultural wastes are not fully utilized, and some are burnt without energy recovery (Onwualu 2010).

It has been observed in our localities that the traditional method of disposing palm tree residues by peasant farmers and oil palm processor is by dumping them around their processing centers because there is no awareness or orientation and available technology of processing these residues to useful products like biofuel or chemical feedstock. Henceforth, this disposal method of palm residues was observed in causing environmental pollution (USEPA 2006) such as land, water and air pollution. The palm fruit fibre and other wastes are not easily decomposed and found litter the environment, and this has disrupted the landscapes where soil percolation and water infiltration are difficult and resulted in low crop cultivation. This also creates hidden spaces for insect pest and disease vectors that are harmful to rural dwellers. Another observation is that during the raining season, rain erodes some of these residues to nearby water bodies (stream and river) which villagers use for domestic consumption. And some residues obstruct waterways (or flow) and cause flood that affects other cities. Air pollution occurs as the residues are sometimes set ablaze after long period of disposal. The release of gas emission from burning is causing health hazard as inhaled by human, and harmful gaseous components of these residues are released to the atmosphere which can cause problem of ozone layer and lead to global warming (UNFCCC 1994).

In view of this, there is need to embark on research development of renewable technology that saves time, and it is usually more environmentally friendly, especially on air emissions (Schroeder 2009) which plays vital roles for future energy and chemical supply. In this study, the main objective is to extract the lignocellulosic content of palm fruit fibre and pyrolyse it for energy (bio-oil) and chemical feedstock productions.

2 Materials and Methods

This section describes the sample preparation, renewable technology and proximate analysis of the residue.

2.1 Sample Preparation

The palm fruit fibre is a waste product of palm kernel fruits after oil processing. The samples were collected from local palm oil processing centre in village near Ogbomoso town (8°07′N, 4°16′E), Nigeria. The samples used for the study were sun-dried and ground into fine particle size. Figures 3, 4, 5 and 6 show the fresh fruits, the palm fruit fibre used, the processing site and residue dump site visited, respectively.

Fig. 3 Fresh palm kernel fruits

Fig. 4 Palm fruit fibre (residue)

The extraction, pyrolysis and proximate analyses of the samples were carried out at the laboratory of the Department of Chemistry, University of Nairobi, Kenya, while ultimate analyses were investigated at the Kenya Agricultural and Livestock Research Organization Laboratory. The samples were then screened to give various fractions using different wire mesh sizes. The diameter of the particle size of palm fruit fibre used ranged from 0.250 to 0.550 mm in form of 0.250, 300, 0.425 and 0.550 mm. Each sample was mixed with sodium anhydrous in order to make it dried and remove totally the moisture in it.

Fig. 5 Oil palm processing site

Fig. 6 Palm tree residues dump site

2.2 *Experimental Procedure*

(a) **Extraction or purification of the lignocellulosic material from the samples**

Lignocellulose was prepared from palm fruit fibre according to the methods reported by Bello et al. (2009). Fifty (50) g of the raw ground palm fruit fibre were measured using an electronic weighing balance Mettler Toledo, with an accuracy of 0.01 g; the sample was put in a 500 mL round bottom flask. Then, 200 mL of ethanol and distil water was measured (1:1 v/v) using a 100 mL measuring cylinder and poured into the sample. The flask containing the sample was put on an electric heating mantle set at 70 °C and allowed to boil for 30 min. Each purification process was maintained for all particle sizes. After the final decantation of the final supernatant, the resultant material (purified sample) was drained and dried in an oven at 105 °C for 48 h and cooled in an airtight desiccator with desiccant. This meant extraction or purification of the sample had been achieved to obtain ligno-celluloses from the samples.

(b) **Pyrolysis Procedure**

The purified samples were pyrolysed in batch-type reactors. Both reactors are the ampoule and the tubular systems. One end-closed narrow cylindrical Pyrex tubes of about 3 cm diameter and 15 cm long was used. The Pyrex ampoules were made by joining bulbs of 5 cm diameter with the cylindrical tubes. 2 g of lignocelluloses was weighed in each case and introduced into the reactor by means of narrow plastic funnel. In order to ensure quantitative transfer of the charge, a small Teflon rubber tubing about 2 cm long was used in the joining of the funnel and reactor outlet, and by gently tapping and varying the amount of the charging materials until the exact weight of 2 g as required was introduced into the reactor.

A small constriction was made a little above the cellulose level before inserting 0.2 g of glass wool into the reactor. The wool was held fixed by another con-striction above so as to prevent the loss of charge during the evacuation. Pirant vacuum gauge was used for the evacuation and the glass was sealed with a hand torch burner when the pressure reaches 0.1 mmHg. The duration of the evacuation was 5–30 min for the tubular and ampoule reactors. A thermostat oven furnace was required as shown in Fig. 7; the isothermal temperature was allowed to run for 10 min before introducing reactor and was maintained throughout the run. Each pyrolysis reaction lasted for 120 min at each temperature selected ranging from 200 to 400 °C. To terminate a run, the reactor was removed from the furnace and allowed to cool to laboratory temperature in a desiccator.

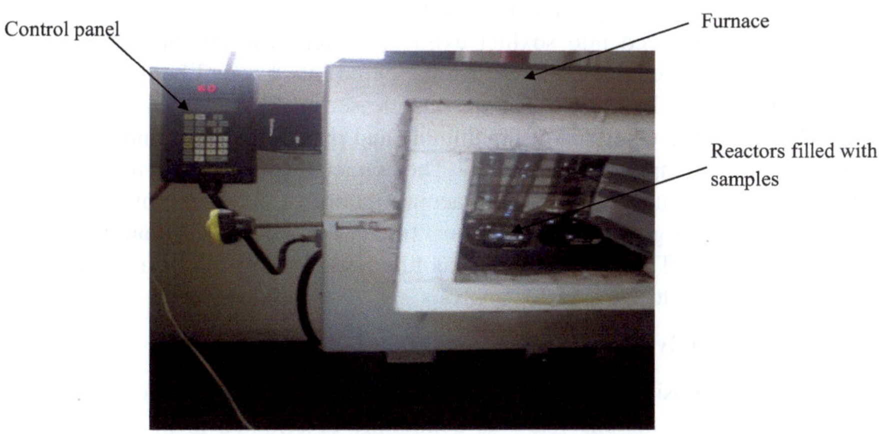

Fig. 7 Pyrolysed samples inside the furnace

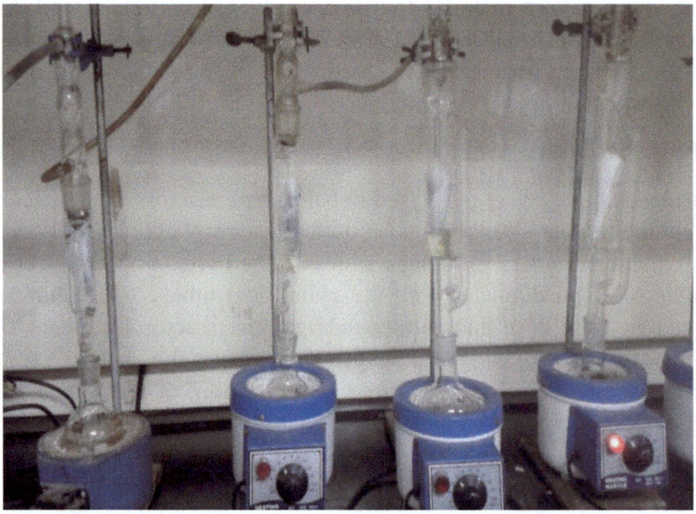

Fig. 8 Soxhlet extraction experiment for palm fruit fibre

The lignocellulose was pyrolysed, and various amount of product of the sample were determined. The buoyancy factor was considered in the evacuation of the gaseous products by subtracting the weight of the air in the ampoules from the weight of the ampoule before ejecting the air inside the ampoule. That of volatile liquid products would be obtained by subtracting the weight of the gaseous product, the weight of the solvent extract and the weight of the unextracted material from the amount charged into the reactor. The solvent extract was obtained by Soxhlet extraction as shown in Fig. 8, for 72 h of the liquid–solid products of the pyrolysed sample after removing the gases and volatile liquids.

2 grammes of the pyrolysed sample were measured and put into dried thimble container and introduced into soxhlet extractor. Then, 100 mL of methanol and dichloromethane solvent (1:1 v/v) each was poured into a round bottom (250 mL) flask and placed on the electric heating mantle set at 20 °C for 24 h. Then, the extracted sample was taken out of the thimble and put into crucible and dried in an oven for 48 h at a temperature of 105 °C. The extracted solvent in the round bottom flask was then concentrated using the vacuum pump set. The concentrated extract was cleaned using a glass column (10 cm) filled with sodium sulphate to remove water or impurities. The clean sample was filled into 0.2 mL chromatograph vials and then introduced to GC/MS for analysis.

(c) **Proximate Analysis**

Proximate composition was carried out in the Department of Physics, University of Nairobi, Kenya. This analysis was commonly determined to investigate agricultural residues as a biomass to develop renewable energy. The sample was analysed for the volatile matter, ash and fixed carbon contents to determine the fuel property to be obtained (Abnisa et al. 2011).

- ***Determination of the percentage volatile matter content***

The volatile matter content was carried out to determine the fuel property of the dried products. The procedure adopted was in accordance with ASTM Standard D5373-02 (2003). 10 g of the dried sample was put in crucible and heated in a muffle furnace of model CARBOLITE-ELF 11/6B at 250 °C for 7 min. The crucible was retrieved and kept to cool in a desiccator to a room temperature of the original weight to obtain the percentage volatile matter content as determined by Fapetu (2000) using Eq. 4.

$$\%\,V_{\mathrm{m}} = \frac{100(w_1 - w_2)}{w_1} \tag{1}$$

where $\%\ V_{\mathrm{m}}$ = % volatile matter content, w_1 = initial weight of the dried sample, w_2 = final weight of the sample after being subjected to 250 °C for 7 min.

- ***Determination of the ash content***

The ash content experiment was carried out to determine the effect of heat on the dried seeds. This was done in accordance with ASTM Standard D5373-02 (2003). About 10 g of finely ground sample of physic nut was placed in a crucible and heated in a muffle furnace CARBOLITE-ELF 11/6B at 250 °C for 1 h. After cooling it in a desiccator, the final weight was measured. The ratio of the initial weight to the final was expressed as a percentage to obtain the % ash content of the residue samples using Eq. 5 as stated by Fapetu (2000).

$$\%\,\mathrm{Ash} = \frac{100w_4}{w_3} \tag{2}$$

where, w_3 = initial weight of the oven-dried sample (g), w_4 = final weight, g.

- ***Determination of fixed carbon content***

This experiment was carried out to determine the fixed carbon content in the dried seed that would be needed to tackle the growing carbon emissions in the atmosphere with a goal to becoming carbon neutral. The fixed carbon content was obtained by using ASTM Standard D5373-02 (2003), as found by Fapetu (2000). The percentage fixed carbon was deduced from the ash content.

$$\%FC = 100 - (\%\mathrm{Ash} + V_{\mathrm{m}}) \tag{3}$$

where $\%\ C$ = the % amount of the fixed carbon, % Ash = % Ash content of dried sample.

(d) **Experimental design**

The response surface methodology (RSM) was employed in this study, and this is necessary to evaluate the performance of variables in pyrolysis system in other to optimize the bio-oil and biogas in the sample (palm fruit fibre). The effects of tested parameters such as temperatures and particle sizes on the pyrolysis efficiency were determined to identify optimal bio-oil and gas conditions. RSM helps proper design of experimental work, and optimization can either be minimum or maximum variables of design parameters. Design expert 6.0.8 software was used to analyse the tested parameters and response. This was obtained by the response surface regression procedure using the following second-order polynomial equation (Yunardi et al. 2011; Abnisa et al. 2011).

The temperatures chosen were 200, 250, 300, 350 and 400, while the particle sizes were 0.25, 0.300, 0.425 and 0.55 mm, and each experiment was replicated three times. The reaction time and product yield are the responses. The model used for predicting is generally a quadratic equation or second-order model. The model equation can be expressed as follows:

$$y = \beta_0 + \sum_{i=1}^{k} \beta_i x_i + \sum_{i=1}^{k} \beta_{ii} x_i^2 + \sum \sum_{i<j=1}^{k-1} \beta_{ij} x_i x_j \tag{4}$$

where y is the predicted response ($T_{\text{retention}}$, Y_{char}, Y_{gas} and Y_{oil}); x_i and x_j are the coded independent variables corresponding to temperature and particle size and β_0, β_i, β_{ii}, and β_{ij} are intercept, linear, quadratic and interaction constant coefficients, respectively. RSM package was also used for regression analysis and analysis of variance (ANOVA). Response surfaces, normal probability and plots were developed using the fitted quadratic polynomial equation obtained from regression, holding one of the independent variables at a constant value corresponding to the stationary point and changing the other variable. The independent variables being studied were temperature and particle size. The dependent variables were the retention time, yields of oil, gas and char. The design of three levels low, medium and high coded as −1, 0 and +1 was applied to this study. Another step is checking the adequacy of the model and testing of the lack-of-fit which is defined as a measure of a model failure in representing data in the experimental domain (Abnisa et al. 2011).

(e) **Characterization of bio-oil**

The bio-oil obtained from lignocellulose of palm fibre was used for characterization. Litmus papers blue and red were used to check its acidity and alkalinity. A microprocessor pH meter (HANNA pH 211) was used to measure the pH of the oil. Analysis was done at room temperature; the metre was calibrated by measuring pH of buffer solution to be 7.03 at 23.6 °C. The density of oil was conducted at room temperature of 23.6 °C. The density is defined as sample mass divided by a fixed empty volume of pycnometer. Then, a 2 mL pycnometer was used to determine the density; the bio-oil was filled into the flask and weighed the mass. Viscosity of bio-oil was measured using Ostwald Viscometer (a U-shaped glass viscometer). All experiments were carried out three times, and average readings were recorded.

The Fourier transform infrared spectrometer (IRAffinity-1S, SHIMADZU) was used to characterize the bio-oil produced from palm fruit fibre to produce spectra of different bands of component compounds. Also, GC-MS was used to identify the chemical compositions of the bio-oil. The analysis was performed with Agilent HP 6890 (version 4.10) N gas chromatograph equipped with Agilent HP 5075 mass-selective detector (mass spectrometer), using a 30 m by 0.25 mm DB-5 ms capillary column (0.25 μm film thickness) with temperature capacity of −60 to 325 °C. The GC oven was heated to 30 °C for 3 min and then to 290 °C at a rate of 3.5 °C/min, while injection port and detector were set at 300 °C. The carrier gas was helium with flow rate of 28 cm^3/min, and the effluent was monitored using a flame ionization detector (FID). Nitrogen was the makeup gas, while hydrogen and compressed air are the lightning flame for FID.

A separate constituent of the pyrolysis mixture together with helium carrier gas goes to the flame ionization detector. It was at this zone that the separated organic compounds ionized at a high temperature in contact with hydrogen or air flame. A polarized electric grid captures the resultant ions which generate a current that will be recorded as a chromatogram. Peak areas of individual gases were measured, and the abundance was determined from the prepared calibration. The calibration which is an external standardization has a technique involving essentially the injection of known amounts of pure compounds as reference substance at the same conditions with that of the sample (pyrolytic products). The sample peak was compared with that of the standard. Also, individual compound was identified by matching their elution times along the baseline with that of the standard. So, different components in the liquid solvent and their retention time were displayed on the screen of the GC-MS monitor.

(f) **Energy used in drying the samples**

To determine energy used in pyrolizing the samples, Boltzmann's equation could be used to translate temperature into energy and vice versa. As temperature goes up, so does the kinetic energy do. Energy used in pyrolizing was determined using Eq. 5:

$$E = K_B T \tag{5}$$

E energy, J
T temperature, K
K_B Boltzmann's constant

At low-temperature 158 °C, $E = 1.38 \times 10^{-23} \times 431$ J = 594.78×10^{-23} J
At high-temperature 442 °C, $E = 1.38 \times 10^{-23} \times 715$ J = 986.7×10^{-23} J

Power dissipated is the amount of energy consumed per time to drying the sample of average weight, 2 g, or it is a measure of how quickly energy is transferred in the system, and this was determined using Eq. 6:

$$P = \text{Energy}/\text{time} \tag{6}$$

Time = 2 h

At low-temperature 158 °C, $P = 594.78 \times 10^{-23}$ J/3600 × 2 s = 8.26×10^{-25} J/s
At high-temperature 442 °C, $P = 986.7 \times 10^{-23}$ J/3600 × 2 s = 1.37×10^{-24} J/s

3 Results and Discussions

3.1 Main Properties of Palm Fibre

The results of the main properties of the agricultural residue (palm fruit fibre) are presented in Table 1. The proximate analyses are the physical characteristics such as moisture, volatile, ash and fixed carbon contents. The ultimate analysis is the composition of the materials in weight percentage such as carbon, oxygen, nitrogen, magnesium, phosphorus and zinc, while the lignocellulosic values are based on the hemicellulose, lignin and cellulose of the materials.

Table 1 presented the main composition of the sample which includes the estimated value for cellulose, lignin and hemicellulose contents. The results revealed that the cellulose, lignin and hemicellulose values of palm fibre are 28.99, 60.36 and 38.96%, respectively. The respective values of other materials are such that mango pod had 22.96, 50.59, 26.45% (Olaleye 2013), wheat straw had 38, 15, 29% (Hongzhang and Liying 2007), while palm shell contained 27.7, 44, 21.6% (Abnisa et al. 2011), respectively.

From Table 1, the moisture content and dry matter of palm fruit fibre are 5.93 and 94.09%, respectively. Agricultural residues have different values like that of mango pod is 7.31% (Olaleye 2013), coconut shell is 6.0% (Ojha and Michael 2006). Several works had been done on proximate analysis of agricultural materials: the value of ash content for mango pod is 2.1% (Olaleye 2013); coconut is 0.6% (Sundaram and Natarajan 2009); rice straw contains 7.56% (Ahmad et al. 2011); palm shell is 2.1% (Abnisa et al. 2011). In this study, the ash content value of palm fibre is 61.30%; this value is higher than that obtained from other materials, and it might be because palm fibre is a product of oil crop and high ash content indicates quality energy source and high fuel property (Onifade and Jekayinfa 2015).

Moreover, the volatile content of palm fibre is 49.90%, while it was reported that volatile content of rice straw is 66.89% (Ahmad et al. 2011), and palm shell

Table 1 Main properties of the palm fruit fibre

Property	Value	Unit
Lignocellulosic analysis		
Cellulose	28.99 ± 0.029	%
Hemicellulose	38.96 ± 0.0897	%
Lignin	60.36 ± 0.41	%
Proximate analysis		
Moisture	5.93	%
Dry matter	94.09	%
Fat	20.13 ± 0.12	%
Ash	61.30 ± 1.15	%
Volatile	49.89 ± 3.12	%
Fixed carbon	38.70 ± 1.15	%
Ultimate analysis		
Nitrogen	1.05	%
Phosphorus	0.24	%
Potassium	0.29	%
Calcium	3.22	%
Magnesium	0.08	%
Iron	20.17	wt%
Copper	2.17	wt%
Manganese	4.33	wt%
Zinc	1.83	wt%

contains 67.2% (Abnisa et al. 2011). This indicates that the particles of others are less dense than the former which makes its component to escape into the air. The fixed carbon content of palm fibre is 38.70%. The values reported by other researchers were lower; rice straw contained 14.56% (Ahmad et al. 2011), and palm shell had 19.7% (Abnisa et al. 2011). There is higher value of fixed carbon content of physic nutshell, 86.40%. It was observed from previous research that fresh physic nut could absorb more carbon content from the atmosphere; i.e., more carbon can be emitted with fresh fruits (Onifade and Jekayinfa 2015), but burning its residues can do more harm to human because of its high volatility and ability to release gases into the atmosphere. This is one of the most promising solutions for tackling the growing carbon emissions from atmosphere (JWT 2010), hence reducing the problem of ozone layer. This implies that physic nut has greater potential to absorb carbon emission, thereby making the atmosphere carbon neutral. This factor can be a strategy for climate change adaptation or mitigation, if more cultivation practice of physic nut is established or encouraged in different countries of the world.

Table 2 Experimental design matrix and results

Actual level of factors			Responses		
Run	Temp. (°C)	Particle size (mm)	Gas (g)	Char (g)	Liquid (g)
1	300	0.55	0.413	1.59	10.758
2	400	0.25	0.679	1.33	10.473
3	200	0.25	0.235	1.98	9.130
4	300	0.42	0.499	1.51	8.273
5	300	0.25	0.489	1.53	9.839
6	300	0.42	0.499	1.51	8.273
7	200	0.55	0.044	1.97	9.023
8	300	0.42	0.499	1.51	8.273
9	158.8	0.42	0.019	1.99	8.253
10	300	0.42	0.499	1.51	8.273
11	300	0.42	0.499	1.51	8.273
12	441.42	0.42	0.994	1.03	6.657
13	400	0.55	0.943	1.09	7.364

3.2 Response Surface Methodology

Table 2 presented design matrixes and the results obtained from each experiment; twenty (20) experiments were carried with three replicates at each level, but thirteen (13) experiments were designed based on the number indicated by the statistical software. Temperature and particle size were the parameters selected to study the optimization of pyrolysis process of palm fruit fibre. The experimental design shows the effects of parameters (temperatures and particle size) on pyrolysis responses (the product yield) using a standard of RSM design based on central composite design (CCD) and two-level factorial design which is commonly used to fit a second other model.

Table 3 shows complete results obtained from analysis of variance (ANOVA) using two-level factorial design of Design Expert (6.8) version which includes experimental design and data analysis. It also expressed quadratic models and graphs in form of three-dimensional response surface or probability plot. It also helps to estimate the coefficients in a mathematical model and predict the response.

Effects of Process Parameters on Product Yield Figures 9, 10 and 11 presented the effect of temperature and particle size on gas, char and liquid yields of palm fibre. It was observed from Fig. 9 that gas yield increased with increase in temperature but slightly decrease with particle sizes. The increase in temperature leads to formation of more gaseous molecules released, the higher the temperature the higher the gases released to the atmosphere. This means if the residues are burnt in an open space, some will be inhaled by human while some components are released to the atmosphere which affects the climate causing global warming (UNFCCC

Table 3 Analysis of variance (ANOVA) for various responses

Source	Sum of square	Degree of freedom	Mean square	*F* value	Prob > *F*	Remarks
For gas model	0.98	3	0.33	332.40	<0.0001	Significant
A:Temperature	0.93	1	0.93	944.28	<0.0001	
B:Particle size	1.486E−004	1	1.486E−004	0.15	0.7061	
AB	0.052	1	0.052	52.77	<0.0001	
Residual	8.826E−003	9	9.807E−004			
Lack of fit	8.826E−003	5	1.765E−003			
Pure error	0.000	4	0.000			
Cor. total	0.99	12				
R^2 = *0.9911*	*Adjusted* R^2 = *0.9881*	*Predicted* R^2 = *0.9725*	*Adeq. precision* = *55.398*			
For char model	1.05	2	0.52	106.80	<0.0001	Significant
A	1.04	1	1.04	212.91	<0.0001	
B	3.409E−003	1	3.409E−003	0.70	0.4235	
Residual	0.049	10	4.896E−003			
Lack of fit	0.049	6	8.159E−003			
Pure error	0.000	4	0.000			
Cor. total	1.09	12				
R^2 = *0.9553*	*Adjusted* R^2 = *0.9463*	*Predicted* R^2 = *0.9064*	*Adeq. precision* = *30.374*			
For liquid model	12.98	5	2.60	5.96	0.0183	Significant
A:Temperature	0.83	1	0.83	1.90	0.2105	
B:Particle size	0.46	1	0.46	1.05	0.3388	
A^2	1.00	1	1.00	2.29	0.1738	
B^2	7.57	1	7.57	17.37	0.0042	
AB	2.25	1	2.25	5.17	0.0571	

(continued)

Table 3 (continued)

Source	Sum of square	Degree of freedom	Mean square	F value	Prob > F	Remarks
Residual	3.05	7	0.44			
Lack of fit	3.05	3	1.02			
Pure error	0.000	4	0.000			
Cor. total	16.03	12				
$R^2 = 0.8098$	*Adjusted* $R^2 = 0.6740$	*Predicted* $R^2 = -0.3525$	*Adeq. precision* = 8.112			

Fig. 9 Effect of temperature and particle size on gas and char yield, respectively

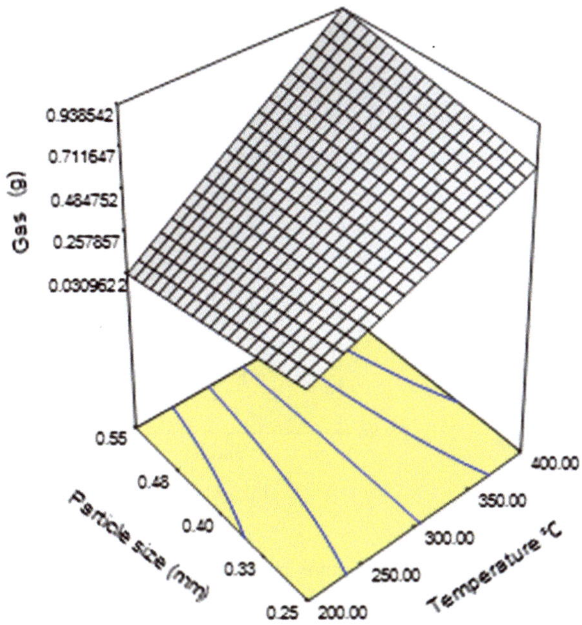

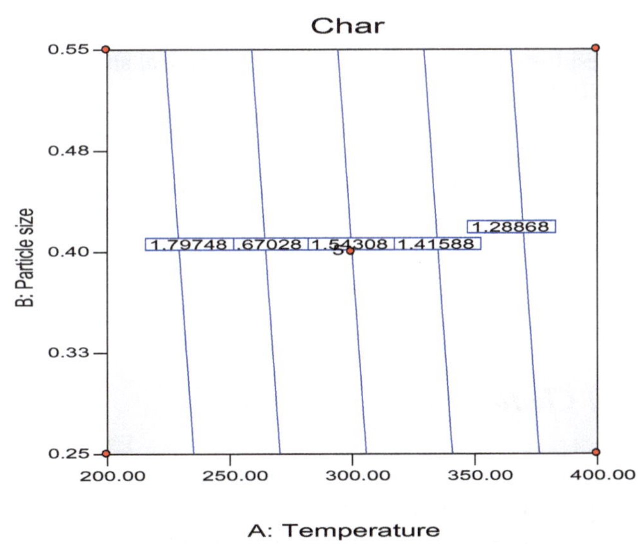

Fig. 10 Effect of temperature and particle size on gas and char yield, respectively

1994). The gaseous component can also be useful as chemical feedstock in industry (Bello et al. 2009).

Figure 10 shows the contour graph depicted the effect of process parameters on the char yield. It was observed from the figure that increases in temperature and

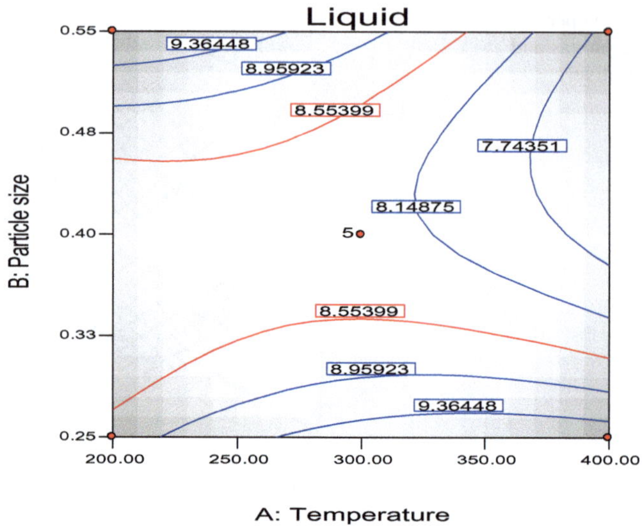

Fig. 11 Effect of temperature and particle size on liquid yield, **a** 3D surface, **b** contour

particle sizes lead to decrease in quantity of char produced. The char has potential soil-quality benefits for farming activities when it is incorporated into the soil. Application of residue char into the soil improves physical and chemical properties of farmland (UNL 2016).

From Fig. 11, it was observed that as temperature and particle sizes increased, the liquid yield increased to a point of 300 °C, 0.55 mm. And there was a fold in Fig. 11a; this means that the bio-oil produced after this point decreased as temperature increased. Figure 11b shows the contour graph of the effect of the processing parameters on the yield; this reveals the optimum process condition is achieved at 300 °C, 0.55 mm.

3.3 *Physical Characterization of Bio-Oil*

The physical properties of bio-oil produced from palm fruit fibre are presented in Table 4. The liquid extracts were acidic in nature because it turned blue litmus paper to red. This indicated that the samples had been burnt under inert condition. The pH values of oil, which ranged from 4.64 to 6.43, increased with increase in temperatures; this meant that oil obtained at high temperatures are slightly acidic while oil obtained at low pyrolitic temperatures indicates low pH readings which are more acidic. The density of the oil is 0.8319 g/mL (831.99 kg/m^3) at 23.6 °C. It was observed that all density values at different temperatures and particle sizes gave similar readings. The density and viscosity are related to phenomenon of liquid

Table 4 Physical properties of bio-oil

Properties	Value	Unit
Viscosity	0.695	cP
pH	4.64–6.43	Acidic
Density at 23.6 °C	831.99	kg/m^3
Calorific value	22.33	kJ/g
Water content	5.39	%
C	65.24	%
H	5.07	%
N	1.10	%
O	28.60	%

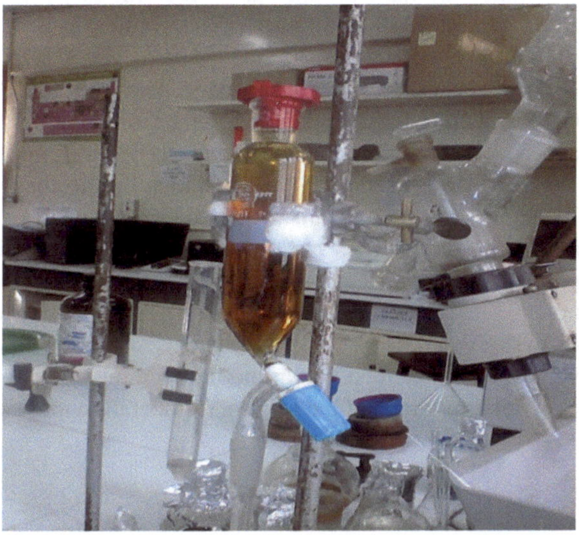

Fig. 12 Bio-oil produced from palm fruit fibre

floatation which can have significant effect on fluid atomizers. The bio-oil produced in this study gave viscosity of 0.695 cPa room temperature; this value was less than that of water used as standard. This indicated that the viscous period of bio-oil is less than that of water. Hence, this could be due to low value of water content (5.93%) which caused it to be less viscous. The presence of water content in bio-oil shows the presence of lignin in the raw material which is 60.36%. The calorific value of the bio-oil is 22.33 kJ/g as shown in Table 4, this shows the amount of energy produced by the complete combustion of 0.291 g of the oil. The calorific value is an important factor to determine the energy content of the fuel. Compared to other common fuel, gasoline (47 kJ/g), diesel (45 kJ/g), ethanol (29.7 kJ/g), wood (15 kJ/g) coal (15 kJ/g) and natural gas (54 kJ/g) (NIST), this indicates that bio-oil from palm fruit fibre is a potential source of energy and can be upgraded first before using it as fuel. Figure 12 presented the bio-oil produced from palm fruit fibre.

3.4 Chemical Characterization of Bio-Oil

Figure 13 presented the results of the spectra obtained from the Fourier transform infrared spectrometer which was used to characterize the bio-oil produced from palm fruit fibre. The O–H stretching vibrations between 3200 and 3400 cm^{-1} show the presence of phenols and alcohols. The C–H stretching vibrations between 2800 and 3000 cm^{-1} and C–H deformation vibrations between 1350 and 1450 cm $^{-1}$ show the presence of alkane groups. The C=O stretching vibrations between 1680 and 1750 cm^{-1} are compatible with the presence of ketones, quinones, aldehyde groups. The peaks between 1500 and 1645 cm^{-1} represented as C=C stretching vibrations indicated the presence of alkenes. Besides, mono- and polycyclic and substituted aromatic groups can be identified by the absorption peaks between 690 and 900 and 1350–1450 cm^{-1}. Then, ethers can be identified by a strong C–O stretching band near 1100 cm^{-1} due to the C–O–C linkage in the compound. Aromatic ethers show a strong band near 1250 cm^{-1}, while cyclic ethers show a C–O stretching band in the range of 1250–900 cm^{-1}. A broad N–H wagging band appears at 750–650 cm^{-1} represents secondary amides.

GC-MS was used to analyse and identify the chemical components in the liquid. Table 5 presents the details results of GC-MS analysis. The most abundant products and highest peak area are achieved by hexadecanoic (81.3%). Other prominent products are pentadecanoic acid (1.47–14.5%), octadecanoic (2.6–70.1%), eicosanoic (3.5–11.3%), 2-2-hydroxyethoxy (2.71–12.3%), Ascorbic 2, 6-dihexadecanoic (7.1–14.3) and Isopropyl palmitate (7.31–41.0). It was observed that different values were obtained at various temperatures and particle sizes. This shows the effect of experimental parameters (temperature and particles sizes) on the chemical

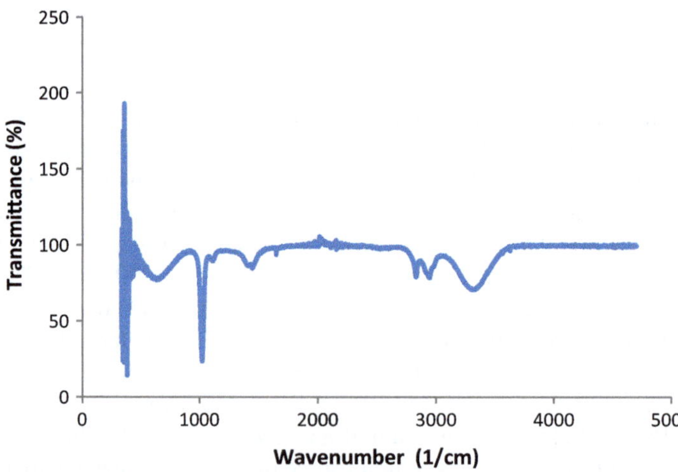

Fig. 13 FT-IR spectra bio-oil derived from palm fruit fibre

Table 5 Identification and quantification of chemical compounds in bio-oil by GC-MS analysis

Chemical compounds	Molecular formula	Molecular weight (g/mol)	Peak probability (%)
3-methyl Pentadecanoic acid	$C_{16}\ H_{32}O_2$	256	1.47–2.85
2,16-methyl Hexadecanoic acid	$C_{18}\ H_{34}O_2$	284	1.30–10.15
Methyl Hexadecanoic acid	$C_{17}\ H_{34}O_2$	270	81.3
14-methyl Pentadecanoic acid	$C_{17}\ H_{34}O_2$	270	9.92–28.7
Tridecanoic acid	$C_{17}\ H_{34}O_2$	270	2.58
Ethyl Pentadecanoic acid	$C_{17}\ H_{34}O_2$	270	6.53
Ethyl Eicosanoic acid	$C_{22}\ H_{44}O_2$	340	3.53–11.3
Ethyl Hexadecanoic acid	$C_{19}\ H_{38}O_2$	298	2.84
Ethyl Tridecanoate	$C_{15}\ H_{30}O_2$	242	1.46
Methyl Octadecanoic	$C_{19}\ H_{38}O_2$	298	4.04–66.6
16-methyl heptadecanoate	$C_{19}\ H_{38}O_2$	298	10.5
Ethyl Octadecanoic acid	$C_{20}\ H_{40}O_2$	312	60.7
Nonadecanoic acid	$C_{21}\ H_{42}O_2$	326	0.93
Cyclopropanepentanoic	$C_{20}\ H_{38}O_2$	310	3.73
Tetradecanoic	$C_{16}\ H_{32}O_2$	256	2.7– 4.63
Ethyl Heptadecanoic	$C_{19}\ H_{38}O_2$	298	5.48
2-2-hydroxyethoxy	$C_{22}\ H_{44}O_4$	372	2.71–12.3
Hexadecanoic acid	$C_{18}\ H_{36}O_2$	284	49.4
Isopropyl palmitate	$C_{19}H_{38}O_2$	298	7.31–41.0
Ascorbic 2,6-dihexadecanoic	$C_{38}\ H_{68}O_8$	652	7.1–14.3
Ascorbic acid, 6-octadecanoate	$C_{24}\ H_{42}O_7$	442	2.8

compounds produced from palm fruit fibre. For instance, highest peak of methyl-hexadecanoic, 81.3% was obtained at 158.8 °C, 0.42 mm, methyl-pentadecanoic acid was high, 14.5%, at 441.42 °C, 0.42 mm, and eicosanoic had value of 11.3% at 300 °C, 0.25 mm. Heptadecanoic has highest peak at 200 °C, 0.55 mm. Only 300 °C 0.25 mm and 400 °C 0.25 mm contained Ascorbic 2,6-dihexadecanoic. There are a great number of other compounds, but their peak areas are low, so this study did not examine them further. Every compound in Table 5 is classified as aromatic oxygenated and hydrocarbon compounds which are dominant compounds in the palm fruit fibre oil. Oxygenated content is favourable to be used for fuels, while other compounds can be useful for chemical productions. The results obtained in this study are different, and values are higher than those obtained from palm kernel shell by Abnisa et al. (2011).

4 Conclusions

The pyrolysis of palm fruit fibre had been examined; experimental factors (temperatures and particle sizes) had significant effect on the bio-oil and chemical compounds yielded from the residues. Investigation was done to determine optimum factors that can maximize bio-oil production by generating mathematical equations from response surface methodology approach. The optimum process condition of bio-oil was produced by 300 °C and 0.55 mm, while highest peaks were attained at the retention time between 14 and 18 min. Power used in pyrolysing the residues inside furnace at temperatures 158 and 442 °C is 442 8.26×10^{-25} J/s and 1.37×10^{-24} J/s, respectively. This paper has outlined how renewable technology is used to break agricultural residues and obtains useful products like bio-oil for biofuel production and chemicals components for industrial purpose. But their usefulness will provide valuable umbrella terms for the guiding principles of climate change policy. Climate change has become one of the major challenges for mankind and the natural environment. The gas emissions released into the atmosphere in ever rapidly growing volumes are recognized to be responsible for this change (JWT 2010). It is reflected from the findings that when the residues are burnt, these bio-fuels and chemical compounds are released into the atmosphere causing air pollution and resulted in depletion of ozone layer which causes global warming and has negative effects on climate. It is profitable and highly beneficial if energy and chemicals can be produced from residues rather causing environmental pollution. Opportunities for success would be enhanced by ensuring a collaborative process which starts at the grassroots (from the rural farmers) to higher educationist and government.

In this framework, climate stewardship becomes a partnership between the public, higher educationist and the government. To implement the findings, it is important that representatives of the relevant stakeholder groups separately deliberate on new system of proper disposal of residues and renewable technology in relevant to climate policy. This could then move to symposium and workshops which bring various ideas from different groups together to embark on technology that is environmentally friendly and use as strategy or policy for climate change.

Acknowledgements The authors would like to acknowledge the financial support of the Department of International Development (DFID) and ACU, UK, and the African Academy of Sciences AAS, Kenya. The research was done at Institute of Climate Change and Adaptation, Kenya.

References

Abnisa, F., Wan Daud, W. M. A., & Sahu, J. N. (2011). Optimization and characterization studies on bio-oil production from palm shell by pyrolysis using response surface methodology. *Journal of Biomass and Bioenergy, 35*, 3604–3616.

Adejuwon, J. O., Odekunle, T. O. (2006). Variability and the severity of the "little dry season" in Southwestern Nigeria. *American Journal of Meteorogical Society*, 483–493.

Ahmad, J. U., Saidur, R., Masjuki, H. H., Mekhilef, S., Ali, M. B., & Furqon, M. H. (2011). An application of energy and exergy analysis in agricultural sector of Malaysia. *Journal of Energy Research, Energy Policy, 39*, 7922–7929. Available at: https://onlinelibrary.wiley.com/doi/doi.org/10.1002/er.1023/fulljohnWiley&Sons.

Balat, M., Kirtay, E., & Balat, H. (2009). Main routes for the thermo-conversion of biomass into fuels and chemicals. Part 1: Pyrolysis systems. *Energy Conversion Management, 50*, 3147–3157. Available at: https://books.google.com/books?isbn=111856040X.

Bello, I. A., Giwa, A. A., Bello, O. S., Oladipo, M. A. & Amuda, O. S. (2009). Study of pyrolysis of lignocellulosic material from mahogany at low temperature and pressure. *International Journal of Biological and Physical Sciences. Science Focus, 14*(2), 146–154. Available at: https://mail.google.com/mail/u/0/#inbox.

Bridgewater, A. V. (1996). Production of high grade fuels and chemicals from catalytic pyrolysis of biomass. *Catalysis Today, 29*, 285–295. Available at: https://books.google.com/books?id=owiVwfMGPKcC.

Bridgewater, A. V., Toft, A. J., & Brammer, J. G. (2002). A techno-economic comparison of power production by biomass fast pyrolysis with gasification and combustion. *Renewable and Sustainable Energy Reviews, 6*, 181–248. Available at: https://eclass.duth.gr/…/Economic%20c.

Fapetu, O. P. (2000). Production of charcoal from tropical biomass for industrial and metallurgical process. *Nigeria Journal of Engineering Management, 1*(2), 34–37.

Glossary–Climate Change. (2001). *Arctic climatology and meteorology education center*. National snow and ice data center, NSIDC. Glossary, in IPCC TAR WG1 2001, http://nsidc.org/arcticmet/glossary/climate_change.html.

Goyal, H. B., Seal, D., & Saxena, R. C. (2006). Bio-fuels from thermo-chemical conversion of renewable resources: A review. Indian Institute of Petroleum, Dehradun 248005, India. *Renewable and Sustainable Energy Reviews 12*, 504–517

Hongzhang, C., & Liying, L. (2007). Unpolluted fractionation of wheat straw by steam explosion and ethanol extraction. *Bioresource Technology, 98*, 666–676. Available at: https://wenku.baidu.com/view/2e907a52f01dc281e53af048.

Jekayinfa, S. O., & Omisakin, O. O. (2005). The energy potentials of some agricultural wastes as local fuel materials in Nigeria. *Agricultural Engineering International, the CIGR E-Journals of Scientific Research and Development, Manuscript EE 05*(3), 10. Available at: www.cigrjournal.org https://ecommons.cornell.edu/…/EE%2005%20003%20Jekayinfa%20fin.

Jekayinfa, S. O., & Scholz, V. (2007). Assessment of availability and cost of energetically usable crop residues in Nigeria. In *Conference on International Agricultural Research for Development*, Tropentag, 2007. Available at: https://citeseerx.ist.psu.edu/viewdoc/download?doi=10.1.1.603.3404.

JWT. (2010). *Jatropha World Team, Centre for Jatropha Promotion and Biodiesel*. Available at: info@jatrophaworld.origin@jatrophaworld.org.

NASA. (2011). *Global warming and climate change*. Available at: http://www.nasa.gov/topics/earth/features/climate_by_any_other_name.html.

Ojha, T. P., & Michael, A. M. (2006). Principles of agricultural engineering. Jain Brothers, New Delhi India, Vol. 1, pp. 205–215. Available at: www.thejainbrothers.com/.

Olaleye, T. O. (2013). *Pyrolysis of mango pod*. An unpublished B. Tech. dissertation at Department of Agricultural Engineering, Ladoke Akintola University of Technology, Ogbomoso, Nigeria. Available at: https://mail.google.com/mail/u/0/#inbox.

Onifade, T. B., Jekayinfa, S. O. (2015). Application of response surface methodology for energy analysis of thin layer drying of physic nut (*Jatropha curcas*). *Journal on Agricultural Engineering, Global Science and Technology Forum, 2*, 1–11.

Onwualu, A. P. (2010). Industrial raw materials from agricultural residues; status, challenges and opportunities. In *Humboldt Kolleg International Conference*, Ogbomoso 11–14th, January, 2010. pp. 367

Osagie, O. I. (2002). Cereal output to plunge next year. *AgroBusiness. THISDAY Online.*

Ren, X., Jinsheng, G., Wenliang, W., Qiang, L., Jianmin, C., & Ben, L. (2013). Optimization of bark fast pyrolysis for the production of phenol-rich bio-oil. *Journal of Bio Resources, 8*(4), 6481–6492. Available at: https://www.researchgate.net/.../273221176ojs.cnr.ncsu.edu‹Home›.

Schroeder, M. (2009). Utilizing the clean development mechanism for the deployment of renewable energies in China. *Journal of Applied Energy, 86*, 237–242. Available at: https://www.google.com/search?q=econpapers.repec.org/RePEc:eee:appene:v:86:y:2009:i:2:p:237-242.

Sundaram, E. G., & Natarajan, E. (2009). Pyrolysis of coconut shell: An experimental investigation. *The Journal of Engineering Research, 6*(3), 33–39. Available at: https://tjer.net/site/issue6-2/p4.pd.

Tsai, W. T., Lee, M. K., & Chang, Y. M. (2006). Fast pyrolysis of rice husk, product yields and composition. *Bioresources Technology, 10*, 22–28. Available at: https://www.google.com/search?q=www.ncbi.nlm.nih.gov/.

UKEA. (2015). Government policy and environmental agency; climate change and adaptation 2010–2015, Department of Environmental Food and Rural Affairs. https://www.gov.uk/government/policies/adapting-to-climate-change.

UNFCCC. (1994). *The United Nations Framework Convention on Climate Change*. Retrieved September 13, 2015. Available at: http://unfccc.int/essential_background/items/1349.php.

UNL. (2016). Panhandle Research and Extension Center & Institute of Agriculture and Natural Resources, University of Nebraska, Lincoln. Available at: www.hpj.com/crops/unl-studyingsugarbeetcharstoimprovesoil.

USEPA. (2006). United States Environment Protection Agency, Municipal Solid Waste in the United States: 2005 Facts and Figures (EPA A530-R-06011, Office of Solid Waste, US Environmental Protection Agency, Washington DC. Available at: https://www.google.com/search?q=https://www3.epa.gov/epaw.

Van de Velden, M., Baeyens, J., Brems, A., Janssens, B., & Dewil, R. (2010). Fundamentals, Kinetics and endothermicity of biomass pyrolysis reaction. *Renewable Energy, 35*, 232–242. Available at: https://www.google.com/search?q=wrap.warwick.ac.uk/5562/.

Yunardi, Z., & Masrianto. (2011). Response Surface Methodology Approach to optimizing process variables for densification of rice straw as a rural alternative solid fuel. *Journal of Applied Sciences, 7*, 1192–1198.

Onifade Tawakalitu Bola is an Agricultural Engineer, a Lecturer in the Department of Agricultural Engineering, Ladoke Akintola University of Technology, Ogbomoso, Nigeria. Her qualifications are B.Eng., M.Sc. and Ph.D. in agricultural engineering. She is specialized in crop processing and storage, renewable energy and agricultural residue utilization, climate change and environmental management for livelihood benefits.

Wandiga Shem Oyoo is a Professor at University of Nairobi, Kenya, and he is also presently the Director of Institute of Climate Change and Adaptation. He is a trained chemist and has supervised or co-supervised dozens of doctoral theses; he had led a variety of international projects.

Bello Isah Adewale is a Professor of Chemistry in the Department of Pure and Applied Chemistry, Ladoke Akintola University of Technology, Ogbomoso, Nigeria. He is a trained chemist and has supervised or co-supervised some doctoral theses.

Jekayinfa Simeon Olatayo is a Professor of Agricultural Engineering in the Department of Agricultural Engineering, Ladoke Akintola University of Technology, Ogbomoso, Nigeria. He is DAAD, Germany alumnus and a fellow of the Alexander von-Humboldt Stiftung Foundation. He has supervised or co-supervised dozens of doctoral theses.

Harvey J. Patricia is a Professor of Biochemistry, at Faculty of Engineering and Science, University of Greenwich Central Avenue, Kent, UK. She has supervised or co-supervised dozens of doctoral theses.

Maize Yields in Varying Rainfall Regimes and Cropping Systems Across Southern Africa: A Modelling Assessment

Siyabusa Mkuhlani, Walter Mupangwa and Isaiah Nyagumbo

Abstract Rainfall variability, which ultimately leads to climate change, is a major threat to smallholder agriculture. It affects time of sowing time and productivity, amongst other challenges. There is therefore need to evaluate the different strategies for their effectiveness in managing climate variability. This study assessed the effects of different strategies on sowing date, season length and maize yields under variable rainfall conditions. Maize (*Zea mays* L.) yield simulations for Southern Africa were conducted using the DSSAT model. Simulated conservation agriculture (CA)-based cropping systems included basins prepared early (*CA-Basins early*) and late (*CA-Basins late*), draught powered planter (*CA-Direct seeder*), ripper (*CA-Ripper*) and Dibble stick (*CA-Dibble*). Conventional systems were mouldboard ploughing early (*CMP-early*) and late (*CMP-late*). Rainfall seasons were classified into *low, medium* and *high* based on the total rainfall amount. Results showed that *high*-rainfall seasons were seeded earlier and had a greater season length compared to *low* rainfall seasons in drier agro-ecologies, translating to higher yields and *vice versa*. Reduced labour requirements and use of draught power, enabled early seeding of *CA-ripper, direct seeder, basins early and CMP-early* systems compared to *CA-Basins late, Dibble stick and CMP-late* systems. However, performance of cropping systems did not vary across season types suggesting that there was thus no evidence of higher yield

---

S. Mkuhlani (✉) · W. Mupangwa · I. Nyagumbo
CIMMYT, International Maize and Wheat Improvement Centre,
Mt. Pleasant, Box MP163, Harare, Zimbabwe
e-mail: siyabusa@gmail.com

W. Mupangwa
e-mail: W.Mupangwa@cgiar.org

I. Nyagumbo
e-mail: i.nyagumbo@cgiar.org

S. Mkuhlani
Climate Systems Analysis Group, Department of Geography
and Environmental Science, University of Cape Town,
Private Bag X3, Rondebosch, Cape Town 7701, South Africa

© Springer International Publishing AG, part of Springer Nature 2019
W. Leal Filho and R. Leal-Arcas (eds.), *University Initiatives in Climate Change Mitigation and Adaptation*, https://doi.org/10.1007/978-3-319-89590-1_12

advantages from CA technologies even during low rainfall seasons. This puts the merits of drought mitigation by CA technologies into doubt despite enabling early planting.

Keywords Conservation agriculture · Conventional agriculture
Semi-arid · Planting date · Season length

1 Introduction

Limited access to capital, unstable political environments, poor access to extension services and low-soil fertility are some of the challenges affecting the productivity of smallholder farming systems in Southern Africa (Mpandeli and Maponya 2014). Smallholder farmers in rain-fed systems have highlighted increased intra- and interseasonal climate variability as the major threat to their livelihood (Thomas et al. 2007). There is therefore need for in-depth evaluation of available adaptation options available to sustainably improve productivity amongst smallholder farmers.

The Southern African region has experienced gradual changes in climate, as manifested by a steady increase in temperature, dry spells, droughts and rainfall variability. In the semi-arid agro-ecologies, rainfall variability is expected to increase and the impacts are expected to surpass those of climate change (Bouba et al. 2013). Increased climate variability has been attributed to increased global warming and the *El Nino* phenomenon. This has led to at least 50% increase in extremely wet and dry seasons in the last 10 years. On the contrary, sub-humid to humid agro-ecologies of Southern Africa have relatively low rainfall variability of less than 10% per annum, despite the notable increases in unpredictability of precipitation within the last three decades. High rainfall is also correlated to the greater number of rainy days per season which is a characteristic of wetter agro-ecologies (Fallis 2013). Climate variability is predicted to be severe amongst communities in developing countries compared, hence there is need to analyse the effects of rainfall variability on smallholder farm productivity (Bouba et al. 2013). The increased seasonality and decreased reliability of rainfall have made it difficult for farmers to determine when to plant, which also has consequences on the length of the growing season and productivity. This has resulted in farmers making repeated planting with the aim of capitalising on the increasingly variable conducive rainfall conditions. The impacts of increased rainfall variability necessitate a better understanding of the impacts of seasonal rainfall variability on crop production (Bouba et al. 2013).

Several strategies for coping with and adapting to rainfall variability have been proposed and explored in smallholder farming systems of Southern Africa. Traditionally smallholder farmers, whose farming systems are centred on mixing crop and livestock, have used livestock as a buffer in times of total crop failure (Valdivia and Quiroz 2003). In response to increased rainfall variability and variation in the commencement and cessation of the rainy season, change in time of planting has been used as a coping strategy (Cooper et al. 2008). Cultivation of drought-tolerant crops, such as sorghum (*Sorghum bicolor* (L.) Moench), and multi-purpose grain legumes has also been adopted by smallholder farmers to cushion them against water stresses associated with rainfall variability. Planting basins, mulch ripping and infiltration pits, are some of the research recommended and widely used soil water and rainwater harvesting techniques for coping with rainfall variability within the Southern African region (Cooper et al. 2008).

Conservation agriculture (CA) is a sustainable option for improving agricultural water management under water limiting conditions. CA has the potential to enable early sowing due to the timeous and reduced land preparation labour needs. CA improves soil structure and fertility, increases rainwater infiltration and retention, reduces soil evaporation, resulting in extended soil water supply to crops during dry spells (Thierfelder et al. 2016). Maize yields under CA systems have been found to increase by 10–100% relative to conventional agriculture under rainfall regimes of less than 600 mm. Yield benefits are, however, suppressed by 5–20% under high-rainfall regimes beyond 1000 mm rainfall per annum (Rusinamhodzi et al. 2011). Various CA cropping systems have been tested and promoted in Malawi, Mozambique, Zambia and Zimbabwe using either manual or animal traction-based tillage techniques. Manual techniques promoted involve the hoe prepared planting stations commonly known as conservation farming basins whose preparation is labour intensive (Thierfelder et al. 2016). To alleviate the high-labour requirements of basins, farmers are recommended to prepare basins during the dry season and are ready for planting at the onset of the rains. However, some farmers sometimes wait for the onset of the rains before preparing the basins and usually fail to plant with the first rains (Mugabe and Banga 2001). Animal traction CA options used in the region include the Magoye ripper and the Brazilian-made Fitarelli direct seeder (Fitarelli Maquinas Agricolas, Brazil; www.fitarelli.com.br) and these have been welcomed by smallholder farmers due to their low labour requirements (Siziba 2007).

Prior limited research has been conducted in evaluating the different strategies for managing climate variability risk in Eastern Southern Africa. The research has been limited to specific agro-ecologies, where previous studies in high potential regions of Zimbabwe have shown that the time of planting is a more important yield determining factor than type of cropping system employed. Use of CA along with draught powered conventional cropping promotes early sowing which tends to lead to high yields in sub-humid to semi-arid areas of Zimbabwe (Nyagumbo et al. 2017). Variations in rainfall often lead to significant maize yield differences. However, no research has been conducted yet to assess how variations in rainfall affect sowing time, season length and the subsequent yields of maize. This study therefore sought to evaluate how climate variability management strategies as depicted by the different cropping systems and climate variability as depicted by the varying seasonal rainfall amounts affects time of planting, season length and subsequent maize yields in contrasting agro-ecologies of Malawi, Mozambique and Zimbabwe.

2 Materials and Methods

The study was based on previous research on sustainable intensification of cropping systems undertaken by ICRISAT and CIMMYT-Southern Africa in Malawi, Mozambique and Zimbabwe (2004–2011). The previous research undertook on-station and on-farm field trials of the performance of maize-based different cropping systems in Eastern Southern Africa in response to seasonal weather (Nyagumbo et al. 2015; Thierfelder et al. 2016). It is, however, complex and expensive to undertake on-farm research on climate change on on-farm trials hence the need to use crop models. Crop models extrapolate outputs to agro-ecological conditions which the research might not have been initially based on. The data was therefore used to calibrate DSSAT process-based crop models and then extrapolated over a period of 30 years (1981–2011).

2.1 Site Descriptions

This study was based on 30-year historical weather data from sub-humid and semi-arid agro-ecological regions of Malawi, Mozambique and Zimbabwe.

2.1.1 Chitala in Malawi

Chitala lies at 13.25°S; 34.30°E and 606 m above sea level (m.a.s.l.) and located in the Salima Agricultural Development Division, Malawi. The location receives an average in-crop rainfall of 1178 mm per annum from October to March. The mean minimum and maximum annual temperatures are 16 and 28 °C, respectively. The soils are predominantly sandy clay loam. The average farm sizes are 1.4 ha with each household size averaging 2.5 adults (Kanyama-Phiri et al. 2000). The commonly grown crops are maize (*Zea mays* L.) and groundnuts (*Arachis hypogaea* L.), and the peak period of labour demand on smallholder farms is in November (Mulwa et al. 2013).

2.1.2 Chitedze in Malawi

Chitedze lies at 13.97°S; 33.65°E and 1150 m.a.s.l., and is located in the mid-altitude plateau of Malawi. The area receives 1300 mm in-crop rainfall per season which often begins in November and ends in April. The area experiences a relatively low seasonal rainfall variability of less than 11.5% (Ngongondo et al. 2011). The soils are *Ferruginous latosol* derived sandy clay loams. The average farm size is 2.6 ha with 3.4 adults per household with the highest labour demand experienced in December. The most commonly cultivated crops are maize, tobacco

(*Nicotiana tabacum* L.), groundnut, soya bean (*Glycine max* (L) Merr), Cassava and common bean. Other legumes are pigeon peas (*Cajanus cajan* L.) and cowpeas (*Vigna unguiculata* (L) Walp) and are mostly cultivated as intercrops in maize-based cropping systems (Kanyama-Phiri et al. 2000).

2.1.3 Chimoio in Mozambique

Chimoio is located at 19.12°S; 33.47°E and 750 m.a.s.l., in Manica Province, Mozambique. The soils are dominated by fine textured red soils derived from *Gneiss* and *Migmatite* which are metamorphic acid rocks. The region experiences rainfall variability of 20%, and mean minimum and maximum temperatures of 19 and 28 °C, respectively (Famba et al. 2011). The average household size in Chimoio is 5 people. The most commonly cultivated crops are maize, common beans (*Phaseolus vulgaris* L.), soya beans, pigeon pea, groundnuts and cowpea on an average land holding of 1.92 ha (Woldemariam et al. 2012).

2.1.4 Angonia in Mozambique

Angonia is located at 14.72°S; 34.37°E and 1200 m.a.s.l. in the Tete Province, Mozambique. The area experiences humid savannah conditions and receives 1200 mm rainfall per annum. Minimum and maximum temperatures are average of 18 and 24 °C, respectively. The average length of the growing season is 180 days (Thierfelder et al. 2016). *Lixisols* and *Luvisols* are the most common soil types in the region. The most commonly cultivated crops are maize, common bean, potatoes (*Solanum tuberosum* L), soya bean and tobacco. Farm size in the area is 0.75–5 ha with about 5 persons per household. Chicken, cattle, goat and pigs are the most common reared livestock species (Franke et al. 2011).

2.1.5 Marondera in Zimbabwe

Marondera lies at 18.18°S, 31.55°E and 1689 m.a.s.l. in agro-ecological region IIa in Mashonaland East Province, Zimbabwe. The area receives rainfall of 750–1050 mm per season and is relatively cooler where the mean monthly minimum temperature ranges from 5.2 to 15.2 °C in July and January, respectively. The maximum daily temperatures range from 18.4 to 26 °C in June and October, respectively (Vincent and Thomas 1960). The average household size in Marondera is 3.9 people (ZIMSTAT 2012). The average household farm size ranges from 2.5 to 3.5 ha but can be as low as 0.4 ha when household heads at times share with their children. The most commonly cultivated crops are maize, soya beans, groundnuts (*Arachis hypogea* L.), bambara nuts (*Voandzeia subterranea* (L) Thouars), round nuts and sweet potatoes (*Ipomea batata* L.) and sunflower (*Helianthus annuus* L) (Saruchera and Matsungo 2003).

2.1.6 Matopos in Zimbabwe

Matopos is located in semi-arid south-western Zimbabwe and lies at 20.38°S; 28.51°E and 1365 m.a.s.l. It lies in the agro-ecological region IV and experiences mean annual rainfall of 557 mm. The mean minimum and maximum temperatures are 10.7 and 25.5 °C, respectively. Matopos is prone to seasonal droughts, mid-season dry spells especially in January and high-rainfall variability of 250–1400 mm per season (Vincent and Thomas 1960). The smallholder farming areas around Matopos are predominated by light and coarse sandy loam soils. The average land holding is 1.34 ha and a household of 6.4 people. At least 50% of the region's cropped land is under maize. Other commonly cultivated crops include sorghum, pearl millet, bambara nut, groundnut and cowpea (Nyamangara et al. 2014).

2.2 Cropping Systems Simulated

The simulations compared seven cropping systems across three rainfall regimes (*low, medium* and *high*) (Sect. 2.4) in sub-humid and semi-arid environments of Malawi, Mozambique and Zimbabwe. Conventional cropping system was subjected to tillage up to 15 cm depth using the draught powered mouldboard plough (Thierfelder et al. 2016). Crop residues were not retained as they are normally consumed by livestock, burnt or taken off the field for use as dry season feed for livestock or bedding in cattle pens (ZCATF 2009). *Early conventional mouldboard ploughing* (*CMP-early*) assumed that land preparation would commence after receiving the first rains when the soil is easy to work. *Late conventional ploughing* (*CMP-late*) cropping system assumed that land preparation would commence at least two weeks after receiving the first rains as the farmer would have to wait for their peers with draught power to finish land preparation and then hire the draught power. Smallholder farmers without access to draught power face delays in planting date of up to three weeks (Waddington et al. 1991).

Basins with dimensions of 15 cm × 15 cm × 15 cm and spaced at 75 cm 60 cm and 85 cm × 70 cm in sub-humid and semi-arid agro-ecologies, respectively, are manually prepared using hand hoes. The *early Basin* (*CA-basins early*) cropping system assumed that farmers would prepare basins after harvesting the previous crop and would plant on the first available opportunity. The *late Basin* (*CA-basins late*) cropping system assumed digging of basins would commence on receiving the first rains.

This is so as some farmers prepare basins after the first rains as the moist soil would be much easier to dig and thus sowing would only commence thereafter (Twomlow et al. 2008). The draught powered ripper tine (*CA-Ripper*) involved a system in which planting furrows at 5–10 cm depth are made with an animal-drawn ripper. The Dibble stick system (*CA-Dibble stick*) makes use of hand-held sticks to make planting stations. The sharpened end of the stick is pushed into the ground to a depth of 5 cm. Seeds are sown into the holes 30 cm apart within rows and 90 cm between the rows. The direct seeding (*CA-Direct seeder*) system makes use of an animal-drawn planter which marks planting stations, places seed and basal fertilizer and covers in one pass.

2.3 Labour Requirements of Different Cropping Systems to Determine When to Plant

The labour hours required for planting a hectare of arable land using the six different CA-based cropping systems were compiled from on-farm research trials managed by CIMMYT and ICRISAT in Malawi, Mozambique and Zimbabwe (Siziba 2007). For each cropping system, the time required to prepare and plant 1 ha was used to determine when planting would be practically achieved for each cropping system, given the average household labour available in that location.

2.4 Rainfall Season Type

Every season in each of the six different agro-ecologies was classified into either of the three rainfall season types as: *Low*, *Medium* and *High*. For each location, the total seasonal rainfall, mean and the coefficient of variation (CV) of the annual rainfall were computed for the 30-year period (Eq. 1). Season classification boundaries were calculated on the basis of the calculated mean seasonal total rainfall plus/minus a product of the CV and mean total rainfall of that location (Eq. 2) (Adamgbe and Ujoh 2013).

$$\text{Coefficient of variation(CV)} = \frac{\text{standard deviation of 30 year rainfall}}{\text{Mean of 30 year rainfall}} \quad (1)$$

$$\text{Season classification boundary} = \text{Mean 30 Year rainfall} \pm \left(\frac{\text{CV}}{100} \times \text{Mean 30 Year rainfall}\right) \quad (2)$$

This enabled three classes of seasons for each location i.e. *low*, *medium* or *high* rainfall. Using a conditional algorithm in an Excel® spreadsheet, each season was thus classified into each of the three classes depending on whether the total seasonal rainfall was below the lower rainfall boundary (*low*), within (*medium*) or above the upper boundary (*high*).

2.5 Planting Date, End of Season and Season Length

The planting date for each season was determined by receipt of at least 30 mm of rainfall within 3 days (ZCATF 2009) during the period 1 October–31 March. End of season was determined using Instat® a meteorological statistical tool (Stern et al. 2005), that has been previously used in *start* and *end of rainy season* determination and rainfall risk analysis (Mupangwa et al. 2016). End of season was regarded as the date when evapotranspiration exceeded precipitation during the cessation period of 1 January–30 June (Mugalavai et al. 2008).

2.6 DSSAT Ceres Maize Model

The DSSAT CERES 4.5 model (Jones et al. 2003) is a dynamic crop simulation model that simulates maize growth and development under different agro-ecologies and management conditions. The CERES model makes use of daily solar radiation, rainfall, maximum and minimum temperatures. These climate parameters interact with soil and cultivar-specific genetic coefficients to compute phenological development, grain and stover yields. The model computes growth by accounting for photosynthetically active radiation (PAR), degree of interception of radiation, inter- and in-row spacing, plant density and the efficiency of radiation conversion into plant biomass. Cultivar-specific ecological coefficients govern the physiological process chief amongst them photosynthesis and growth. Phenological coefficients determine plant root, stem and leaf growth and development (Jones et al. 2003). The model, though built in North America, has been used extensively in Africa for research on soil water dynamics (Dzotsi et al. 2010), crop yield forecasting (Manuela et al. 2007), agro-meteorology and climate change assessment (White et al. 2011).

2.7 Calibration of the Model

The DSSAT model was calibrated based on the measured maize phenological and biophysical data from on-farm trials maize yield data from each of the 6 locations in the three countries (Nyagumbo et al. 2015; Thierfelder et al. 2016). The root mean square error (RMSE) computed using Eq. (1) was used to determine the successfulness of model calibration.

$$\mathrm{RMSE} = \sqrt{\sum_{i=1}^{n} (P_i - O_i)^2 / n \cdot \frac{100}{O}} \tag{3}$$

A RMSE value of less than 30% shows the model's ability to correctly predict the parameters under study whereas a value greater than 30% was an indication of the model's inability to correctly predict the parameters under study (Moriasi et al. 2007).

Upon successful calibration, maize yields were simulated under different cropping systems and agro-ecological conditions for each of the six locations across the three rainfall regimes (Sect. 2.4). Maize yield simulations were conducted based on the earliest sowing date for each cropping system and location. Fertiliser application in this study was based on the recommendations in SIMLESA (2010a, b).

2.8 Statistical Analysis

GENSTAT® version 6 was used to perform an analysis of variance on planting dates, season length and maize grain yields using the REML approach since the data was unbalanced (VSN 2002). Data was unbalanced because for each location the number of seasons belonging to each rainfall regime was not the same, with most belonging to the medium category. Location, rainfall season type and cropping system were considered fixed effects, while the year was considered a random effect. The main response variables were maize grain yield, planting date and season length which were separately run for each location. Separation of means was done through the least significant difference (LSD) test at $p \leq 0.05$.

3 Results

3.1 Rainfall Season Type

Different agro-ecologies received varying amounts of rainfall. The least amount of seasonal rainfall amongst all the locations was 390 mm per season at Matopos. Higher rainfall exceeding 1000 mm per season was received at Chitala, Angonia, Chimoio and Marondera. Matopos and Chimoio received the mean lowest and highest rainfall amounts of 557 and 1035 mm per season, respectively. Rainfall seasons for each location were classified as *low*, *medium* or *high* based on the seasonal rainfall means and variability of that location (Table 1).

3.2 Planting Date

No notable differences in cropping system * season type interactions were observed on maize planting dates across all locations (Table 2). However, planting

Table 1 Classification of seasons based on rainfall received across different agro-ecologies in Southern Africa*

Location	Low (mm)	Medium (mm)	High (mm)	Mean (mm)
Chitala	<649 (8.3)	649–1127 (7.4)	>1127 (6.7)	888 (7.4)
Chitedze	<784 (9.3)	784–986 (8.7)	>986 (7.2)	885 (8.4)
Angonia	<802 (8.8)	802–1092 (8.2)	>1092 (8)	947 (8.3)
Chimoio	<761 (9.4)	761–1309 (8.6)	>1309 (8)	1035 (8.7)
Matopos	<390 (9)	390–725 (8.6)	>725 (7.2)	557 (9.6)
Marondera	<570 (12)	570–1059 (8.6)	>1058 (7.2)	814 (9.3)

*Mean monthly coefficient of variation (%) for each location and rainfall regime is given in the brackets

dates varied significantly between season types in all locations except Angonia, Chitedze and Chitala (Fig. 1). No notable differences were observed in maize planting dates at Chitala where maize was seeded from 13 to 15 December. At Chimoio, Marondera and Chitedze, high-rainfall seasons were generally seeded earlier compared to low rainfall seasons. At Marondera, high-rainfall seasons were seeded earlier around 27 November, whereas low rainfall seasons were seeded a month later around 25 December (Fig. 1). Planting dates also varied significantly amongst the different cropping systems (Tables 2 and 3). At Angonia, *CA-ripper, direct seeder, basins early and CMP-early* systems were seeded earlier in the season around 2 December, whereas *CA-Basins late* and *Dibble stick* were seeded almost 2 weeks later around 13 December. The *CMP-late* cropping system was then seeded much later in the season around 26 December. The pattern was generally similar in the other locations (Tables 2 and 3).

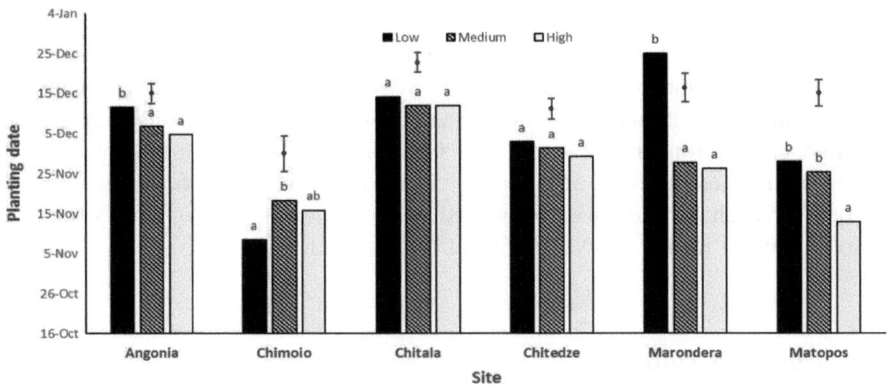

Fig. 1 Changes in maize planting dates across contrasting rainfall season types (*Low, Medium* and *High*) in (a) Angonia, (b) Chimoio, (c) Chitala, (d) Chitedze, (e) Marondera and (f) Matopos. Error bars denote standard error of differences ($p < 0.05$). Means having different letters above the column are significantly different at $p < 0.05$ probability level

Table 2 Summary of the analysis of variance for the maize planting dates across different agro-ecologies in Southern Africa

Source	Angonia_pr.	Chimoio_pr.	Chitala_pr.	Chitedze_pr.	Marondera_pr.	Matopos_pr.
Cropping system	<0.001	<0.001	<0.001	<0.001	<0.001	<0.001
Season type	0.006	0.099	0.651	0.400	<0.001	0.009
Cropping system * season type	0.997	1.000	0.998	0.981	0.702	0.999

Table 3 Earliest maize planting dates (Julian date) in different cropping systems across different agro-ecologies in Southern Africa

Cropping system	Angonia	Chimoio	Chitala	Chitedze	Marondera	Matopos
CA-Basins early	336^{a}	313^{a}	342^{a}	329^{a}	323^{a}	313^{a}
CA-Basins late	348^{b}	330^{b}	352^{b}	343^{b}	344^{bc}	345^{bc}
CMP-late	360^{c}	342^{c}	362^{c}	356^{c}	350^{c}	353^{c}
CMP-early	338^{a}	319^{a}	346^{ab}	332^{a}	342^{b}	322^{a}
CA-Dibble	347^{b}	319^{ab}	346^{a}	334^{a}	343^{b}	342^{a}
CA-Direct seed	336^{a}	313^{a}	342^{a}	329^{a}	315^{a}	313^{a}
CA-Ripper	336^{a}	313^{a}	344^{a}	330^{a}	336^{b}	320^{a}
sed	*3.51*	*6.58*	*3.61*	*4.11*	*5.55*	*4.9*
$Lsd_{0.05}$	*6.93*	*12.97*	*7.12*	*4.05*	*10.94*	*9.66*

N.B. The numbers in Table 2 denote day of year starting from 1 January of each year. Means in the same column followed by different superscript letters are significantly different at $p < 0.05$

3.3 Season Length

No notable cropping system * season type interactions were observed on maize growing season length in all locations (Table 4). However, the results showed that season length significantly varied amongst all the cropping systems in each of the six locations (Tables 4 and 5). At Angonia, *CMP-late* cropping system had the shortest season length of 104 days. The maize crop in the *CA-basins late*, *CA-Dibble* and *CMP-late* cropping systems had similar ($p > 0.05$) season lengths which ranged from 116 to 119 days. On the contrary, *CA-ripper*, *basins early* and *CA-direct seeder* had the greatest season lengths of more than 127 days per season. The pattern was generally similar across all locations.

At Chimoio, *CMP-late* had the shortest season length of 119 days with all the other cropping systems having season lengths of at least 138 days. At Chitedze, the *CMP-late* system also had the shortest season length compared to all cropping systems. At Marondera, *CA-basins late* and *CMP-late* had the shortest season length of about 114 days. *CA-ripper* had 121 days but the greatest length was in the *CA-basins early* and *CA-direct seeder* both with at least 140 days. The pattern was similar for Matopos in Zimbabwe (Tables 4 and 5).

Differences in maize growing season lengths were highly significant amongst the contrasting rainfall season types in Angonia, Chitala, Chitedze and Matopos. Semi-arid Matopos had the shortest season length of 89 days in *low* rainfall seasons, whereas high-rainfall season type had the greatest length of at least 135 days. The trend was uniform across all locations where the low- and high-rainfall season types had short and long season lengths respectively (Fig. 2).

Table 4 Summary of the analysis of variance for the maize growing season length in the 6 locations in Southern Africa

Source	Angonia_pr.	Chimoio_pr.	Chitala_pr.	Chitedze_pr.	Marondera_pr.	Matopos_pr.
Cropping system	<0.001	<0.001	0.023	<0.001	0.014	<0.001
Season type	<0.001	0.072	0.002	<0.001	0.058	<0.001
Cropping system * season type	0.999	1.000	1.000	1.000	1.000	1.000

Table 5 Maize growing season lengths (days) in the different cropping systems across different agro-ecologies in Southern Africa

Cropping system	Angonia	Chimoio	Chitala	Chitedze	Marondera	Matopos
	Days					
CA-Basins early	129^{d}	152^{b}	122^{b}	132^{b}	141^{bc}	128^{b}
CA-Basins late	117^{b}	138^{b}	112^{ab}	121^{b}	114^{a}	97^{a}
CMP-late	104^{a}	119^{a}	102^{a}	107^{a}	113.8^{a}	89^{a}
CMP-early	118^{bc}	147^{b}	118^{b}	128^{b}	120^{ab}	99^{a}
CA-Dibble	119^{bc}	147^{b}	118^{b}	128^{b}	119^{ab}	99^{a}
CA-Direct seed	129^{d}	152^{b}	122^{b}	133^{b}	149^{c}	128^{b}
CA-Ripper	128^{cd}	147^{b}	117^{b}	129^{b}	121^{ab}	120^{b}
sed	*5.41*	*8.21*	*6.37*	*6.13*	*11.7*	*10.16*
$Lsd_{0.05}$	*10.68*	*16.19*	*12.57*	*12.1*	*23.09*	*20.05*

Means in the same column followed by different superscript letters are significantly different at $p < 0.05$

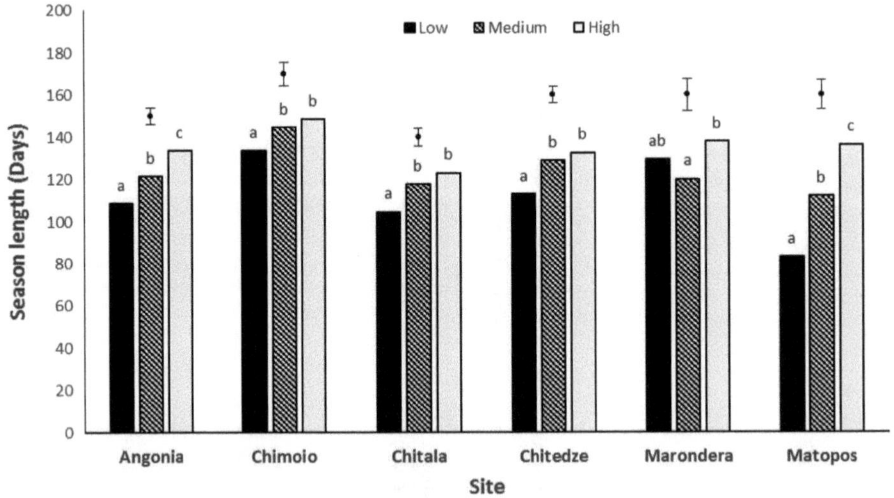

Fig. 2 Maize growing season length in different season types (*Low*, *Medium* and *High*) in (a) Angonia, (b) Chimoio, (c) Chitala, (d) Chitedze, (e) Marondera and (f) Matopos. Error bars denote standard error of differences ($p < 0.05$). Means having different letters above the column are significantly different at $p < 0.05$ probability level

3.4 *Maize Grain Yield*

Simulation results showed no cropping system * season type interaction effect on maize grain yield in all locations except Chitedze (Table 6). At Chitedze, no notable yield differences were noted in all cropping systems except *CA-Basins late* and *CMP-late*. Both *CA-Basins late* and *CMP-late* cropping systems had the

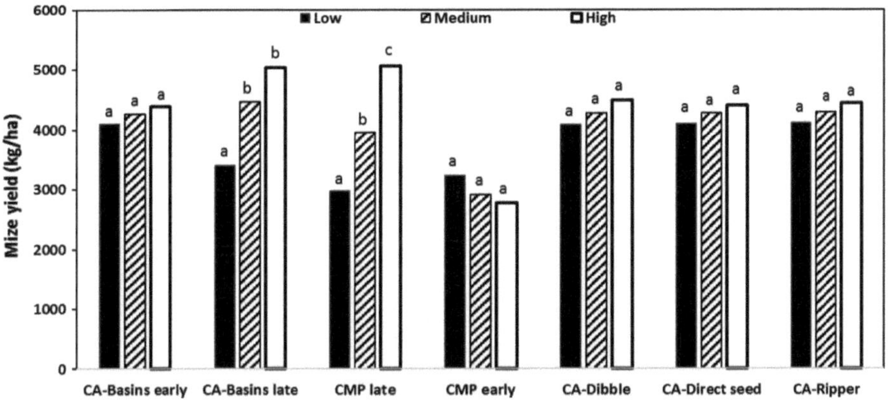

Fig. 3 Maize crop yields (kg ha^{-1}) under different cropping systems and season types at Chitedze. Error bars denote standard error of differences ($p < 0.05$). Means having different letters above the column are significantly different at $p < 0.05$ probability level

highest and lowest yields in *high* and *low* rainfall seasons, respectively. *CA-Basins late* cropping system had the highest yield of 5034 kg ha^{-1} under the *high*-rainfall seasons. On the contrary, *low* rainfall season type had the least yields of 3402 kg ha^{-1} (Fig. 3).

Maize grain yields significantly differed between the contrasting rainfall season types (Tables 6 and 7). Generally *low*, *medium* and *high*-rainfall season types led to corresponding maize grain yield patterns. The pattern was more pronounced at Matopos, where the lowest yield of 438 kg ha^{-1} was realised in the *low* rainfall season. The highest yields were then realised in the *high*-rainfall season type at 4981 kg ha^{-1}. The pattern was similarly pronounced at Marondera. At Chitala, *low* and *medium* rainfall season types attained no notable yield differences at 2987 and 3149 kg ha^{-1}. The highest maize grain yields of 3362 kg ha^{-1} were realised in *higher* rainfall seasons. As an exception, the yield pattern was different at Angonia, where the least yield of 2890 kg ha^{-1} was realised under the *medium* rainfall season and the highest yields of 3080 and 3150 kg ha^{-1} being noted from the *high* and *low* rainfall seasons, respectively (Tables 6 and 7).

Crop yields also differed amongst the different cropping systems with significant differences being realised at Angonia, Chimoio and Chitedze (Fig. 4; Table 6). The lowest maize yields were realised at Angonia, Mozambique in the *CA-basins early, CA-Direct seeder* and *CA-Ripper* cropping systems compared to the relatively higher yields in the *CA-Basins late, CA-Dibble* and *CMP-late* cropping systems. Chimoio, Mozambique had the lowest yields in the *CA-basin early* system compared to the high yields in the other cropping systems. No notable yield differences were realised at Chitala, Malawi. Chitedze, Malawi had notably high yields in all systems except in the *CMP-early* system. Marondera, Zimbabwe had the lowest yields under *CMP-late* compared to the highest yields realised in the *CA-basins* and *ripper* cropping systems. The pattern was similar at Matopos, Zimbabwe (Fig. 4; Table 6).

Table 6 Summary of the analysis of variance of maize grain yields across different agro-ecologies in Southern Africa

Source	Angonia_pr.	Chimoio_pr.	Chitala_pr.	Chitedze_pr.	Marondera_pr.	Matopos_pr.
Cropping system	0.002	<0.001	0.995	<0.001	0.460	0.330
Season type	<0.001	<0.001	0.002	<0.001	<0.001	<0.001
Cropping system * season type	0.854	0.974	0.999	0.001	0.668	0.988

Table 7 Maize grain yields (kg ha^{-1}) in different rainfall season types across different agro-ecologies in Southern Africa

Season type	Angonia	Chimoio	Chitala	Chitedze	Marondera	Matopos
Low	3151^{b}	1962^{a}	2987^{a}	3714^{a}	2218^{a}	438^{a}
Medium	2894^{a}	2908^{b}	3149^{a}	4063^{b}	4186^{b}	2658^{b}
High	3080^{b}	3025^{b}	3362^{b}	4372^{c}	4600^{c}	4981^{c}
sed	*77*	*134.6*	*90.17*	*136.9*	*177.3*	*350.2*
$Lsd_{(0.05)}$	*151.7*	*265.6*	*177.9*	*270.1*	*349.8*	*690.8*

Means in the same column followed by different superscript letters are significantly different at $p < 0.05$

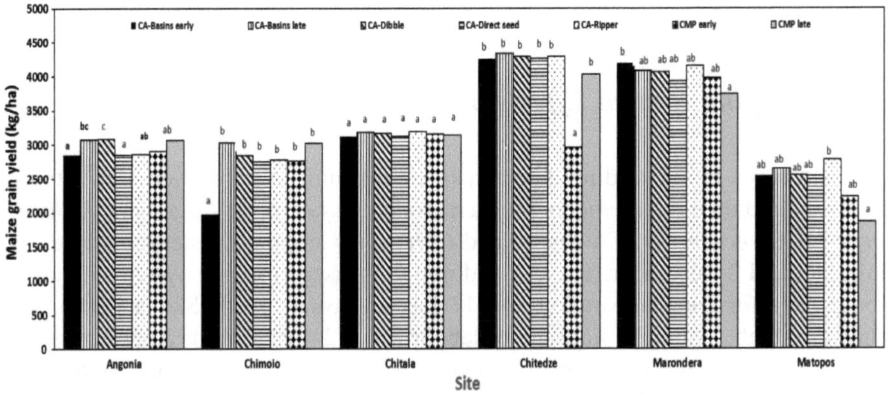

Fig. 4 Maize crop yields under different cropping systems in (a) Angonia, (b) Chimoio, (c) Chitala, (d) Chitedze, (e) Marondera and (f) Matopos. Means having different letters above the column are significantly different at $p < 0.05$ probability level

4 Discussion

4.1 *Effects of Rainfall Season Type on Maize Planting Date*

Early and late sowing was generally related to high and low rainfall seasons in the drier agro-ecologies of Chimoio, Marondera and Matopos. This is attributed to the amount of rainfall received. Low rainfall seasons are often associated with increased rainfall variability and reduced frequency of rainfall events (Masvaya et al. 2007). As a result, the chances of satisfying the sowing requirements of 30 mm rainfall in 3 days are reduced thus delaying sowing opportunities in low rainfall seasons (Raes et al. 2004). At Marondera, low rainfall seasons received 570 mm with a coefficient of variation of 12%. This implies reduced chances of rainfall events which ultimately lead to delayed maize planting. On the contrary, high-rainfall seasons are characterised by increased frequency of rainfall events and relatively low rainfall variability. At Marondera, high-rainfall seasons receive at least 1000 mm per season, which is associated with low rainfall variability (<7%), which increases the chances of early sowing and lower risks of crop failure during such seasons.

In some agro-ecologies such as Angonia, Chitedze and Chitala, there were no notable differences in planting dates despite the contrasting nature of rainfall season types. Lack of significant differences is attributed to the relatively high rainfall and the relatively more stable nature of seasons in these environments. Angonia in Mozambique receives an average rainfall of 900 mm per season (Table 1). The relatively higher rainfall is associated with less rainfall variability of about 8% across all season types, thus the season types are relatively more stable despite the contrasting rainfall amounts. Therefore, the chances of fulfilling the sowing requirements are higher regardless of the season type.

4.2 Cropping System Effects on Planting Date

The differences in the seeding date across the cropping systems were attributed to the labour and time requirements for each cropping system. At Angonia, *CA-Basins early, CA-Direct seeder, CA-Ripper* and *CMP-early* cropping systems were seeded earlier around 2 December. The CA and non-CA-based cropping systems require 7 labour days for land preparation and 11 labour days to seed (Siziba 2007). The relatively high available household labour of 5.2 persons at Angonia would potentially reduce the need for extended periods of land preparation. As a result, *CA-Basins early, CA-Direct seeder, CA-Ripper* and *CMP-early* systems were seeded at almost the same date despite having different labour requirements. The abundance of household labour at Angonia can be a buffer against the high-labour requirements associated with certain cropping systems (Woldemariam et al. 2012).

CA-Direct seeder system was seeded earlier as direct seeding involves preparation of planting stations, seed and fertilizer placement and covering in one pass. This reduces the amount of labour and time needed to plant a crop. This is also supported by studies undertaken in Southern Africa, which show that draught powered direct seeding systems require 11.6 person-h ha^{-1} for preparing planting stations and covering seed, compared to 55.8 person-h ha^{-1} required under conventional agriculture (Wall et al. 2013).

CA-basins late and *CMP-late* cropping systems were seeded later due to high-labour requirements in preparing land. The delay was also worsened by the late preparation of basins, which commences after the first rainfall events of the season and this is commonly practised by some smallholder farmers (Twomlow et al. 2008). The poor access to draught power coupled with the nature of sowing opportunities at the start of seasons resulted in delayed seeding of *CMP-late* cropping systems in most locations (Mugabe and Banga 2001).

4.3 Season Type Effects on the Growing Season Length

Generally, *low* rainfall season types experienced short season lengths and *high*-rainfall seasons corresponded to extended crop growing season. Extended season length under *high* rainfall was attributed to an earlier start of the rainfall season and increased the availability of soil moisture for maize growth (Xu et al. 2012). This promotes the natural morphological and phenological development of the maize crop. Our findings thus confirm other studies in this region, which also suggested that high-rainfall seasons are often characterised by an early start and prolonged season lengths while the opposite is generally true for low rainfall seasons (Masvaya et al. 2007).

On the contrary, reduced rainfall leads to low plant available soil moisture during the growing season (Xu et al. 2012). Soil moisture stress on maize plants hastens phenological development but shortens the flowering period at the same time. Assimilates are thus mobilised towards early seed setting, thus the maize crop has a shorter growing cycle under *low* rainfall seasons. This is typical of *El Nino* or drought years where the rainfall is less than 300 mm, which shortens the crop's growing period (Leary et al. 2013). As consequence, a normal maize crop which receives 1000 mm per season matures in at least 160 days but will be reduced to 90 days or less under moisture stressed conditions (Aslam et al. 2013).

4.4 Cropping System Effects on Growing Season Length

The start of the rainy season which is determined by receipt of at least 30 mm of rainfall in 3 days also determines when both CA and non-CA based cropping systems can be seeded (ZCATF 2009). *CA-Basins late, CMP-early* and *CMP-late* had a shorter growing season due to delayed planting at Marondera. In contrast, the other *CA*-based cropping systems had an extended growing season, which was also attributed to early planting. Early planting was attributed to the reduced labour demands in land preparation of CA-based and draught animal powered systems. Early seeded systems extended the crop growing season ensuring the maize crop would grow until available soil moisture becomes limiting when the frequency of rainfall events decreases later in the season. Early planting also ensures that the maize crop develops early such that the crop can endure mid-season dry spells which late seeded relatively young plants would not be able to withstand (Tsimba et al. 2013). Late planting reduces the growing season as this shortens the time the maize crop would be exposed to favourable soil moisture conditions for crop growth and development. Basins collect rain and runoff water thus storing water for crop growth and development. This extends the crop growing season. However, late sown basins shorten the growing season which was attributed to the time of planting. These findings also confirm other studies suggesting that the time of seeding supersedes the cropping system employed (Nyagumbo et al. 2017).

4.5 Season Type Effects on Maize Yield

Maize grain yields were generally higher in the *high*-rainfall seasons across all agro-ecologies. This is because high rainfall increases soil moisture availability and plant available water. This in turn increases photosynthetic productivity and ultimately maize grain yield (Xu and Zhou 2011). High-rainfall agro-ecologies normally attain high yields due to increased available soil moisture. The pattern is more pronounced at Matopos and Marondera which receive low rainfall of high variability (Vincent and Thomas 1960). Low rainfall therefore led to severe soil moisture deficiency which resulted in crop water stress leading to poor yields (Table 1). This led to poor maize yields of about 438 kg ha^{-1} compared to at least 4500 kg ha^{-1} achieved under the high-rainfall seasons. In addition, arable lands surrounding Matopos and Marondera are dominated by light-textured sands with at least 50% sand content and low water holding capacity (Nyamapfene 1991) which reduces available soil water during low rainfall seasons. The impacts of low rainfall and light textured soils are worsened by the late seeding associated with low rainfall seasons. Low rainfall seasons are also associated with reduced season length thus further reducing crop yields.

Despite the low rainfall in semi-arid agro-ecologies, medium to high rainfall totals exceeding 700 mm can be experienced (Table 1) leading to occasional and opportunistic high maize grain yields exceeding 4500 kg ha^{-1} in such rare high-rainfall seasons. High maize grain yields (>2600 kg ha^{-1}) were also observed in semi-arid Gwanda district of Zimbabwe similar to Matopos, after receiving some 700 mm in 2008/9 (Mashingaidze et al. 2012). Yet in the following season which could be classified as *low* rainfall season type (receiving 265 mm), maize yields dropped to 500 kg ha^{-1} (Mupangwa et al. 2015) at the same location.

At Angonia, *low* and *high*-rainfall seasons had the highest maize grain yield which is in contrast to the normal pattern on other sites where *low* rainfall season types would normally lead to poor yields. Here, both the *low*- and *high*-rainfall season type had the same sowing date of 28 November. Thus, *low* rainfall season types benefitted from early planting which led to higher yields which were equal to those attained under *high* rainfall (Shumba et al. 1992). The similar yields under *low* and *high*-rainfall season types may have been due to the yield penalties suffered by high-rainfall amounts due to leaching of nutrients or waterlogging caused by the excessive rains. This may have been further exacerbated by the heavy textured soils at Angonia which have at least 30% clay content which tends to promote waterlogging under high-rainfall conditions (Akhtar and Nazir 2013). This is also supported by Thierfelder et al. (2016) and Nyagumbo et al. (2015) who observed that CA basin system depresses maize yields by at least 3% under higher rainfall agro-ecologies.

4.6 Cropping System Effects on Maize Yield

The absence of interaction between cropping systems and season type at all locations except Chitedze suggests a consistent behaviour of cropping systems across different rainfall regimes. This means there are no yield advantages from CA-based systems during dry seasons. This is in contrast to meta-analysis results from field experimental studies which suggest more positive yield merits from CA in dry seasons compared to wet ones (Rusinamhodzi et al. 2011). This is also in contrast to Thierfelder and Wall (2009) who highlighted that CA practices obtain high yields in drought conditions. The findings of this research partially correspond to Nyamangara et al. (2014) who realised non-significant yield differences amongst non-CA and non-Basin CA systems. Nyamangara et al. (2014), however, contrast the findings of this study where CA-based basin systems attain high yields under low rainfall conditions compared to non-CA cropping systems.

In addition, soil moisture measurements from CA systems also confirm the yield benefits attributed to moisture conservation; however, the benefits of such moisture-saving measures are not apparent from this study during the period. This result could be attributed to the DSSAT model failing to accurately predict or capture the soil water balance and dynamics in rainfall seasons which had poor rainfall distribution. Models that use the curve number approach have been observed to fail to accurately predict the soil water balance in some rainfall seasons, resulting in poor prediction of crop yields which are directly influenced by soil moisture content (Mupangwa and Jewitt 2011).

The interaction between cropping system and rainfall season type at Chitedze led to significant maize yield differences in *CA-basins late* and *CMP-late* systems (Fig. 3). Chitedze is a generally high rainfall receiving agro-ecology receiving about 1000 mm per annum. This agro-ecology is dominated by heavy textured sandy loam soils with a high water-holding capacity. *CA-basins early*, *CA-Dibble*, *CA-Direct seeder*, *CA-ripper* and *CMP-early* did not attain significant yield differences amongst the different rainfall season types. This is attributed to early sowing associated with these cropping systems. As a result of early sowing, the yield advantages of specific cropping systems are diminished (Nyagumbo et al. 2017). *CA-basins* and *CMP-late* are seeded later in the season where moisture becomes limiting. However, high-rainfall seasons will have considerable soil moisture available for crops despite being late in the season, thus high yields were attained. On the contrary, low rainfall will have limited available soil moisture leading to reduced maize grain yields. Time of planting and cropping system therefore have a more significant effect on maize grain yields.

Agro-ecologies receiving high rainfall, which include Angonia, attained low maize grain yields in early seeded *CA-basins early, CA-Direct seeder* and *CA-Ripper* cropping systems around 2 December. This is in contrast to delayed and manually prepared *CA-Basins late, CA-Dibble* and *CMP-late* cropping systems seeded later than 13 December which attained higher yields. Shumba et al. (1992) suggest that a week's delay in sowing leads to at least 5% decrease in crop yields.

This was, however, not the case at Angonia, where the late seeded systems had higher yields compared to early seeded systems. Angonia has heavy textured soils which are prone to waterlogging in high-rainfall seasons (Nyagumbo et al. 2015) which lead to lower yields despite promoting early planting. Delayed sowing led to the escape of waterlogging damage on the maize crop. The most sensitive phenological stages of maize growth may not have coincided with high-rainfall periods (Ramamasy and Baas 2007).

5 Conclusion

The study realised that generally high-rainfall seasons are associated with early sowing which leads to greater season lengths and ultimately higher maize yields. On the contrary, low-rainfall seasons lead to reduced season lengths and ultimately low grain yields. The pattern is, however, significant in semi-arid agro-ecologies, growing seasons which receive low and high rainfall are associated with delayed and early planting, respectively, as planting conditions were satisfied earlier in the season in high than in low rainfall seasons. Similarly, low and high-rainfall season types experience reduced and extended growing seasons, respectively. On the contrary, high-rainfall agro-ecologies do not experience significant differences in sowing dates between high and low rainfall seasons, an effect attributed to low rainfall variability in those agro-ecologies. CA and non-CA cropping systems that enable early land preparation as well lead to early sowing, with consequences in increased season length and yields.

Draught powered CA-*Direct seeder* and *Ripper* systems and winter prepared CA basins that enable early planting, increase season length and in some locations result in higher yields. Labour intensive and late prepared CA and non-CA based basins and conventional cropping systems had the shortest seasons due to late seeding and the subsequent reduced season length leading to low maize yields. Maize grain yields were lower in early seeded CA and non-CA cropping systems in high-rainfall agro-ecologies compared to late prepared systems. This suggests that the use of CA systems that are powered by draught animals or prepared before the onset of the season can help to prolong season lengths but their capacity to mitigate droughts in terms of yields was not evident from this study. In addition, the absence of notable maize yield differences within the individual locations puts to doubt the theory that use of different cropping systems, such as CA, has the potential to mitigate against droughts. This study therefore puts to doubt the potential of some CA technologies in mitigating droughts at least in terms of yields during low rainfall seasons. We, however, recommend further studies to be carried out from field experiments to further validate the drought mitigation merits of CA practices during low rainfall season. The realisation of notable yields under different rainfall regimes within each location highlights the importance of use of seasonal forecasts for farmers to prepare for the upcoming season.

Acknowledgements The authors of this paper would like to acknowledge funding received from the Australian Centre for International Agricultural Research through the projects *Integrating crop and livestock production for improved food security and livelihoods in rural Zimbabwe (ZimCLIFS)* project number CSE/2010/022 and the *Sustainable Intensification of Maize–Legume Systems in Eastern and Southern Africa (SIMLESA)* project number CSE/2009/024. Further, we also acknowledge the financial support received through the CGIAR Research Program on Climate Change, Agriculture and Food Security (CCAFS) through our CIMMYT colleagues Drs Clare Stirling and Santiago Ridaura. The authors also acknowledge Rumbidzai Matemba-Mutasa for advising on data analysis.

References

Adamgbe, E. M., & Ujoh, F. (2013). Effect of variability in rainfall characteristics on maize yield in Gboko, Nigeria. *Journal of Environmental Protection, 4*(9), 881–887.

Akhtar, I., & Nazir, N. (2013). Effect of waterlogging and drought stress in plants. *International Journal of Water Resources and Environmental Sciences, 2*(2), 34–40.

Aslam, M., Zamir, I., Afzal, I., Yaseen, M., Mubeen, M., & Shoaib, A. (2013). Drought tolerance in maize through Potassium: Drought stress, its effect on maize production and development of drought tolerance through potassium application. *Cercetări Agronomice în Moldova, XLVI,* 16.

Bouba, T., et al. (2013). Effects of climate variability and climate change on crop production in southern Mali. *European Journal of Agronomy, 49,* 115–125.

Cooper, P. J. M., et al. (2008). Coping better with current climatic variability in the rain-fed farming systems of sub-Saharan Africa: An essential first step in adapting to future climate change? *Agriculture, Ecosystems & Environment, 126*(1–2), 24–35.

Dzotsi, K. A., et al. (2010). Modeling soil and plant phosphorus within DSSAT. *Ecological Modelling, 221*(23), 2839–2849. Available at: http://dx.doi.org/10.1016/j.ecolmodel.2010.08.023.

Fallis, A. (2013). Changes in extreme rainfall events in South Africa. *Journal of Chemical Information and Modeling, 53*(9), 1689–1699.

Famba, S. I., et al. (2011). Conservation agriculture for increasing maize yield in vulnerable production systems in central Mozambique. In *African Crop Science Proceedings* (pp. 255–262). Kampala, Uganda: African Crop Science Society Conservation.

Franke, A. C., Rufino, M. C., & Farrow, A. (2011). Characterisation of the impact zones and mandate areas in the N2Africa project. *Report N2Africa Project*, (1), 50.

Jones, J. W., et al. (2003). The DSSAT cropping system model. *European Journal of Agronomy, 18* (3–4), 235–265. Available at: http://linkinghub.elsevier.com/retrieve/pii/S1161030102001077.

Kanyama-Phiri, G., et al. (2000). Towards integrated soil fertility management in Malawi: Incorporating participatory approaches in agricultural research. *Managing Africa's Soils*, Working Paper 11.

Leary, N., et al. (2013). Climate change and vulnerability and adaptation. J. P. Neil Leary, C. Conde, J. Kulkarni, A. Nyong, J. Adejuwon, V. Barros, I. Burton, & R. Lasco (Eds.), p. 922.

Manuela, C., Soler, T., & Paulo, C. (2007). Application of the CSM-CERES-Maize model for planting date evaluation and yield forecasting for maize grown off-season in a subtropical environment. *European Journal of Agronomy, 27,* 165–177.

Mashingaidze, N., et al. (2012). Crop yield and weed growth under conservation agriculture in semi-arid Zimbabwe. *Soil & Tillage Research, 124,* 102–110.

Masvaya, E., Mupangwa, W., & Twomlow, S. (2007). Rainfall variability impacts on farmers' crop management strategies. *Waternetonline.Ihe.Nl.* Available at: http://www.waternetonline.ihe.nl/challengeprogram/ir34masvayarainfallstrategy.pdf.

Moriasi, D. N., et al. (2007). Model evaluation guidelines for systematic quantification of accuracy in watershed simulations. *American Society of Agricultural and Biological Engineers, 50*(3), 885–900.

Mpandeli, S., & Maponya, P. (2014). Constraints and challenges facing the small scale farmers in Limpopo Province, South Africa. *Journal of Agricultural Science, 6*(4), 135–143. Available at: http://dx.doi.org/10.5539/jas.v6n4p135.

Mugabe, F. T., & Banga, D. J. (2001). Assessment of the nitrogen requirements of planted maize (*Zea mays* L.) in the semi-arid areas of Zimbabwe. *UNISWA Journals of Agriculture, 10,* 5–11.

Mugalavai, E. M., et al. (2008). Analysis of rainfall onset, cessation and length of growing season for western Kenya. *Agricultural and Forest Meteorology, 148*(6–7), 1123–1135.

Mulwa, C., et al. (2013). Characterization of Maize-Legume farming systems and household socio-economic and risk profiles in Malawi: Baseline survey report for the project: Sustainable Intensification of maize-legume Farming Systems for Food Security in Eastern and Southern Africa, p. 125.

Mupangwa, W., et al. (2016). Rainfall risk and the potential of reduced tillage systems to conserve soil water in semi-arid cropping systems of Southern Africa. *AIMS Agriculture and Food, 1*(1), 85–101.

Mupangwa, W., & Jewitt, G. P. W. (2011). Simulating the impact of no-till systems on field water fluxes and maize productivity under semi-arid conditions. *Physics and Chemistry of the Earth, 36*(14–15), 1004–1011. Available at: http://linkinghub.elsevier.com/retrieve/pii/S1474706511001859. Accessed March 10, 2014.

Mupangwa, W., Twomlow, S., & Walker, S. (2015). Reduced tillage and nitrogen effects on soil water dynamics and maize (*Zea mays* L.) yield under semi-arid conditions. *International Journal of Agricultural Sustainability*, (February), 1–18. Available at: http://www.tandfonline.com/doi/abs/10.1080/14735903.2015.1011393.

Ngongondo, C., et al. (2011). Evaluation of spatial and temporal characteristics of rainfall in Malawi: A case of data scarce region. *Theoretical and Applied Climatology, 106*(1–2), 79–93.

Nyagumbo, I., et al. (2015). Maize yield effects of conservation agriculture based maize–legume cropping systems in contrasting agro-ecologies of Malawi and Mozambique. *Nutrient Cycling in Agroecosystems*, 10705.

Nyagumbo, I., et al. (2017). Labour and yield benefits from conservation agriculture practices across Southern Africa. *Agriculture, Ecosystems & Environment, 150*, 21–33. Available at: http://dx.doi.org/10.1016/j.agsy.2016.09.016.

Nyamangara, J., et al. (2014). Effect of conservation agriculture on maize yield in the semi-arid areas of Zimbabwe. *Experimental Agriculture, 50*, 159–177. Available at: http://www.journals.cambridge.org/abstract_S0014479713000562.

Nyamapfene, K. (1991). *Soils of Zimbabwe*. Harare, Zimbabwe: Nehanda Publishers (Pvt) Ltd.

Raes, D., et al. (2004). Evaluation of first sowing dates recommended by criteria currently used in Zimbabwe. *Agricultural and Forest Meteorology, 125,* 177–185.

Ramamasy, S., & Baas, S. (2007). *Climate variability and change : Adaptation to drought in Bangladesh—A resource book and training guide*. Rome, Italy.

Rusinamhodzi, L., et al. (2011). A meta-analysis of long-term effects of conservation agriculture on maize grain yield under rain-fed conditions. *Agronomy for Sustainable Development, 31*(4), 657–673.

Saruchera, M., & Matsungo, O. (2003). Understanding local perspectives: Participation of resource poor farmers in biotechnology-The Case of Wedza District of Zimnbabwe. *Biotechnology and the Policy Process in Developing Countries, Background*, p. 45. Available at: www.ids.ac.uk/biotech.

Shumba, E. M., Waddington, S. R., & Rukuni, M. (1992). Use of tine-tillage, with atrazine weed control, to permit earlier planting of maize by smallholder farmers in Zimbabwe. *Experimental Agriculture, 28*(4), 443.

SIMLESA. (2010a). *Manica: Sustainable intensification of maize-legume cropping systems for food security in eastern and southern Africa (SIMLESA)*. Exploratory Trial Protocol and Data Collection Sheets, Manica, Mozambique.

SIMLESA. (2010b). *Malawi Highlands: Sustainable intensification of maize-legume cropping systems for food security in eastern and southern Africa (SIMLESA)*. Exploratory Trial Protocol and Data Collection Sheets, Malawi.

Siziba, S. (2007). *Institute of Agricultural Economics and Social Sciences in the Tropics*. University of Hohenheim.

Stern, R., et al. (2005). *Instat Tutorial*, p. 40. Available at: http://www.reading.ac.uk/ssc/resource-packs/ICRAF_2007-11-15/InStat/Instat.html.

Thierfelder, C., et al. (2016). Evaluating manual conservation agriculture systems in Southern Africa. *Agriculture, Ecosystems & Environment, 222*, 112–124. Available at: http://linkinghub.elsevier.com/retrieve/pii/S0167880916300810.

Thierfelder, C., & Wall, P. (2009). Investigating conservation agriculture systems in Zambia and Zimbabwe to mitigate future effects of climate change. In *African Crop Science Conference*. African Crop Science Society, pp. 303–307.

Thomas, D. S. G., et al. (2007). Adaptation to climate change and variability: Farmer responses to intra-seasonal precipitation trends in South Africa. *Climate Change, 83*(3), 301–322.

Tsimba, R., et al. (2013). The effect of planting date on maize grain yields and yield components. *Field Crops Research, 150*, 135–144. Available at: http://dx.doi.org/10.1016/j.fcr.2013.05.028 .

Twomlow, S., et al. (2008). Precision conservation agriculture for vulnerable farmers in low-potential zones. In *9th WaterNet/WARFSA/GWP-SA Annual Symposium, Johannesburg, South Africa*, 29–31 October 2008. Amsterdam, Netherlands: WaterNet. Johannesburg, South Africa: WaterNet.

Valdivia, C., & Quiroz, R. (2003). Coping and adapting to increased climate variability in the Andes. In *American Agricultural Economics Association Annual Meeting*, pp. 1–28.

Vincent, V., & Thomas, R. G. (1960). *An agricultural survey of Southern Rhodesia: Part I: The agroecological survey*. Salisbury, Zimbabwe: Government Printers.

VSN. (2002). GenStat for Windows (6th edn.). Available at: http://www.vsni.co.uk.

Waddington, S., et al. (1991). Extent and causes of low yield in maize planted late by smallholder farmers in sub humid areas of Zimbabwe. *Farming Systems Bulletin Eastern and Southern Africa, No. 9. CIMMYT, Mexico City*, (9), 15–30.

Wall, P. C., et al. (2013). Conservation agriculture in Eastern and Southern Africa. In R. A. Jat & J. G. de Silva (Eds.), *Conservation agriculture: Global prospects and challenges* (pp. 263–292). Cambridge, USA: CABI publishing.

White, J. W., et al. (2011). Methodologies for simulating impacts of climate change on crop production. *Field Crops Research, 124*, 357–368.

Woldemariam, H. H., et al. (2012). *Characterization of maize-legume farming systems and farm households in Mozambique: Analysis of technology choice, resource use, gender, risk management, food security and poverty profiles*. Zimbabwe: Harare.

Xu, Q., et al. (2012). Effects of rainfall on soil moisture and water movement in a subalpine dark coniferous forest in Southwestern China. *Hydrological Processes, 26*(25), 3800–3809.

Xu, Z., & Zhou, G. (2011). Responses of photosynthetic capacity to soil moisture gradient in perennial rhizome grass and perennial bunchgrass. *BMC Plant Biology, 11*(1), 21. Available at: http://www.pubmedcentral.nih.gov/articlerender.fcgi?artid=3037845&tool=pmcentrez&rendertype=abstract.

ZCATF. (2009). *Farming for the future: A guide to conservation agriculture in Zimbabwe*. Harare: Zimbabwe Conservation Agriculture Task Force.

ZIMSTAT. (2012). *Census 2012: Preliminary report*. Zimbabwe: Harare.

Mr. Siyabusa Mkuhlani graduated with a Masters Degree in Crop science at the University of Zimbabwe, Zimbabwe. He is currently a Ph.D. fellow at the Climate Systems Analysis Group, University of Cape Town, South Africa. His current research interests include small-scale farming systems, sustainable agriculture, cropping systems and climate modelling.

Dr. Walter Mupangwa graduated with a Ph.D. in Soil Science and Agro-meteorology at the University of the Free State, South Africa. He is currently working with CIMMYT as a systems agronomist within the Sustainable Intensification programme with CIMMYT Ethiopia. His interests include farming and cropping systems design, management and modelling of sub-Saharan African smallholder farmers.

Dr. Isaiah Nyagumbo graduated with a Ph.D. from the Soil Science and Agricultural Engineering Department, University of Zimbabwe, Zimbabwe in 2002. He is currently working with CIMMYT-Zimbabwe as a cropping systems agronomist in the central and eastern South African region. His research interests include farming systems design and sustainable intensification.

Climate Change Education Trends in Canadian Post-secondary Educational Institutions (PSIs)

Ana Maria Martinez and Steve Alsop

Abstract Taking into consideration (1) the extent to which climate change impacts humans (and non-humans) globally; (2) the continuous political discussions around this issue; and (3) claims that post-secondary educational institutions (PSIs) make through their mission statements in regards to their roles within societies, this research sought to explore climate change curricula responses in Canadian PSIs. Specifically, this paper presents an analysis of climate change curricula in 225 Canadian PSIs, focusing on the academic year (2014–15). Three key terms are used to identify course syllabi: "climate change," "global warming," and "greenhouse gases or GHGs." Our analysis codes syllabi ($N = 562$) into *Tiers-of-Concentration* and *Streams* according to area/discipline of concentration (technical/scientific, managerial, environmental/sociological, economics, policy/governance). Gathered data reveals a series of trends, including: (i) a relative paucity of courses with climate change focus; (ii) an emphasis on scientific/technical aspects of climate change, and (iii) complex relationships between provincial and institutional policies and climate change curricula practices. We conclude discussing implications of these trends for higher education theory and practice.

Keywords Climate change · Curriculum · Post-secondary educational institutions
Canada

A. M. Martinez (✉)
Faculty of Environmental Studies, York University, 4700 Keele Street,
Toronto, ON M3J 1P3, Canada
e-mail: ana_martinez@edu.yorku.ca

S. Alsop
Faculty of Education, York University, 4700 Keele Street,
Toronto, ON M3J 1P3, Canada
e-mail: salsop@edu.yorku.ca

© Springer International Publishing AG, part of Springer Nature 2019

W. Leal Filho and R. Leal-Arcas (eds.), *University Initiatives in Climate Change Mitigation and Adaptation*, https://doi.org/10.1007/978-3-319-89590-1_13

1 Introduction

Tipping points, thresholds, points of no return, projections, and future scenarios are some of the terms frequently associated with the current global climate. Although scientists established the influence of greenhouse gases (GHGs) on the Earth's temperatures in the late 1800s and relationships between the industrialization and changing climate systems in the late 1950s, climate change and climate science remain controversial.[1] Indeed it is now labeled and treated as a "hoax" by some high-profile political leaders.

In the introduction to *This Changes Everything*, Klein (2014) outlines how she suffered for many years from a form of climate denial—a propensity to "look away" (p. 3). Reflecting on this statement, we wonder if PSIs are guilty of enacting a form of climate denial? Are they looking away? How are PSI curricula responses addressing the urgency to shift current societal paradigms in regards to climate change perceptions, ideas, but most importantly toward an understanding that enable global systems to reach or rather remain within the 1.5 °C target?

There is an abundance of statements calling for the reorientation of higher education in response to pressing environmental and social challenges. Many of these focus on climate change, persuasively arguing that post-secondary institutions need to take a lead in responses to what is indisputably one of the most pressing issues of our time. Following our belief in the importance of higher education responding to climate change, our chapter has two aims. Firstly, it presents results of an audit of climate change curricula. We document how universities and colleges in Canada respond to climate change, focusing on course syllabi in the academic year 2014–15. Secondly, it explores trends in course syllabi, analyzing content, course distribution, and institutional policies. It is part of a broader study of climate change in Canadian post-secondary institutions (Martinez 2017).

Environmental sustainability continues to be a focus of reform in higher education. A series of high-profile institutional surveys now exist (including the Sustainability Tracking and Rating System—STARS developed by the Association for the Advancement of Sustainability in Higher Education—AASHE; the Sustainability Assessment Questionnaire (SAQ) developed by the Association of

[1]The literature on "manufacturing doubt" (this term is used by John Cook from the University of Queensland in Australia to group papers on climate change denial) argues that obscuring knowledge on climate change is purposely done to maintain the *status quo*. Read "*Merchants of doubt: how a handful of scientists obscured the truth on issues from tobacco smoke to global warming*" by Oreskes, N., & Conway, E. M., 2010; "*Lost in translation: Climate denial and the return of the political*" by Goeminne, G., 2012; "*Leading voices in the denier choir: Conservative columnists' dismissal of global warming and denigration of climate science*," by Elsasser, S. W., & Dunlap, R. E., 2012; "*Climate change prediction: Erring on the side of least drama?*" by Brysse, K., Oreskes, N., O'Reilly, J., & Oppenheimer, M., 2013; "*The Subterranean War on Science*" by Lewandowsky, S., Mann, M. E., Bauld, L., Hastings, G., & Loftus, E. F., 2013; "*Toxic seepage: Climate denial and its corrosive effect on the scientific community*" by Lewandowsky, S., Oreskes, N., Risbey, J. S., Newell, B. R., & Smithson, M., 2015.

University Leaders for a Sustainable Future—ULSF; Sustainable Development on Campus: Tools for campus decision-makers developed by the International Institute for Sustainable Development (IISD); the Platform for Sustainability Performance in Education developed by the Environmental Association for Universities and Colleges—EAUC). These surveys document considerable progress in many universities, especially in campus energy and waste reduction (Leal Filho et al. 2015; Dyer and Duer 2017; Rappaport and Hammond Creighton 2007). There are also inspirational case studies of pedagogical and curriculum innovation (see, for example, Cebrián in-press; Kricsfalusy et al. 2016). Our study draws from and adds to this work in three distinctive ways.

First, we explicitly focus on climate change curriculum (in the form of course syllabi). We recognize the inseparable nature of climate change and sustainability, but also note that studies of "environmental sustainability" in PSIs are much more abundant that studies explicitly focusing on climate change (Henderson et al. 2017; Leal Filho 2010; Vaughter et al. 2013, 2016). As Henderson et al. (2017) note, post-secondary institutional change often focuses on "changing energy consumption behaviors" (p. 20), largely associated with improving campus sustainability. Second, we offer a pan-Canadian analysis of selected course syllabi. Some inventories of climate change courses exist particularly within sustainability frameworks of institutional assessments. However, there is only report that has been published in Canada and this focuses on courses in British Columbia during the 2010–2011 calendar year (PICS 2011). Third, we provide an analysis of curricula content, course distribution and climate change policy and politics, exploring where particular courses are located (provincially and institutionally) and how this connects with local and regional policies.

2 Methodology

Our study sought to explore how PSIs are responding to climate change. Viewed from the outside, curricula appear as content structured, and somewhat systematically inspired by, guidelines and high-level directives of what areas of study a PSI should explore. We offer a course audit exploring the question: where are climate change courses found and what do they contain? Then, we reflect on some factors influencing the formation of these courses.

Our research consists of a survey of university and college course syllabi for the academic year 2014–15. We identified 310 PSIs (colleges and universities) in Canada. We focus our analysis solely on English and Bilingual speaking institutions, a total of 225. The other institutions are French-speaking.

To isolate climate change courses, a *quantitative census analysis* (Dal 2011) was performed comprising of an online key term Boolean search of published course program materials. These materials include academic calendars, lists courses of every faculty/department, and central databases of registrar's systems. Faculties and departments were condensed into clusters according to the area of study/discipline, in order to facilitate the final analysis. Three key terms were used to identify courses: “climate change,” “global warming,” and “greenhouse gases or GHGs.” Syllabi were selected containing these terms in either the title or course description. Where syllabi were not available, requests by email (n = 702) were made directly to professors or the contact persons in the relevant department or faculty where the course was being offered.

Our choice of these particular key terms was informed by our intended focus (climate change) and also practicalities of isolating a revealing and manageable data set. We reasoned that these terms are so readily equated with climate change that they should serve as reliable indicators of course emphasis. Our demarcation methods, of course, have some weakness. For instance, they exclude courses not explicitly mentioning our chosen terms but cover climate change content under broader labels, such biodiversity, sustainability, and ecology. We recognize these limitations but reasoned that explicitly mentioning these terms in the title or course description is revealing indicator of climate change commitment. The inclusion and omission of content are revealing, as course titles and descriptions appear widely in university program information for students.

An early reflection on the suitability of our search criteria was conducted after we analyzed three institutions. It suggested that our selection criteria were gathering together a diverse range of courses with a firm content focus and commitment. Some courses emerged that did not address climate change (emerging from combinations of key words but with different meanings). These were discarded. This resulted in a total of 562 course syllabi.

A grounded content analysis (Creswell 2013) was then performed on these syllabi (extending the focus to the entire course syllabi). Initially, each outline was coded into *Tiers* (as an indication of course emphasis on climate change) and then according to the areas of content and disciplinary knowledge (into what we refer to as “*Streams*”).

Our classification into tiers was adopted from the 2011 Pacific Institute for Climate Solutions’ (PICS) paper and modified for this study. Courses were allocated into three tiers using PICS (2011: 5–6) definitions:

Tier 1. Climate change is the main focus of the course;
Tier 2. Climate change is a core topic (A module or an important section); or
Tier 3. Climate change is included to a lesser extent: This includes courses where climate change is mentioned in the description [or syllabus], but not listed as a core module.

Our analysis then focused on the disciplinary/subject nature of the courses. Through an iterative process, guided by grounded theory methods (Glasser and Strauss 1967), we sought to categorize syllabi into various content streams. Our choice of streams was inspired by Sauvé’s (2005) analysis of environmental education literature, seeking to convey a sense of fluidity, recognizing that syllabi often flow in and out of particular content groups. In this sense, our coding was not intended to be categorical, in the sense of allocating each syllabus definitively to a singular stream, but sought to identify multiple streams with given syllabi.

Five streams were generated. The following descriptors emerged and acted as guides.

Technical/Scientific: This stream contains courses that focus on explaining the science behind climate change. This includes the study of climate systems, data, models, projections, scenarios, monitoring local and global climate systems.
Managerial: This stream includes course content that proposes strategies and "formulat[ing] appropriate responses to global climate change" (IPCC 1990, xxv). Two areas were clearly identified: adaptation and mitigation. Adaptation referring to the ability to adjust to the impacts produced by changes in the climate, while mitigation indicating the mechanisms to correct and minimize such impacts. In both cases, management strategies are part of the language, as well as vulnerability, scale (local, coastal, national, regional, global), the sector (water, agriculture, health, energy, tourism, urban, planning, transportation), the target group (humans, biodiversity, resources), or the solution (guiding principles/rules, management, financial, economics, legal, community-based, place-based, etc.). One common feature of this stream is the link between administrative actions and technological shifts, which has encompassed a significant number of courses in "green technology," "capacity building," and "energy efficiency."
Environmental–Sociological: This stream refers to social, cultural, philosophical, and environmental practices associated with climate change, including understanding, coping, access to and sources of information, controversy, behavior, historical, or current events that have or continue defining climate change, morality, ethics, faith, and attitudes. This stream also includes content exploring questions of equity and equality and challenges lifestyles, (un)sustainable practices and value-centered ideologies.
Economics: This stream is highly tied to the managerial stream in the sense that many of the administrative strategies convey a "financial responsibility". The language of this course content is economic, characterized by the primary goal of transitioning to a cleaner, most effective green economy, but also at a low cost. This stream includes course discussions of how local, regional, national or global markets perform, and the procedures and implications for the use of economic mechanisms and tools for transitioning efforts, the commodification of resources, marketing and markets, accounting and being accountable.
Politics and Governance: Policies, regulations, and policy-makers provide the foundation for the context and content of the courses. This stream is concerned with governance, authority, democracy, organizational structures, and public opinion at the local, regional, national, international and global levels.

Course data was uploaded into NVivo (for Mac, ver 11.3.2). Saldaña (2009) guided our analysis, informed by fields of study and disciplines, which were identified as common terms in course syllabi. Saldaña (2009) calls this process "coding for patterns: similarity, difference, frequency" (p. 16). Each syllabus was carefully allocated. A quantification system was developed to determine if a syllabus covered one or more streams. If it covered one stream only, it was given a value of 1. If judged as covering two or more streams, 1 was divided accordingly ($S = 1/(S_x + S_y\ldots)$). Through this process, an approximate weight was given to each stream for a given course.

3 Results

(i) ***Course Numbers***

The first point to note is that the number of courses selected is extremely modest, especially when compared with the total number of institutions and the total number of courses offered in an academic year. Indeed, based on our sampling methods, the majority of institutions surveyed (54%) did not offer any identifiable climate change course. 104 PSIs (43 colleges and 61 universities) offered a total of 849 climate change courses, of which universities offered approximately four-fifths and colleges one-fifth. Almost three-quarters of these courses were offered at the undergraduate level; eleven out of the 104 PSIs offered 20+ courses. These trends are summarised in Fig. 1.

In terms of faculty and department breakdown, 179 faculties and 169 departments across Canada offered climate change courses in the 2014–15 academic calendar year. The faculties that offered the most climate change courses, perhaps unsurprisingly were: Science and Technology (37%) and the Arts (28%). Geography Departments offered two-fifth (22%) of the total of climate change courses, followed by Biological and Applied Sciences (11%), Atmospheric and Oceanic Sciences (9%) and Environmental Sciences (9%). Perhaps is important to note here that most Geography Departments are located in the Faculties of Art.

Eleven PSIs offered more than 20 climate change courses each and these are located in ON ($n = 5$), BC ($n = 5$), and NS ($n = 1$). All of these institutions are publicly funded. A breakdown of course distribution and student enrollment was performed. Student numbers were obtained from University Canada statistics (www.univcan.ca). Table 1 summarizes these data in terms of the total number of courses per institution.

An analysis per capita of these eleven institutions suggests that although the size of the institution clearly affects the number of climate change courses per student, it is not a single determining factor. Climate change courses are unevenly distributed with particular universities and colleges having relatively high concentrations of courses for their size (University of British Colombia, Lakehead University, and Douglas College are noteworthy in this regard).

(ii) ***Content Trends, Part 1—Tiers***

The majority (57%) of the courses were categorized at the Tier 3 level, meaning that climate change was considered to a lesser extent in these courses. Twenty-three percent of courses were judged as Tier 1, and 20% at Tier 2. These trends are summarised in Fig. 2.

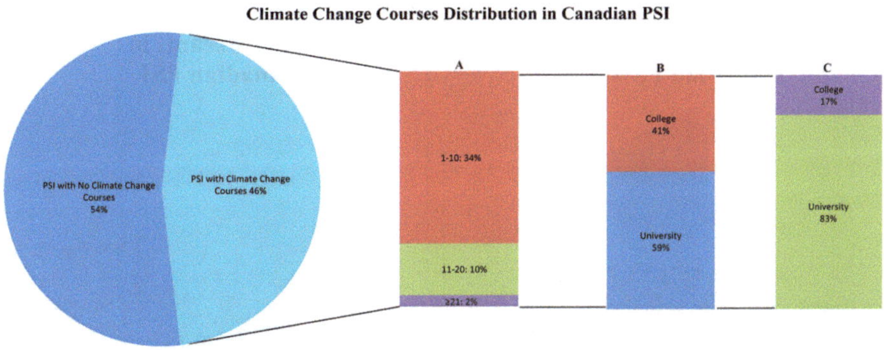

Fig. 1 Climate change courses distribution in Canadian PSIs. Key—A: Range of number of courses offered per PSI; B: College and universities (%) with climate change courses; C: Percentage of climate change courses by college and university

Table 1 Canadian PSIs that offered 20+ climate change courses

PSI Name	Province	Total climate change courses	Student enrollment	Climate change courses per capita (1000)
University of Waterloo	ON	51	35,900*	1.4206
York University	ON	45	53,900*	0.8348
University of Toronto	ON	35	84,400*	0.4146
Dalhousie University	NS	29	18,560**	1.5625
Lakehead University	ON	22	8,510*	2.5851
Douglas College	BC	20	10,464**	1.9113
University of British Columbia	BC	20	58,030*	0.3446
University of Northern British Columbia	BC	20	3,420*	5.8479
University of Ottawa	ON	20	43,100*	0.4640
University of the Fraser Valley	BC	20	8,190*	2.4420
University of Victoria	BC	20	20,910*	0.9564

Enrollment numbers were extracted from *Universities Canada and **each PSI's Web site (Universities Canada 2016)

(iii) ***Content trends, Part 2—Streams***

Around three-fifths of the courses were judged as containing technical/scientific and managerial streams (54% and 12%, respectively) (See Fig. 2). Other streams were less influential. For example, a minority of identified courses included discussion of sociological and political aspects of climate change. This observation echoes recent discussions suggesting that some disciplinary areas have been

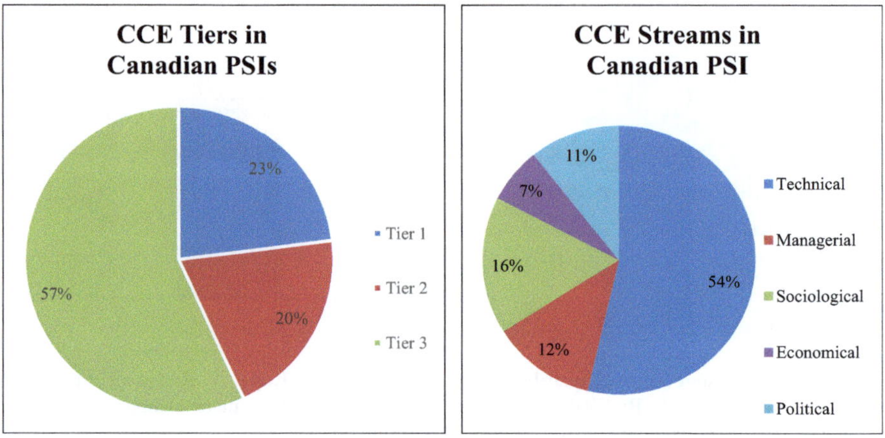

Fig. 2 Distribution of climate change courses by Tiers of Concentration and Streams departments in Canadian PSIs (n = 104)

comparatively slower in incorporate climate change into their research and pedagogical practices. Dunlap and Brulle, for instance, argue; "from the very beginning of the climate change research effort, the social sciences have been and continue to be marginalized" (2015, p. 3). The general point is that some content areas are pedagogically responding to climate change more than others. The contrasting cultures of academic curriculum practices have been a topic of sustained reflection (see, for example, Snow 1959).

In this study, almost two-thirds of the courses at the national level focused on the technical (science) and managerial (adaptation and mitigation) aspects of climate change. This distribution is broadly in line with recent IPCC reports where one volume is dedicated to science, one to mitigation, and one to adaptation. In subsequent work, we conclude that the emergence of climate change curricula is deeply connected to disciplinary structures and individual climate change curriculum leaders (Martinez 2017). It is much more loosely connected to institutional-wide initiatives and policies.

(iv) ***Jurisdictional Trends***

In Canada, provinces hold constitutional authority for higher education. As a consequence, distinctive systems of education (including guiding policies and associated funding) have emerged in each of the ten provinces. The three territories (Nunavut, Northwest Territories, and Yukon) also have separate higher systems reflecting the territorial history, geography, and governance. Different provinces currently have quite a different climate change stance—for example, Alberta and Saskatchewan are oil-rich provinces (associated with the Tar Sands) and have opposed recent discussions of carbon taxation. In contrast, Ontario has moved toward the production of renewable energies and mechanisms to reduce its GHG emissions.

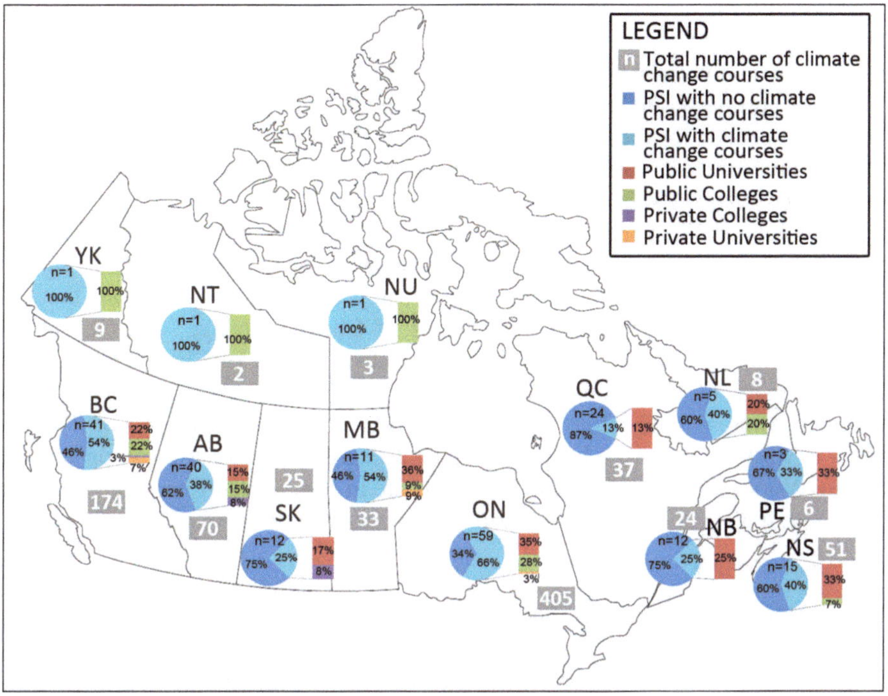

Fig. 3 Percentage of PSIs offering climate change courses per Canadian Province and Territory (n = number of PSIs)

Given this, we thought that a jurisdictional breakdown of courses and content might be revealing. In terms of Tiers, the largest number of Tier 1 courses were offered in Ontario (ON) (53%), British Columbia (BC) (16%), and Alberta (AB) (9%). Undergraduate (UG) courses are predominant in New Brunswick (NB) (89%), and Nova Scotia (NS) (88%). The largest percentage of graduate (GR) courses was found in Newfoundland and Labrador (NL) (38%). Institutions in Northwest Territories (NT), Nunavut (NU), and Yukon (YK) did not contain Tier 1 courses (based on our selection criteria). Manitoba (MN), NB, NL, NS, and Saskatchewan (SK) did not offer any continuing education, distance education or online (DE) courses, whereas NU offered all its courses through DE programs.

Two provinces offered the majority of the courses ON (48%) and BC (21%). These trends are summarised in Fig. 3. This also includes a breakdown of public and private institutions. In Canada, the vast majority of higher educational institutions are publicly funded (89.5%). Climate change courses predominately occur in public education institutions. This study found that publicly funded PSIs offered nine-tenths of the climate change courses at the national level.

A regression analysis reveals that the total number of courses, as one might expect, is positively correlated to the number of PSIs ($r = 0.88$). However, a comparison between British Colombia and Alberta seems revealing. These are neighboring provinces of similar size, with a similar number of higher education institutions and students. However, the number of climate change courses varies quite dramatically. British Columbia offers twice as many courses as Alberta. These provinces approach climate change from strikingly different angles. Our study has sought to account for these political differences through a more grounded analysis of macro/meso/micro-politics of curricula formation within particular institutions (see Martinez and Alsop, in preparation).

As of July 2017, there are three UG (undergraduate) and three GR (graduate) programs in climate change in Canada. These are offered in BC and ON. Two of these programs which grant degrees specifically on climate change are the UG minor program called "Human Dimensions of Climate Change" (HDCC) at the University of Victoria and the GR program called "Master of Climate Change" at the University of Waterloo. The University of Toronto and the University of Waterloo offer specializations at the UG level, while the University of Toronto offers a Master's and a Ph.D. with a general degree but with a focus on climate change. York University offers a UG Certificate in Sustainable Energy and starting in Winter 2017, the University of Victoria offers a Certificate in HDCC.

(v) ***Policy Trends***

In higher education, much attention is placed on institutional climate declarations, plans, and policies. Our analysis provided an opportunity to compare and contrast numbers of climate courses with these commitments. Table 2 represents the relative number of courses mapped alongside side whether the institution was: a signatory of The University and College Presidents' Climate Change Statement of Action (UCPCCSAC); has Climate Action Plans (CAP); or a central institutional policy/strategy for addressing sustainability and/or climate change. This data focuses on the year 2014–15.

Our representation suggests that relationships between institutional climate change policies and curriculum practices are far from straightforward. The University of Waterloo, is a case in point in this regard, as it contained more climate change courses than any other institution, and yet it was not a signatory of UCPCCSAC, did not have an institutional climate action plan, or a central climate change policy. In contrast, McMaster University and Sheridan College have a comparatively modest number of courses and yet have signed the UCPCCSA declaration and have climate change policies and plans in place.

4 Conclusions

This paper offers an analysis of climate change courses in Canadian post-secondary education in a given year (2014–15). Using particular selection criteria, we sought to isolate and analyze course syllabi with climate change keywords. Our analysis reveals four observations.

Table 2 Canadian PSIs with climate change courses who are signatories of UCPCCSAC have Climate Action Plans (CAP), or policies/strategies addressing sustainability and/or climate change

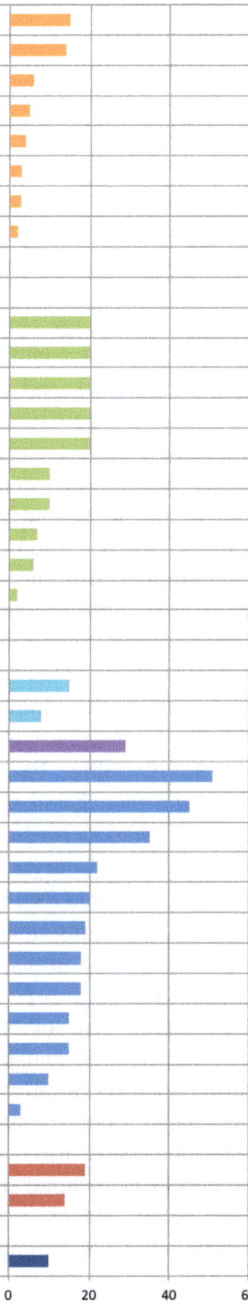

		UCPCCSAC	CAP	POLICY/STRAT.
AB	University of Alberta		✓	✓
AB	University of Calgary	✓	✓	✓
AB	Athabasca University	✓		
AB	Grant MacEwan University	✓		✓
AB	King's University College Edmonton	✓		
AB	St. Mary's University College	✓		
AB	Northern Alberta Institute of Technology	✓		✓
AB	Red Deer College	✓		
AB	Alberta College of Art and Design	✓		
AB	Bow Valley College	✓		
BC	University of the Fraser Valley	✓		
BC	University of Victoria	✓		✓
BC	Douglas College			✓
BC	University of British Columbia	✓	✓	✓
BC	University of Northern British Columbia	✓		✓
BC	Simon Fraser University	✓		✓
BC	Vancouver Island University	✓		✓
BC	Thompson Rivers University	✓		✓
BC	Kwantlen Polytechnic University	✓		✓
BC	Royal Roads University	✓		✓
BC	Emily Carr University of Art and Design	✓		
BC	Capilano University	✓		✓
MB	University of Winnipeg	✓	✓	✓
MB	University of Manitoba	✓		✓
NS	Dalhousie University	✓	✓	✓
ON	University of Waterloo			
ON	York University			✓
ON	University of Toronto			✓
ON	Lakehead University			✓
ON	University of Ottawa		✓	✓
ON	Queen's University	✓	✓	✓
ON	Trent University			✓
ON	University of Western Ontario			✓
ON	Wilfrid Laurier University			✓
ON	University of Guelph			✓
ON	McMaster University	✓	✓	✓
ON	Sheridan College	✓	✓	✓
ON	University of Trinity College	✓		
QC	McGill University			✓
QC	Concordia University			✓
QC	Université de Sherbrooke	✓		
SK	University of Saskatchewan	✓	✓	✓
	n=42 (65%)	29	10	31

Firstly, perhaps most significant observation is the relatively small number of syllabi containing our key terms. This suggests that climate change was not a central subject of sustained study in Canadian higher education (during the year 2014–15). We acknowledge the innovative courses and programs that did (and continue to) exist, but stress that the number of courses identified was modest when compared with the total number of university course offerings in general, or in a given institution. Indeed, if you average the total number of climate change courses across all institutions, the average is approximately four per institution. Moreover, courses are far from evenly distributed, the majority of courses are found in specific institutions. This suggests that only limited number of students had access to courses with a clearly specified climate change emphasis. Furthermore, the majority of available courses were judged to be at Tier 3, the lowest level of concentration of climate change content.

Secondly, our analysis identifies and maps a series of content streams. The majority of the courses identified contained technical, scientific, and managerial content streams. Environmental, sociological, and political discussions of climate change were in a minority of climate change courses. This suggests that different subject areas placed greater emphasis on climate change education.

Thirdly, our data suggests some jurisdictional trends. While the number and size of institutions is statistically a predictor of the number of courses, a comparison of provinces we suggest is revealing. British Colombia and Alberta are bordering provinces with roughly the same number of higher education institutions and students, and yet the number of courses in these provinces is markedly different. This seems informative given that the climate change policy and political contexts of these provinces contrast along similar lines. This forms the focus of our ongoing research on politics of higher education curriculum formation.

Ontario is the province with the largest Tier 1 number of courses (n = 104), which is probably related to the fact that three PSIs in this province offer climate change programs, combined with the fact that it is the province with the largest student population. An aspect to emphasize in this analysis is that the technical aspects of climate change dominate Tier 1 courses in all three provinces with the larger number of Tier 1 courses (AB, BC, and ON). However, a closer look shows a significant difference between them. Each province has rather a distinctive curriculum focus, which has, we speculate, to do future employability and the capacity to make courses and programs attractive to prospective students. For instance, in AB most of the technical Tier 1 courses were focused on carbon management and technologies associated with site remediation and bio-solutions. While in BC and ON, technical courses were mostly focused on physics, models, and broad aspects of the science behind climate change. For the latter province, topics such as renewable energy and energy efficiency were also evident. While this could suggest provincial trends, it is necessary to keep in mind that individual PSIs are significantly shaping provincial curriculum development. To be more concrete, three-fourths of Tier 1 courses offered in Ontario are found in the three PSIs with climate change programs.

Our data also compares policy and curricular practices. It seems to complicate an assumption that institutional policy is a reliable indicator of institutional curriculum practices. In a number of other studies, the presence of institutional-wide climate declarations and policy statements are taken as direct indicators of an educational institution's climate change allegiances and priorities (Vaughter et al. 2015, 2016). Indeed, climate change education policy has garnered considerable attention in both practice and research (Sauvé et al. 2007; Henderson et al. 2017), based on the logic of getting the "right policy" is the first stage in education reform. Somewhat contrary to this notion, our study reveals institutions without central declarations and policies but with significant concentrations of climate change courses, and institutions with the reverse trend (pro-climate institutional statements but no courses). In the case of *curriculum*, it seems that policy is not always a reliable indicator of curriculum. It should be stressed that curriculum, of course, is just one of the ways that higher education institutions are responding to climate change. We should also stress associated limitations of this empirical study. It draws data from a single calendar year and is based on particular search criteria, the presence of combinations of keywords in course syllabi titles and descriptions. It offers an analysis of the intended curriculum (as outlined in available course syllabi). Future work is needed to explore the taught and assessed curricula. Future work might also study different pedagogical approaches and how these relate (or not) to content tiers and streams.

We wish to leave these discussions on a positive note. Throughout this analysis we were struck by innovative climate change courses. We remain committed to PSIs playing a leading role in climate change. In this respect, we acknowledge the leadership of those involved with these innovations. Our ongoing research seeks to gather data concerning the micro-politics of climate curricula formations. We are interested in better understanding the political conditions underpinning curriculum formation (see Martinez 2017). When, why, and how do climate change curriculum innovations take place?

In Canada, 52% of the Canadian population between 25 and 64 years of age holds some sort of degree, certificate, or diploma acquired through a PSI (Statistics Canada 2011). This means that PSIs are distinctively positioned—especially at this particular moment in time—to explore broader questions of what it might mean to be part of a climate just society. We hope that this chapter raises critical questions about course content and brings greater attention to the climate change courses offered by Canadian PSIs. We suggest that more courses and programs are needed that make explicit reference to climate change. Ideally, these courses are made available for all students, regardless of their program or faculty affiliation. There are also a series of related questions concerning geographic location and culture. There is a noted absence of institutions and climate change courses in the North. It should be acknowledged that these institutions are grappling with a dramatically changing Arctic climate. This needs to be brought to a much wider attention. Canadian First Nations and governments in the North have emphasized their concern in regards to climate change, highlighting the importance of local knowledge and ongoing climate change adaptations. As many persuasively argue (Barnhardt 2005; Beckford et al. 2010), traditional indigenous knowledge should be more extensively incorporated into climate change education curricula (and vice versa), not only in the North but also across Canadian PSIs' climate change curricula. This has profound implications for our ongoing work.

References

Barnhardt, R. (2005). Indigenous knowledge systems and Alaska Native ways of knowing. *Anthropology & Education, 36*(1), 8–23.

Beckford, C. L., Jacobs, C., Williams, N., & Nahdee, R. (2010). Aboriginal environmental wisdom, stewardship, and sustainability: Lessons from the Walpole Island First Nations, Ontario, Canada. *The Journal of Environmental Education, 41*(4), 239–248.

Cebrián, G. (in press). A collaborative action research project towards embedding ESD within the higher education curriculum. *International Journal of Sustainability in Higher Education.*

Creswell, J. (2013). Qualitative inquiry & research design. In *Choosing among five approaches* (3rd ed.). Los Angeles, London, New Delhi, Singapore, Washington D.C: Sage.

Dal, M. (2011). Online data collection and data analysis using emergent technologies. In S. Hesse-Biber (Ed.), *The handbook of emergent technologies in social research* (p. 687). New York: Oxford University Press, Inc.

Dunlap, R., & Brulle, R. (Eds.). (2015). *Climate change and society. Sociological perspectives.* New York: Oxford University Press.

Dyer, G., & Duer, M. (2017). Strategic leadership for sustainability by higher education: The American College & University Presidents' Climate Commitment. *Journal of Cleaner Production, 140*(1), 111–116. (Elsevier).

Glasser, B. G., & Strauss, A. L. (1967). *The discovery of grounded theory. Strategies for qualitative research.* New Brunswick (USA) and London (UK): Aldine Transaction. A Division of Transaction Publishers, Rutgers.

Henderson, J., Bieler, A., & McKenzie, M. (2017). Climate change and the Canadian Higher Education System: An institutional policy analysis. *Canadian Journal of Higher Education, 47* (1), 1–26.

IPCC. (1990). Climate change: The IPCC response strategies. *Report of the Working Group III of the Intergovernmental Panel on Climate Change*, 332. WMO/UNEP. UK: Cambridge University Press.

Klein, N. (2014). *This changes everyhting*. Toronto: Penguin Random House Knopf Canada.

Kricsfalusy, V., George, C., & Reed, M. G. (2016). Integrating problem- and project-based learning opportunities: Assessing outcomes of a field course in environment and sustainability. *Environmental Education Research*, 1–18 (Taylor & Francis).

Leal Filho, W. (2010). Climate change at Universities: Results of a World Survey. In W. Leal Filho (Ed.), Universities and climate change. Introducing climate change to university programmes. Heidelberg, Germany: Springer.

Leal Filho, W., Shiel, C., do Paco, A., & Brandi, L. (2015). Putting sustainable development in practice: Campus greening as a tool for institutional sustainability efforts. In J. P. Davim (Ed.), *Sustainability in higher education* (pp. 1–19). Waltham/Kidlington, MA/OX: Chandon Publiching/Elsevier.

Martinez, A.M. (2017). *Climate change education: An exploration of curriculum enactments in Canadian post-secondary educational institutions* (Doctoral dissertation). York University, Toronto, Canada.

Martinez, A. M., & Alsop, S. (in preparation). Influencing factors in the formation of climate change curriculum in Canadian Post-Secondary Educational Institutions.

PICS. (2011). *Climate change courses in British Columbia postsecondary institutions. A scoping review*. Pacific Institute for Climate Solutions. Victoria: University of Victoria.

Rappaport, A., & Hammond Creighton, S. (2007). *Degrees that matter. Climate change and the university*. Cambridge, London: MIT Press.

Saldaña, J. (2009). *The coding manual for qualitative researchers*. Thousand Oaks, California, USA: Sage.

Sauvé, L. (2005). Currents in environmental education: Mapping a complex and evolving pedagogical field. *Canadian Journal of Environmental Education, 10*(1), 11–37.

Sauvé, L., Berryman, T., & Brunelle, R. (2007). Three decades of international guidelines for environment-related education: A critical hermeneutic of the United Nations discourse. *Canadian Journal of Environmental Education, 12,* 33–54.

Snow, C. (1959). *The two cultures and the scientific revolution*. The Rede Lecture. New York: Cambridge University Press.

Statistics Canada. (2011). Distribution of the population aged 25 to 64, by highest certificate, diploma or degree and age group, Canada, provinces and territories, 2011. Retrieved November 6, 2015, from 2011 National Household Survey (NHS), 2011, Statistics Canada: http://www.statcan.gc.ca/pub/81-582-x/2015002/tbl/tbld6.4-eng.htm.

Universities Canada. (2016). *Enrolment by University*. Retrieved December 22, 2016, from http://www.univcan.ca/universities/facts-and-stats/enrolment-by-university/.

Vaughter, P., McKenzie, M., Lidstone, L., & Wright, T. (2016). Campus sustainability governance in Canada. A content analysis of post-secondary institutions' sustainability policies. *International Journal of Sustainability in Higher Education, 17*(1), 16–39.

Vaughter, P., Wright, T., & Herbert, Y. (2015). 50 Shades of green: An examination of sustainability policy on Canadian campuses. *Canadian Journal of Higher Education, 45*(4), 81–100.

Vaughter, P., Wright, T., McKenzie, M., & Lidstone, L. (2013). Greening the Ivory Tower: A review of educational research on sustainability in post-secondary education. *Sustainability, 5* (5), 2252–2271.

Ana Maria Martinez is Research Associate at the Las Nubes Project in the Faculty of Environmental Studies at York University. She is a marine biologist with a specialization on environmental law, a Master's in Environmental Studies, and a doctorate in Education.

Steve Alsop is a Professor at the Faculty of Education and Department of Science and Technology Studies at York University, Canada. He has published widely on the politics and practices of education within scientific and environmental sustainability contexts.

Part II
Experiences and Lessons From Climate Change and Related Projects

Geospatial Analysis of Rainfall and Temperature Variations Effect on Maize (Zea Mays) Yield

Olumuyiwa Idowu Ojo, Johnson Olawale Olonnu and Masengo Francois Ilunga

Abstract Climatic variation results from the gradual warming of our planet, which accounted for the variability of the weather globally. This variation in turn affects agricultural crop production negatively; this often necessitates price increase and hence food scarcity. Maize is an important food in Nigeria, especially in the south-west Nigeria. Maize farmers are in need of data to better plan crops planting and monitor rate of yield. Thus, the study examined the relationship between the inter-annual climate variability (temperature and rainfall) and the maize yield in south-western states of Nigeria. Annual maize yield data for years 1991–2007 were collected from the Federal Ministry of Agriculture, Nigeria, while temperature and rainfall data were obtained from the Nigerian Meteorological Station. Analysis of the relationship between climate variation of temperature, precipitation on maize yield and maps of the respective parameters was produced using GIS for selected years. The results of the work include production of geospatial variation maps of the rainfall, temperature and maize yields and the correlation of the variables in study area. Ekiti State had the highest rainfall value of 182.86 cm, while Ogun State had the least rainfall value of 110.81 cm. Projecting for the year 2017, Ondo and Ekiti states would have a joint highest rainfall value of 190.13 cm while Ogun State may have rainfall value of 110.25 cm. For temperature, Ogun State had the highest temperature value of 32.21 °C while Lagos State had the lowest temperature value of 31.39 °C. By prediction, Ogun State will have mean annual temperature of 32.08 °C by the year 2017, while Lagos State would have the least temperature value of 31.43 °C. Mean annual maize yield on temperature-based is highest for the year 2012 at Osun State (1.81 tonne value) and lowest at Ogun (1.43 tonne).

O. I. Ojo (✉) · M. F. Ilunga
Department of Civil Engineering, University of South Africa, Pretoria, South Africa
e-mail: olumuyiwaojo@gmail.com

O. I. Ojo · J. O. Olonnu
Department of Agricultural Engineering, Ladoke Akintola University of Technology, Ogbomoso, Nigeria

© Springer International Publishing AG, part of Springer Nature 2019
W. Leal Filho and R. Leal-Arcas (eds.), *University Initiatives in Climate Change Mitigation and Adaptation*, https://doi.org/10.1007/978-3-319-89590-1_14

Mean annual maize yield, however, varied from state to state. By the year 2017, mean annual maize yield would decline to 1.46 tonne for Osun State and least in Ogun State with 1.16 tonne. The study depicted a general pattern, but the results change from state to state. The findings from the study will help governments and farmers to better plan crops production activities in order to address the problem of increasing food security.

Keywords Maize yield · Weather variability · Ecological · South-western Nigeria

1 Introduction

Crop agricultural production depends largely on weather as the nature of soil formed in an area in the process of denudation also depends on the weather factors. Favourable weather condition fosters crop agriculture and vice versa. The weather of Nigeria has undergone changes during the last 10 years, and this did affect the subsistence farmers who depend mainly on rainfed agriculture (Adejuwon 2004). Weather variability over a period of instrumental record is important to understand the nature of different climate systems and their impact on the environment and society (Oguntunde et al. 2012). Ayanlade and Odekunle (2006) observed that crop yield does not depend on the climatic zone but on the behaviour of weather elements. Based on the characteristics of weather elements and the type of vegetation/crop that can they support, the whole Nigeria landmass (about 923,766 km^2 in total mass area) is divided into seven ecological zones (according to Federal Government of Nigeria report on drought management (FGN 1999)). They are mangrove swamp, rainforest, montane forest/grassland, derived savannah, Guinea savannah, Sudan savannah and the Sahel savannah (Sowunmi and Akintola 2009). The mangrove swamp and rainforest zones, and part of derived savannah zone are found in the south-western part of the country (Sowunmi and Akintola 2009). They reported that the zones are characterized by high rainfall intensity, long wet season, dense vegetation, rugged topography and temperature range of 25–28 °C and small farm holdings. A general study on crop yield variability as influenced by climate (Chen et al. 2004) submitted that precipitation and temperature are found to have opposite effects on yield levels and variability of maize (Zea mays). Hence, at higher annual precipitation (majorly rainfall), maize yield escalates into greater tonnes. Weather variability and changes have a direct, often adverse, influence on the quantity and quality of most crop agricultural produce (Sowunmi and Akintola 2009). They also concluded that the weather of an area is highly correlated to the vegetation and by extension the type of crop that can be cultivated. Temperature, rainfall, humidity, sunshine (day length) are the important weather elements that influence cropping production.

Maize is the most widespread crop, with a distribution ranging from about 50°N to about 45°S about the equator, and it is the second most popular crop after wheat (Leff et al. 2004). This therefore justifies maize as tropical crop; since the distribution ranges along the latitudinal lines of the Tropics of Cancer and Tropics of Capricorn, maize (Z. mays) is adapted to a warm, dry subtropical climate with plentiful sun, occasionally abundant rain during the night and limited rain during the daytime, low wind speed and loamy soils (Anderson 2000; Odgarrd 2001). During germination, the soil temperature should not move below 8–10 °C. These ecological requirements constrain its global geographical distribution. It is restricted to tropical and the temperate climatic zones with more than forty per cent (40%) of the total annual production supplied by the USA, primarily the so-called corn belt (Leff et al. 2004). Maize (Z. mays), one of the principal cereal grains (Okaka 1999), a fruit of cultivated grasses belonging to the monocotyledonous family of Gramineae is also commonly referred to as corn. The growing season temperatures in south-western Nigeria are warm enough to allow maize to be grown to maturity for human consumption. At higher annual rainfall, maize yield is usually improved. Maize is not only a culturally important food but also inevitable in modern days processed food for both infants and adults of the vast-growing population of the south-western Nigeria. Sowunmi and Akintola (2009) observed that virtually all tribes in Nigeria consume maize, either fresh or processed. It is consumed as pap (Ogi) and Tuwo in virtually all parts of the country. Maize is a major ingredient in infant food, brewery and poultry feed industries, and it is fermented to produce hydrolyzed sugars and syrup. According to Sowunmi and Akintola (2009), maize is a culturally important food in Nigeria, especially in the south-west Nigeria and certain level of effect on the mental health could result in a typical Oyo (most especially Osun or Ogun) man in the case of an abrupt shortage of maize in the market. This is owing to his everyday early morning consumption of "pap". Maize yield variability was observed in the south-west to be attributed to weather variation (Ayanlade and Odekunle 2006).

1.1 The Weather in the South-West

Metrological consideration in the south-west tends to be more on rainfall characteristics, such as rainfall amount, duration and intensity (Omogbai 2010). The various rainfall characteristics have been studied in terms of their distributions over space and time, trends, periodicities, onset, retreat, probabilities (IPCC 2001). The highly variable nature of the rainfall in the region as compared with the relatively stable nature of the temperature appears to have imbued more relevance to the former as the major component in the study of climate in the region (Gbuyiro et al. 2002). One established factor, which promotes or inhibits rainfall precipitation in south-west Nigeria in particular, is the progression or regression of the zone of influence of the tropical maritime air mass (MT) and its associated zone of discontinuity (Omogbai 2010). The intertropical discontinuity (ITD) is the most popularly accepted medium that influences rainfall distribution in Nigeria (Adejuwon et al. 1990). It is established that to the southern part of ITD, varying degrees of convective activity and precipitation take place with little or no cloud. In other words, rain falls mostly when an area is overlain by the MT air mass and ceases when the area is overlain by the continental tropical air mass. This makes the position of ITD a great determinant of most rainfall attributes in the region. Prominent among these attributes are the length of the rainy season, as it appears as if the ITD is a rain-producing phenomenon in itself, but in reality, weather zones occur in a latitudinal spatial relationship to it. However, weather change is attributed to the phenomenon of global warming influence and variability over space and time in Nigeria and through West Africa (Omogbai 2010). Rainfall in each state of the south-west is largely seasonal and varies from year to year. Two distinct seasons are observed, which are dry and wet season. The dry season extends over a period of about six to seven months from October to middle of March (or till early April). The wet season extends over a period of about five to six months (from May to September). Subsistence and mechanized farming are being majorly practiced in the states, but the former (farming system) being highly practiced than the later. This is probably because the area comprises mostly of smallholders farmers who use simple techniques of maize production.

1.2 Maize Production in South-West Nigeria

Maize production in south-west Nigeria is now on increased trend in the recent years than other part of Nigeria owing to the favourable soil type and the prevalent climatic condition of the area (Ayanlade and Odekunle 2006). According to Iken et al. (2002), maize (Z. mays) is a major important cereal, which thrives well in the rainforest and the derived savannah zones of Nigeria. Maize has been in the diet of Nigerians for centuries; it started as a subsistence crop and has gradually become more important crop, and maize has now risen to a commercial crop upon which many agro-based industries depend as raw materials.

1.3 Weather Variability of the South-West

Just as in the case of other parts of the country, the weather of south-western Nigeria is dominated by the influence of three major wind currents, namely the maritime tropical air mass, the continental tropical air mass and the equatorial easterlies (Omogbai 2010). The maritime tropical air mass originates from the southern high-pressure belt located off the coast of Namibia, and in its trajectory, picks up moisture from over the Atlantic Ocean, crosses the Equator and enters south-western Nigeria (Omogbai 2010). The continental tropic air mass originates from the high-pressure belt north of the Tropic of Cancer. It picks up much little moisture along its path and is thus dry. The two masses meet along a slanting surface called the intertropical discontinuity (ITD). The equatorial easterlies are rather erratic cool air masses, which come from the East and flow in the upper atmosphere along the ITD. Occasionally, however, the air mass dives down, undercuts the air mass and gives rise to line squalls or dust devils (Omogbai 2010). Rainfall of the study area mostly results from the saturation after the subsequent cooling and accumulation of warm moist air (cloud) at some higher layers of the atmosphere (many a time, in the ionosphere layer). At hot noonday, cosmic ray supplies heat to earth surface, which in turn supplies heat to the moist earth surface. When the heat reaches an approximate threshold, earth moisture vapour pressure reaches and exceeds the prevailing atmospheric pressure. The air, warmer front, rises while less dense air moves downward. The rest heat supplied (at continuous sunshine) is converted to vertically moving energy by the moist air. This results in formation of cloud often time with a high vertical extent (stratus, stratocumulus, alto-cumulous, cumulonimbus are some of these cloud type). Continuous lifting of the (warm) moist air results in saturation of the cloud (IPCC 2002) and hence results into rainfall. They also concluded that owing to global warming, there is a considerable difference in the cosmic ray heat being incidental on the earth surface. This alters heat reception and hence evaporation, evapotranspiration as well as the lifting described above and thus causes rainfall occurrence as well as relative humidity (weather condition in general) to show variations.

1.4 Geospatial Analysis of Maize (Z. Mays) Yield

Spatial analysis is one of the remarkable aspects of ArcGIS as using spatial analysis; ArcGIS (version 9.2) can combine information from many independent sources and derive entirely new sets of information, the results (ESRI 2006). ArcGIS (version 9.2) accesses and run analysis models and display their results as a new map layer using the previously recorded statistical data of the feature in consideration (Omogbai 2010). The geographic information system (GIS) database creation and mapping techniques were used to create interactive maps of the maize production areas, areas harvested and the yield (Odekunle et al. 2007). GIS is a computer-based technology that helps in management, manipulation, analysis, display (mapped) data and decision-making process. GIS offers a relatively efficient way for weather elements maize yields modelling. GIS technology has the ability to integrate spatially explicit information into different formats from multiple sources.

GIS interpolation makes it easy to work with spatial data, to see patterns, which could not be seen originally, and revealing hidden trends and distributions, thus gaining new insights (Ayanlade and Odekunle 2006). Building spatial-dominant models through GIS tools is frequently a time-consuming process, and a common requirement of much rainfall/maize yields research is the assimilation of physical data and biological data from many sources (Kufoniyi 2004). Therefore, the search for a relatively easy method to integrate this information is ongoing and it is thought that interpolation, quarry, overlying and an object-oriented approach, as mentioned above, are useful systems to assess the impact of rainfall variability on maize yields. The statistical methods that were employed for this study include correlation of crop yield with annual rainfall total, multiple correlation analysis of yield with annual rainfall total and Z-distribution anomalies distribution anomalies are the only valid statistical technique to use when the data are normally distributed. Generally, the rainfall total in Nigeria is normally distributed, especially in a situation where the mean annual rainfall is greater than 750 mm (Adejuwon and Odekunle 2006). Smart maize agriculture would therefore be necessary to combat and further prohibit food insecurity and malnutrition in the south-western Nigeria.

2 Methodology

2.1 Acquisition of Data

The weather data used in this project work were obtained from the Nigeria Metrological Station (NIMET) from 1991 to 2007 and essentially comprised of the temperature and rainfall data of each mentioned state in the south-western states of Nigeria. The maize yield data, however, were obtained from the Federal Ministry of Agriculture Nigeria on annual basis. Maize yields were based on the ratio of expected annual yield and exact annual yield of the crop.

2.2 Study Area

The study area is the south-western eco-region of Nigeria that lies between the semi-arid and the wet southern part. The area lies between longitude 2° 31′ and 6° 00′E and latitude 6° 21′ and 8° 37′N with a total land area of 77,818 km^2 and a projected population of 28,767,752 in 2002. The study area is bounded in the East by Edo and delta states, in the North by Kwara and Kogi states, in the West by the Republic of Benin and in the south by the Gulf of Guinea. Figure 1 shows the position of the study area in the map of Nigeria. The location coordinates of the areas (Ogun, Oyo, Ondo, Osun, Ekiti and Lagos State) in reference to the map of the country; Nigeria is depicted in Table 1 and Fig. 1.

2.3 Data Analysis

ArcGIS (9.2 version) was used to analyse annual temperature, rainfall and maize (Z. mays) yield data geospatial variation using Kringing method of interpolation.

Table 1 Coordinates of the Study Area

S. No.	Name of station	Latitude (degree, minute, seconds)	Longitude (degree, minute, seconds)
1	Ekiti	7° 40′ 00″N	5° 15′00′E
2	Ondo	7° 10′ 00″N	5° 05′ 00″E
3	Osun	7° 30′ 00″N	4° 30′ 00″E
4	Ogun	7° 00′ 00″N	3° 35′ 00′E
5	Oyo	8° 00′ 00″N	4° 00′ 00″E
6	Lagos	6° 35′ 00″N	3° 45′ 00′E

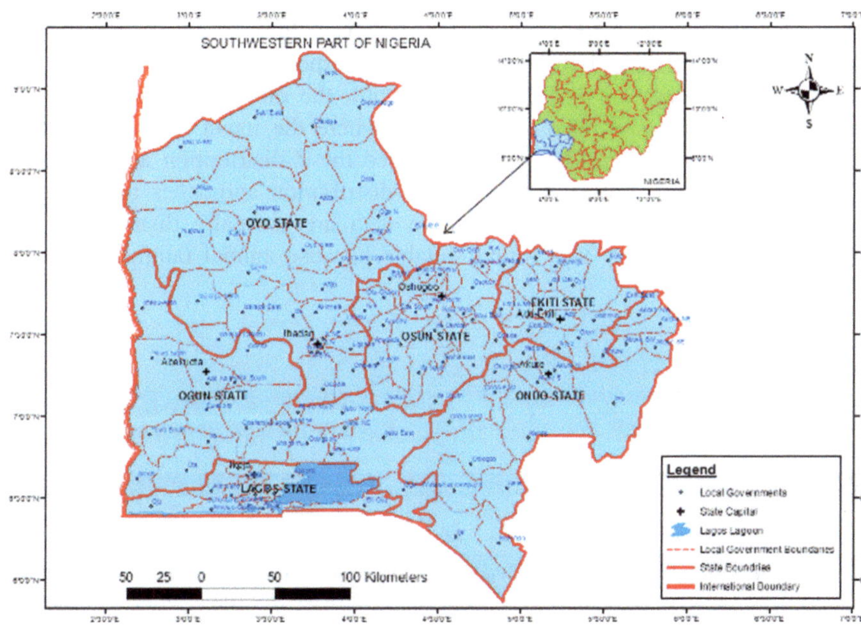

Fig. 1 Map of the study area (south-western, Nigeria) *Source* Ojo and Olawale (2014)

Existing map of Nigeria in JPEG format was imported into the ArcGIS (version 9.2) environment. This was done by the add data icon (a "cross" sign) at the ArcGIS (version 9.2) interface. The Nigeria map was geo-referenced by correcting and updating the map with the correct coordinates of each state of the study area. After importing the Nigeria map into the GIS interface, it was then processed by geo-referencing, rectifying, updating and digitizing. Geo-referencing was achieved by assigning the right coordinates to the (study area) Nigeria map. Rectification is the act of changing the coordinates of a map from old coordinates to the newly assigned correct coordinates to register the map into ArcGIS (version 9.2) internal environment and updating is re-adding the registered map. Digitizing which is the last phase in map pre-processing refers to the act of editing/carving out necessary features from an existing map. The south-western part of Nigeria map, states, states' capitals, local governments and boundaries, is carved out of the existing map. Temperature, rainfall and maize (Z. mays) yield values were inputted in the State Capital's attribute table. ArcGIS (version 9.2) was then used to develop the interpolated maps showing the variation of temperature, rainfall and maize (Z. mays) yields at year intervals within the south-western part of Nigeria by performing Kringing method of geospatial interpolation (at the spatial analyst tool) for the temperature, rainfall and maize (Z. mays) yield values. The interpolated maps that serve as major outputs are then reclassified and post-processed by laying out and exporting them. Layout was done to give legend, scale bar, north arrow, grid lines, title and other map characteristic feature to the map. The interpolated maps were exported in TIFF formats. Figure 2 is a flow chart describing map interpolation process using Kringing process of interpolation.

According to Adejuwon and Odekunle (2006), the rainfall total in Nigeria is normally distributed, especially in a situation where the mean annual rainfall is greater than 750 mm. Therefore, this study used the Z-distribution anomalies for the variability and correlation assessment. Z-distribution anomalies are the only valid statistical technique to use when the data are normally distributed. The measurement of variability was found suitable because annual rainfall total data used.

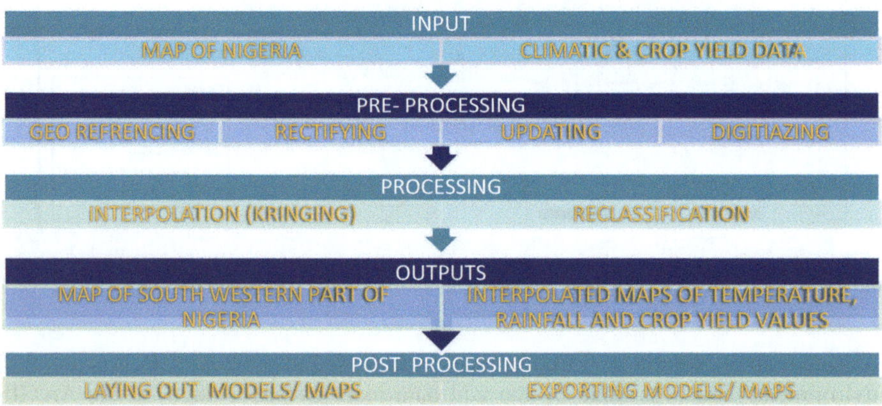

Fig. 2 Kringing method of interpolation flow chart

3 Results and Discussion

3.1 Maps of Mean Annual Maize Yield for the Study Period

The mean maize yield map through the years of study in the south-western states (as obtained from Kringing interpolation through the method described in Fig. 2) is as shown in Fig. 3. The variation in the (blue) colour intensity relays the trend of

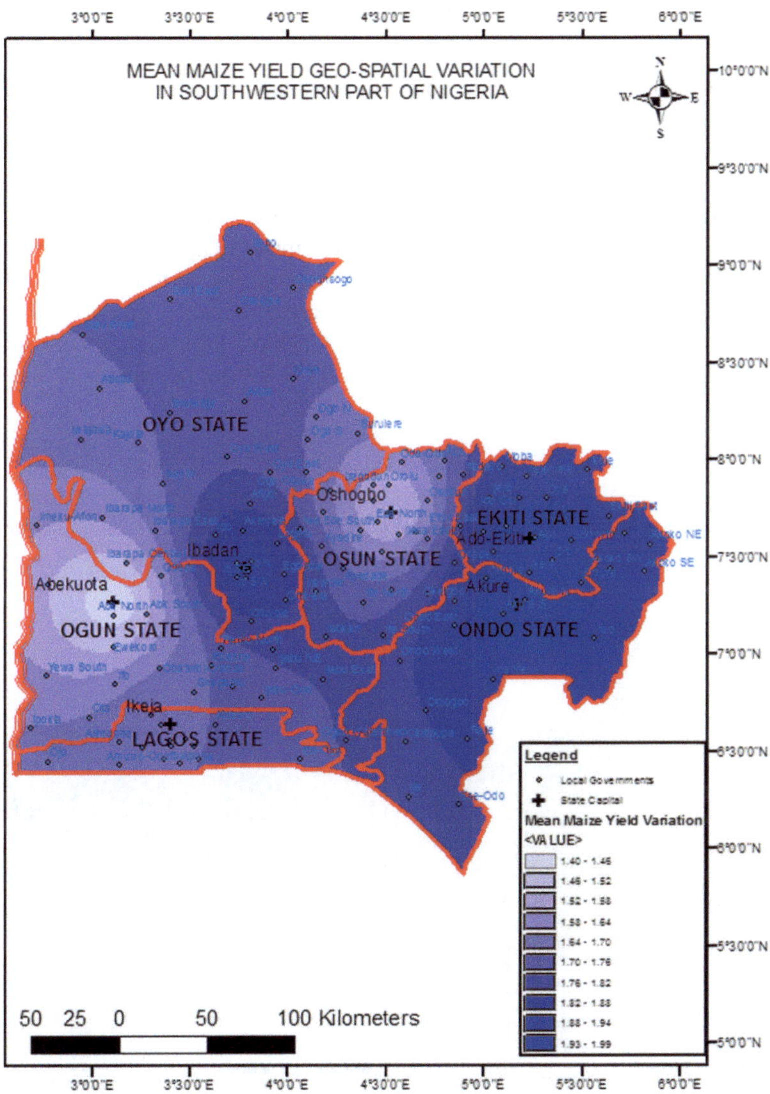

Fig. 3 Map showing mean maize yield geospatial variation in south-west

the maize yield: 1.40–1.48 tonne of maize yield value to 1.93–1.99 tonne. Areas in lighter blue indicate a much lower maize yield than the areas with denser blue. Areas in moderate blue colour intensity depict medium maize yield. The lightest blue areas have the least mean yield value of 1.40–1.45 tonne. The densest blue areas have the highest mean maize yield ranging between 1.93 and 1.99 tonne.

3.2 Geospatial Variation of Mean Annual Temperature and Mean Annual Maize Yield

The mean geo-partial temperature map (as obtained from process illustrated in Fig. 2) in the south-west during the study time range of study is shown in Fig. 4. The variation in the colour relays the trend of temperature. Areas in dense green indicate low-temperature range of about 31.30–31.37 °C. Areas in dense red have highest temperature ranges of about 31.34–32.09 °C. Yellow areas have medium temperature variation ranges. Areas with highest mean temperature as indicated in dense red 31.34–32.09 °C, Abeokuta for instance, correspond to the point of lowest mean maize yield on the yield map, whereas, areas in dense green on the mean temperature map, value of about 31.30–31.37 °C, (Ekiti for instance) correspond to the area of highest maize yield output 1.93–1.99 tonne area on the yield

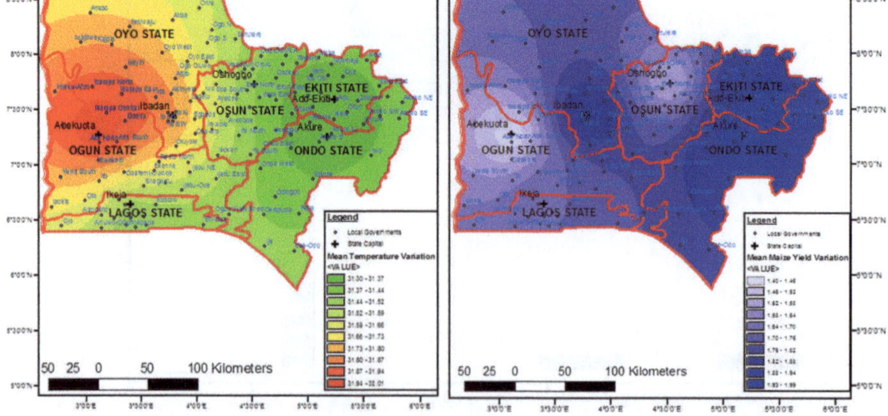

Fig. 4 Mean temperature geospatial variation versus mean maize yield variation

map. There is therefore an inverse relation between the mean annual maize yield and mean annual temperature.

3.3 *Mean Annual Rainfall Variation and Mean Annual Maize Yield Variation*

The mean geo-partial rainfall map in the south-west during the study time range of study is shown in Fig. 5. The variation in the colour indicates the trend of the rainfall intensity (96.20–99.45 and 125.43–128.83 cm) and correlates to the nature of the maize yield on the mean maize yield map. Areas in dense green indicate lowest rainfall depth (areas from Ibadan through Ogun State), values between 94.20 and 99.43 cm. These areas correspond to the least mean maize yield, values ranging from 1.40 to 1.46 tonne in value on the mean maize yield map. The areas in dense red colour have the highest rainfall depth, value ranging between 126.62 and 126.68 cm (which include Ondo and Ekiti States), and correspond to the highest mean maize yield 1.93–1.99 tonne areas thus revealing a direct correlation between mean rainfall intensity and mean maize yield. Light green and light red colours

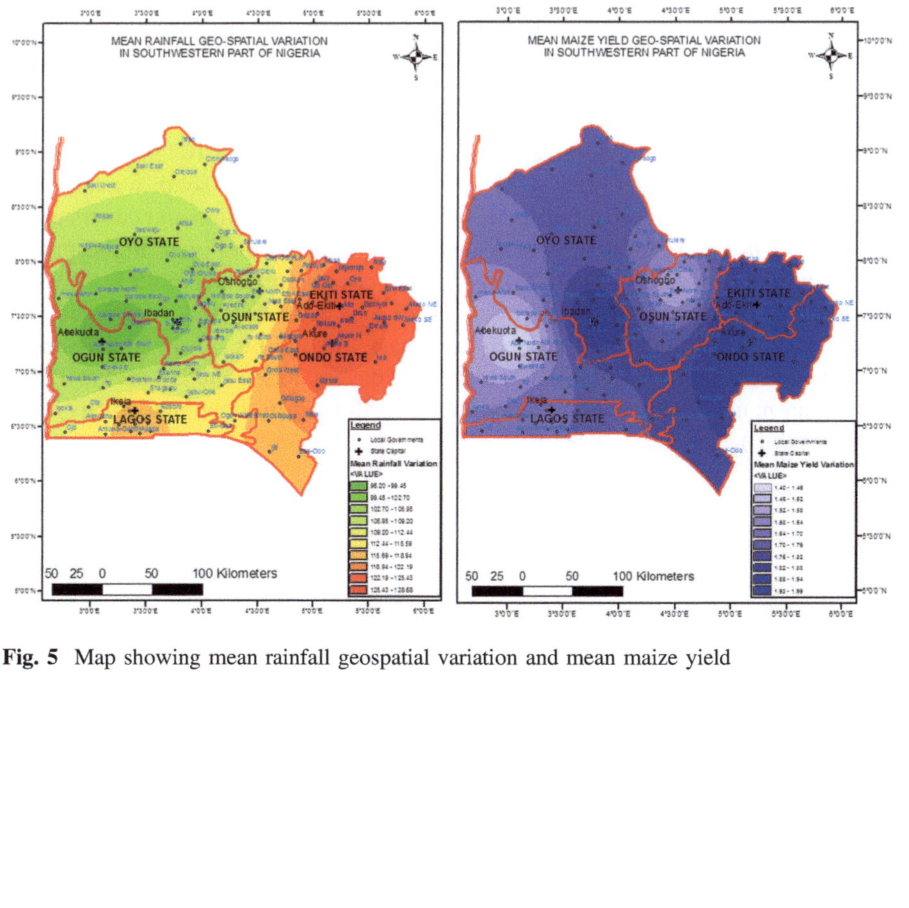

Fig. 5 Map showing mean rainfall geospatial variation and mean maize yield

depict area with relatively low and relatively high rainfall intensity, while the yellow areas have moderate rainfall intensity, which correlates to different yield nature on the mean yield map.

3.4 Correlation and Prediction of Future Maize Yields

Table 2 shows the equation to be used for the future prediction of rainfall and temperature values for each state, and Table 3 shows the values obtained. This was achieved by making use of the linear regression lines' equations of variables. The dependent variable is maize yield represented by (*Y*), while the independent variable is year difference from the datum represented by *x* (i.e. year 1991). Table 4 shows the future predictions of maize yields in the study area states, which were obtained by substituting the values of the predicted year (2012 and 2017). Table 5 shows the correlation.

3.5 Relation of Variables

The area covered by mean rainfall, mean temperature and mean maize yield ranges is as shown in Figs. 6, 7 and 8. The abscissas represent the mean rainfall amount, mean temperature and mean maize yields, respectively, while the ordinates represent the area covered in percentage. That is, percentage proportion occupied in south-west land area (which is about 191,843 km^2, (Omogbai 2010)). In Fig. 6, the highest mean rainfall is within the range of 109.02–112.44 and 105.95–109.20 cm and covers the same land area of 17.5% (33572.525 km^2) of south-western land mass. The least mean rainfall ranges between 96.20 and 99.45 cm and covers about 3.2% (6138.98 km^2) of south-western land mass. Similarly on Fig. 7, highest mean temperature is between 31.37 and 31.44 °C and covers 15% (28776.45 km^2) while the least mean temperature is between 31.94 and 32.01 °C and covers just 2% (3836.86 km^2) from the south-western land mass. Also in Fig. 8, the highest mean maize yield ranges from 1.70 to 1.76 tonne and covers 26.5% (50838.40 km^2). The least mean maize yield coverage ranges from 1.40 to 1.46 tonne and covers 0.7% (1342.90 km^2).

Table 2 Equation for future prediction of rainfall and temperature values for each state

S. No.		Ekiti	Ondo	Osun	Ogun	Oyo	Lagos
1	Rainfall	$Y = 1.455x + 152.3$	$Y = 1.455x + 152.3$	$Y = 0.210x + 115.2$	$Y = 0.667x + 92.91$	$Y = 0.081x + 108.7$	$Y = 0.769x + 120.4$
2	Temperature	$Y = 0.023x + 30.95$	$Y = 0.023x + 30.95$	$Y = 0.006x + 31.42$	$Y = -0.025x + 32.73$	$Y = -0.011x + 31.87$	$Y = 0.009x + 31.20$

where Y is maize yield, and x is year difference, taking 1991 as the datum

Table 3 Future prediction of average rainfall and average temperature values for each state from equations

S. No.	States	Mean temperature (2012) in °C	Mean temperature (2017) in °C	Mean rainfall (2012) in cm	Mean rainfall (2017) in cm
1	Ekiti	31.43	31.55	182.86	190.13
2	Ondo	31.43	31.55	182.85	190.13
3	Osun	31.54	31.58	119.61	120.66
4	Ogun	32.21	32.08	106.92	110.25
5	Oyo	31.64	31.58	110.40	110.81
6	Lagos	31.39	31.43	136.55	140.39

Table 4 Future prediction of maize (Zea mays) yield using the average rainfall and temperature values in each state from equations

S. No.	States	Mean maize yield (temperature-based) (2012) in tones	Mean maize yield (temperature-based) (2017) in tones	mean maize yield (rainfall-based) (2012) in tones	Mean maize yield (rainfall-based) (2017) in tones
1	Ekiti	1.60	1.29	9.43	7.63
2	Ondo	1.60	1.29	9.43	7.63
3	Osun	1.81	1.46	121.83	4.69
4	Ogun	1.43	1.15	5.27	5.26
5	Oyo	1.45	1.18	4.28	4.28
6	Lagos	1.54	1.25	6.88	5.56

Table 5 Correlation of maize yields with growing seasonal rainfall and temperature

S. No.	States	Maize versus rainfall	Maize versus temperature
1	Ekiti	−0.254	−0.099
2	Ondo	−0.278	−0.159
3	Osun	0.304	−0.326
4	Ogun	0.281	0.359
5	Oyo	0.229	−0.056
6	Lagos	0.101	−0.458

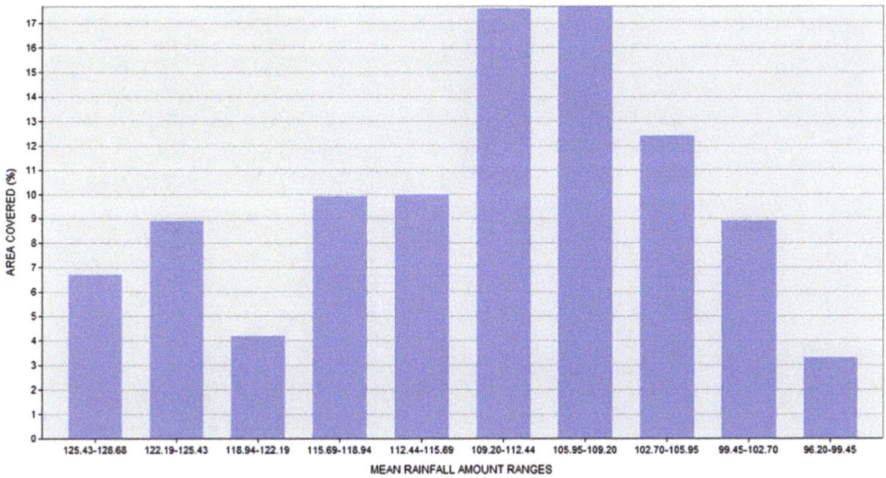

Fig. 6 Bar chart showing area covered by mean rainfall amount ranges

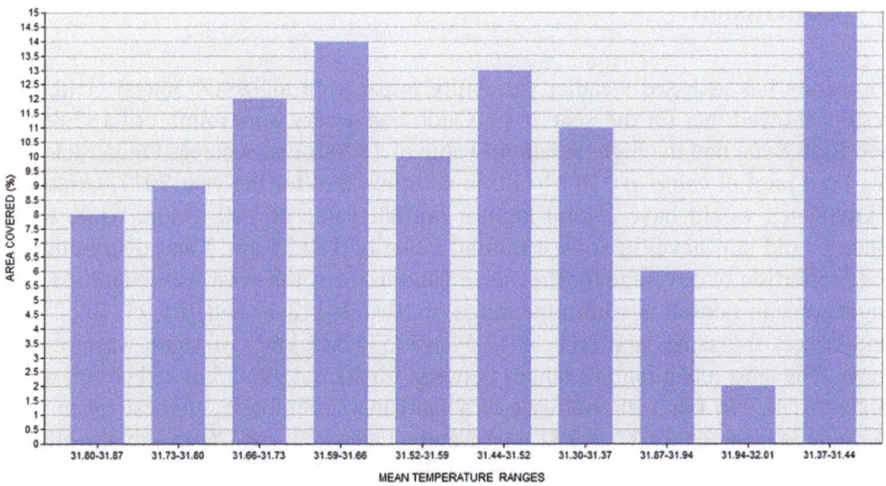

Fig. 7 Bar chart showing area covered by mean temperature ranges

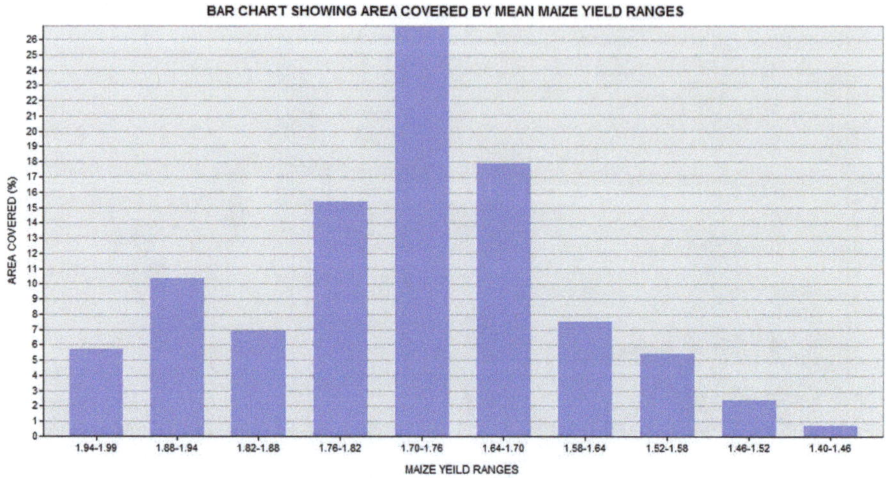

Fig. 8 Bar chart showing area covered by mean maize yield ranges

4 Conclusion

The study has assessed weather variability impacts on maize (Z. mays) yield. The results showed that for the year 2012, Ondo had a very high rainfall of 182.85 cm and Ekiti State had the highest rainfall value of 182.86 cm, whereas Ogun State had the least rainfall value of 110.81 cm. In addition, that for the year 2017, Ondo and Ekiti States would have a joint highest rainfall value of 190.13 cm, while Ogun State would still have the lowest rainfall value of 110.25 cm. Land distribution in the proportion of coverage for the maize yield in the south-west was examined. The highest mean rainfall is within the range of 109.02–112.44 and 105.95–109.20 cm and covers the same land area of 17.5% (33572.525 km^2) of south-western land mass. The least mean rainfall ranges between 96.20 and 99.45 cm and covers about 3.2% (6138.976 km^2) of south-western land mass. Similarly, highest mean temperature is between 31.37 and 31.44 °C and covers 15% (28776.45 km^2), while the least mean temperature is between 31.94 and 32.01 °C and covers just 2% (3836.86 km^2) from the south-western land mass. Mean maize yield, mean temperature and mean rainfall variation through the seventeen years of study for the study area had been known. Mean rainfall and mean temperature are seen to have had inverse relation on the mean maize yield. That is, while areas with increased mean rainfall turned out to have increase maize yield output, areas of higher temperature are of declined mean maize yield output. Results deducted are good for maize farmers as to make better plan on maize (Z. mays) farming in spite of the varying weather condition as to ensure smart agriculture. Geospatial analysis techniques (such as spatial analysis of the 3D analyst tool of ArcGIS 9.2) should be explored further for realistic assessment of the impact of weather variability on farming activities.

References

Adejuwon, S. A. (2004). Impacts of climate variability and climate change on crop yield in Nigeria. In *Proceedings of the Conference Centre Workshop on Assessment and Adaptation of Impacts of Climate Change (ICC'2004)*, Obafemi Awolowo University, Ile-Ife, pp. 271–279.

Adejuwon, J. O., Balogun, E. E., & Adejuwon, S. A. (1990). On the annual and seasonal patterns of rainfall fluctuants in Sub-Sahara West Africa. *International Journal of Climatology, 10,* 839–949.

Adejuwon, J. O., & Odekunle, T. O. (2006). Variability and the severity of the "little Dry Season" in southwestern Nigeria. *Journal of Climate, 19,* 483–493.

Anderson, R. L. (2000). Ecology & interference of Proso Millet in semi-arid corn. *Weed Technology, 14,* 45–50.

Ayanlade, S., & Odekunle, T. (2006). Application of GIS to assess rainfall variability impacts on crop yield in Guinea Savanna part of Nigeria. *Research Journal of Environmental and Earth Sciences, 2*(1), 19–30. (2010).

Chen, C., McCarl, B. A., & Schimmelpfennig, D. E. (2004). Yield variability as influenced by climate: A statistical investigation. *Climatic Change, 66*(2004), 239–261.

ESRI (2006). Introduction to ARC View GIS. Environmental system research institute. Inc. USA.

Gbuyiro, S. O., Lamin, M. T., & Ojo, O. (2002). Observed characteristics of rainfall over Nigeria during ENSO years. *Journal of Nigeria Meteorological Society, 3*(1), 1–17.

Iken, J. E., Amusa, N. A., & Obatobu, V. O. (2002). Nutrient composition and weight evaluation of small newly developed maize varieties in Nigeria. *Journal of Food Technology Africa (Kenya), 7,* 25–35.

IPCC. (2001). Climate change: The scientific basis. In J. T. Houghton, Y. Ding, D. J. Griggs, M. Noguer, P. J. van der Linden, X. Dai, K. Maskell, & C. A. Johnson (Eds.), *Contribution of working Group I to the third assessment report of the intergovernmental panel on climate change.*

IPCC. (2002). Climate change 2001: Synthesis report. In R. T. Watson (Ed.), *A contribution of working Group I, II and III to the third assessment report of the intergovernmental panel on climate change* (pp. 239–397). Cambridge, UK: Cambridge University Press.

Kufoniyi, O. (2004). *Development of geospatial data infrastructure in Nigeria.* A Paper presented at the National Workshop on Satellite Remote Sensing (Nigeria Sat–1) and GIS: A solution to sustainable National Development Challenges, 15th–17th June 2004, Le Meridien Hotel Abuja, p. 4.

Leff, B., Ramankutty, N., & Foley, J. A. (2004). Geographic distribution of major crops across the world. *Global Biogeochemical Cycles, 18.* https://doi.org/10.1029/2003gb002108. ISSN: 0886-6236.

Odekunle, T. O., Orinmoogunje, I. O. O., & Ayanlade, A. (2007). Application of GIS to assess rainfall variability impacts on crop yield in Guinean Savanna part of Nigeria. *African Journal of Biotechnology, 6.*

Odgarrd, M. V. (2001). Climatic and non-climatic drivers of spatiotemporal maize-area dynamics across the northern limit for maize production.

Oguntunde, P. G., Abiodun, B. J., & Gunnar, L. (2012). Spatial and temporal temperature trends in Nigeria, 1901–2000. *Meteorology and Atmospheric Physics, 118,* 95–105.

Ojo, O. I., & Olawale, S. O. (2014). Assessment of weather variability impact on Cassava Yield in South Western Nigeria. *LAUTECH Journal of Engineering & Technology, 9*(2). ISSN: 1597-0000.

Okaka, C. E. (1999). Helminth parasites of some tropical freshwater fish from Osse River in Benin, southern Nigeria. *Tropical Freshwater Biology, 8,* 41–48.

Omogbai, B. E. (2010). Rain days and their predictability in south-western region of Nigeria. *Journal of Human Ecology, 31*(3), 185–195.

Sowunmi, F. A., & Akintola, A. J. O. (2009). Effect of climatic variability on maize production in Nigeria. *Research Journal of Environmental and Earth Sciences, 2*(1), 19–30.

Olumuyiwa Idowu Ojo is a senior lecturer at Ladoke Akintola University of Technology, Ogbomoso, Nigeria. He is a trained agricultural water engineer and researcher, has supervised or co-supervised projects and theses and has in excess of 35 publications to his credit. His research areas include application of remote sensing and GIS for irrigation water management, land use and land suitability/capability evaluation, erosion mapping, waste/wastewater management and climate change impacts on the environment.

Johnson Olawale Olonnu is an agricultural engineer and postgraduate student at Ladoke Akintola University of Technology, Ogbomoso, Nigeria, and works on evaluation of climate change impacts on some crops.

Masengo Francois Ilunga is a professor in Civil and Chemical Engineering Department at University of South Africa (UNISA). He is also the founder and leader of the UNISA Water Research Flagship. He is involved in teaching water engineering-related subjects at UNISA. His current research interests cover artificial intelligence, entropy, statistical techniques and modelling in hydrology and water resources and climate change issues.

Climate Change Adaptation for Russian Cities: A Case Study of the Thermal Comfort Assessment

Natalia Shartova and Pavel Konstantinov

Abstract The aim of this paper is to discuss the possibility of creating software for the automated assessment of bioclimatic comfort in Russian cities. This research is relevant because 74% of the population in Russia live in cities, and these areas require the development of primarily standardized methods for the assessment of bioclimatic comfort. To select the bioclimatic indices that are suitable for use across the territory of Russia, a critical analysis of the existing and applied bioclimatic indices used in Russia and worldwide was carried out. Based on the results of the analysis, the three most relevant indices were selected, and software was developed for their calculation. Using this software, the evaluation of bioclimatic comfort in 112 large Russian cities (with populations over 100,000) was conducted for 2010–2014, and the results are displayed in map view for cartographical analysis. These results will be the basis for further research on categorizing Russian cities according to their level of thermal comfort and developing climate change adaptation measures.

Keywords Climate change · Bioclimatic comfort · Urban centres

1 Introduction

Climatic variables impact all living organisms, the health of the human population and various aspects of population human activity (McMichael et al. 2006; Costello et al. 2009). One way of assessing the impact of climate change on health and human activity is the evaluation of bioclimatic conditions. The most relevant bioclimatic condition is the assessment of the bioclimatic comfort in cities, where a significant percentage of the population lives.

N. Shartova (✉) · P. Konstantinov
Faculty of Geography, Lomonosov Moscow State University, Leninskye Gory-1, Moscow 119991, Russia
e-mail: shartova@yandex.ru

© Springer International Publishing AG, part of Springer Nature 2019
W. Leal Filho and R. Leal-Arcas (eds.), *University Initiatives in Climate Change Mitigation and Adaptation*, https://doi.org/10.1007/978-3-319-89590-1_15

Currently, more than 100 indices have been introduced to assess the bioclimatic comfort of human activity (Blazejczyk et al. 2012); however, most are used sporadically for specific purposes. Many papers have presented detailed studies on the theory and methodology of assessment using various indices (Fanger 1970; Landsberg 1972; Driscoll 1992; Parson 2003; Budd 2008; Frohlich and Matzarakis 2013).

There are numerous approaches to define bioclimatic indices. In Russia, the classification of indices is primarily based on a combination of meteorological variables commonly used for these types of calculations, e.g. temperature and humidity; temperature and wind; and temperature, humidity and wind (Isaev 2003).

Worldwide, bioclimatic index classification is usually based on one of the following methods of index calculation:

1. Indices based on direct measurements of environmental variables and their analysis are called "direct indices". This group of variables is evaluated for a combined effect on the human body, using a number of meteorological values such as air temperature, wind speed and ambient humidity.
2. Indices based on human body heat balance models are predominantly called "rational indices". Heat balance models account for numerous aspects of heat transfer mechanisms.

The first group of indices based on the monitoring of environmental variables is more convenient and more practical to implement than the second group of indices. Indices in the second group are difficult to implement, since they depend on many variables, some of which require invasive measurements; however, their application produce more accurate results.

Russian studies featuring the application of bioclimatic indices cover practically every region in Russia. The following research findings have been presented at the Federal District level: Andreev (2009) for the Southern Federal District; Isaeva and Perevedentsev (2010) and Merkulov et al. (2011) for the Volga Federal District; and Krasnoschekov et al. (2013) for the Central Federal District. Many studies targeting the assessment of bioclimatic comfort for single subjects have been undertaken in recent years and include every region within the Central Federal District (Fedotov et al. 2001), Altay Krai (Arkhipova et al. 2005), the Republics of Adygea (Varshanina and Mitusov 2006), Udmurtia (Isaeva 2009), Tatarstan (Perevedentsev et al. 2016), Tyva (Erdyneeva 2010), the Jewish Autonomous Region (Grigorieva 2010), the Belgorod (Akimov 2012), Voronezh (Dobrynina et al. 2013), Irkutsk (Latysheva et al. 2014) and Murmansk (Vinogradova 2015) regions and Moscow (Shartova et al. 2015). Several studies have assessed the comfort in resort territories (Litvinov et al. 2014; Andreev and Popova 2015). Research into the bioclimatic environment of urban territories, including large cities in Russia, is seldom found in the literature (Regions and Cities…, 2015). However, assessment of bioclimatic comfort in populous cities is currently the most urgent issue.

2 Methodology

Since 74% of the Russian population live in cities, software development for automated assessments of bioclimatic comfort in Russian cities is an urgent task. Usage of this software would allow us to obtain the analytical material for further research into the field of categorizing Russian cities using the level of comfort and allow for the development of adaptation measures for extreme temperature conditions.

The following steps have been specified to attain these goals:

1. The bioclimatic indices that are used worldwide and in Russia are critically reviewed. Indices are selected to assess the urban bioclimatic comfort/discomfort regarding the climatic environment in Russia, based on the availability of relevant meteorological data.
2. Software is developed for the calculation of bioclimatic indices for the cities in Russia.
3. The software is validated and a preliminary cartographic analysis of the results is obtained.

2.1 *Bioclimatic Indices*

The research methodology in this study is based on the calculation of bioclimatic indices that is most common in practice worldwide. A literature review was conducted to select the bioclimatic indices. According to this study and the availability of the methods being developed for various, and in some cases limited, meteorological datasets, three indices have been chosen to define the thermal comfort/discomfort environment. On the one hand, these indices may, with the utmost objectivity, reflect both the discomfort and extreme environment under high temperatures, such as the *Heat Index* (*HI*) and *Humidex* (*HU*), or low temperatures, such as the *Wind Chill Temperature* (*WCT*); on the other hand, they are the most simple and versatile calculation methods. In addition, these indices have become applicable in countries featuring contrasting climatic environments similar to that in Russia, that is, the USA and Canada.

The choice of two thermal indices is stipulated by hot environments exposing a human body to significant stress, especially for those living in temperate climate environments. Therefore, the correlation between the population death rate and heat waves has been observed to be stronger than the correlation between the population death rate and cold waves (Revich et al. 2015). In this respect, it seems practical to test several thermal indices in the territory of the Russian Federation.

The HI is widely used in the USA and combines the effect of air temperature and relative humidity to determine an apparent temperature, i.e. the human body thermal sensation. The HI Eq. (1) (Rothfusz 1990) is derived from a multiple regression analysis of air temperature and relative humidity. At high humidity, the water evaporation rate is reduced and thus heat is removed from the body at a lower rate, causing the body to retain more heat than it would in dry air. The HI is valid for daily average air temperatures above 20 °C. Its values are expressed in degrees Celsius, and environmental discomfort is designated for index values above 27 °C.

$$\begin{aligned}\mathrm{HI} = &-8.784695 + 1.61139411 * T + 2.338549 * \mathrm{RH} - 0.14611605 * T * \mathrm{RH} \\ &-1.2308094 * 10^{-2} * T^2 - 1.6424828 * 10^{-2} * \mathrm{RH}^2 + 2.211732 * 10^{-3} * T^2 * \mathrm{RH} \\ &+7.2546 * 10^{-4} * T * \mathrm{RH}^2 - 3.582 * 10^{-6} * T^2 * \mathrm{RH}^2\end{aligned} \tag{1}$$

T is air temperature in °C, and RH is relative humidity rounded to its integer value in %.

The HU is the heat stress index that has been used in Canada since 1965 but was later revised to its present status (Masterson and Richardson 1979). Its application is identical to the HI, i.e. to determine how an average human feels in hot and humid weather. The HU combines air temperature (T, in °C) and air vapour pressure (vp, in hPa) into one index to reflect the perceived temperature (2). Discomfort is designated for index values above 30 °C.

$$\mathrm{HU} = T + 0.5555 * (\mathrm{vp} - 10) \tag{2}$$

where $\mathrm{vp} = 6.11 * e^{\left(5417.753 * \left(\frac{1}{273.16}\right) * \left(\frac{1}{273.16 + \mathrm{td}}\right)\right)}$ and td is dew point temperature (in °C).

Currently, cold environment indices are less developed than the warm environment indices. The WCT index compares the cooling effect to the wind environment assessment (Brauner and Shacham 1995). This index was developed based on the assessment of human skin dry heat loss during exposure to low temperatures. It is calculated for air temperature below 0 °C, and environmental discomfort is designated for index values below −35 °C. The formula differs for low (3) and high wind speeds (4).

WCT:

$$\begin{aligned}&\text{For low wind speeds (wind speeds} < 5\ \mathrm{km\ h^{-1}}) \\ &13.12 + 0.6215 * T - 11.37 * v_{10}^{0.16} + 0.3965 * T * v_{10}^{0.16};\end{aligned} \tag{3}$$

$$\begin{aligned}&\text{For high wind speeds (wind speeds} > 5\ \mathrm{km\ h^{-1}}) \\ &T + \frac{-1.59 + 0.1342 * T}{5} v_{10}\end{aligned} \tag{4}$$

T is the air temperature from the dry bulb thermometer (°C), and v_{10} is the wind speed recorded at the 10-m standard anemometer height (km h^{-1}).

Thus, in this study, three bioclimatic indices were chosen to evaluate the territory of Russia. The results obtained from these indices can be compared with the results for other cities worldwide, which will further contribute to the development of adaptation measures.

2.2 *Study Area and Data*

A total of 177 Russian cities with a population of more than 100,000 were chosen as the urbanized territories for the bioclimatic comfort/discomfort assessment. The location of the meteorological station providing the meteorological data for a particular city was verified for each city. Verification entailed selecting 115 cities with meteorological stations located within the city area and reporting the urban area temperature regime. For 22 cities, the meteorological stations are located farther away (5–10 km from the city, depending on the settlement size). These cities have also been incorporated into the study, assuming that the nearest meteorological stations report the urban area temperature regime on a conditional basis.

Meteorological data were collected in 137 selected cities. This study used station daily air temperatures (°C), relative humidity (%) and wind speed recorded at the 10-m standard anemometer height (km h^{-1}) from the archive of the All-Russian Research Institute of Hydrometeorological Information—World Data Center, http://meteo.ru, during the period 2010–2014.

For 25 cities, more than 10% of the entire meteorological data were not recorded, and these cities were excluded from the research. Thus, 112 cities in Russia were included in the research, following the initial data evaluation. Moscow is the largest urbanized territory in Russia; therefore, we decided to use the data from two meteorological stations, Balchug (in the city centre) and VDNKh (north of the city).

2.3 *Software and Techniques*

Calculations of the above indices for the selected cities of this study resulted in the establishment of an automated algorithm for the calculation of bioclimatic indices, which was implemented as software, based on DBMS MS Visual FoxPro 9.0 (Fig. 1).

The algorithm operates in stages. During the first stage, the initial meteorological data are loaded, verified and processed. If necessary, the loaded meteorological data may then be deleted. Tables with the calculated index values are created over the subsequent stages:

1. The hourly index value tables are calculated per day of observation for each of the 8 periods of meteorological observation.
2. The daily index value tables are calculated as the mean index value per day of observation. The values are averaged as follows: each day is graded against the maximum index value for a specified day. According to the HI and HU, the daily averaging is introduced separately for daytime and night-time (night-time is 10.00 p.m. to 09.00 a.m.).

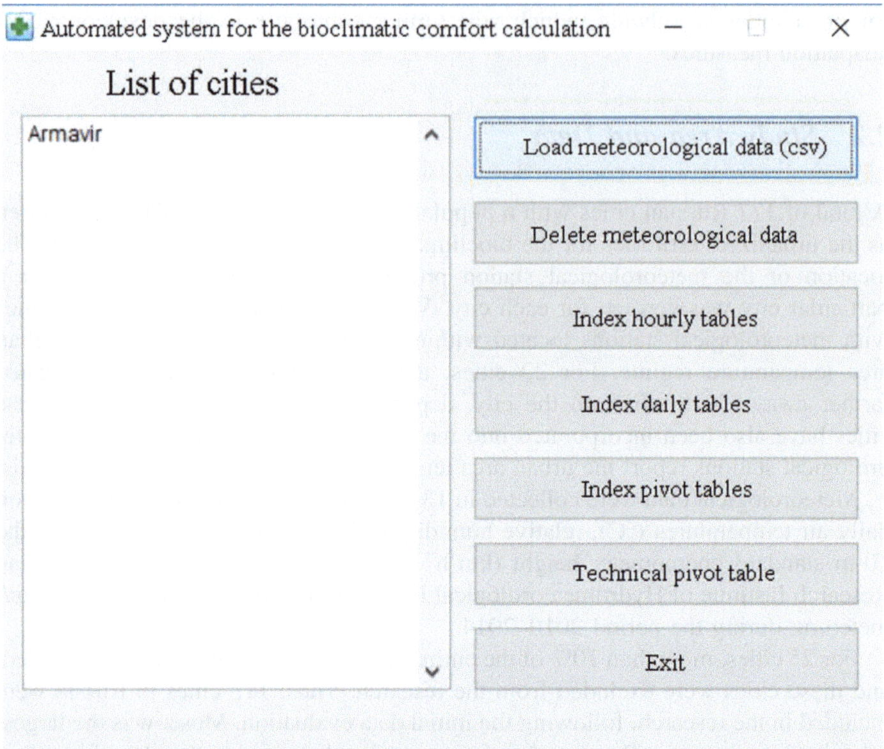

Fig. 1 Interface of the software for the bioclimatic comfort calculation

3. The index pivot tables are then calculated, which indicate the total discomfort days by month, the corresponding index value and their relative values over the entire period of study.
4. The pivot table provides technical information about the years that have been loaded into the database and the days with missing meteorological data.

After processing the data for the entire set of cities included in this study using the automated algorithm, we created 112 index pivot tables over the period 2010–2014. Subsequently, we transformed these tables into a single database to be used in the QGIS Geographic Information System. Legends for the thermal discomfort summary maps for each bioclimatic index analysed in this study have been derived from natural intervals.

3 Results and Discussion

A primary analysis of the overall heat and cold discomfort results is outlined below.

3.1 Assessment Provided by the Heat Index

The maximum number of days of discomfort has been calculated for 12 cities, according to a HI above 27 °C (Fig. 2). This is a predominant index for characterizing the cities in the south of European Russia and the Urals, including Armavir, Astrakhan, Balakovo, Volgograd, Derbent, Kamyshin, Krasnodar, Orenburg, Orsk, Rostov-on-Don, Saratov and Taganrog. In these cities, the number of days of discomfort during the five-year period was 483–256 days, and the average was approximately 100 days of discomfort per year. Comfort analysis results for cities in the North Caucasus are particularly noteworthy. For instance, Kislovodsk and Nevinnomysk most likely have a terrain cooling effect and thus feature only 60 and 45 days of thermal discomfort, respectively.

Moscow clearly stands out among the temperate zone cities, with 182 days of discomfort. Moscow may be comparable to cities located much further south, including Belgorod, Voronezh, Lipetsk, Penza and Tambov. Therefore, the capital of Russia features a severe heat island effect, compared with Saint Petersburg, with only 87 heat discomfort days.

It is worth mentioning here that the number of days of discomfort for any year may vary significantly within a single month. For example, 30 days of discomfort were calculated in Moscow in July of 2010, 24 in 2011, 15 in 2012, seven in 2013, and one in 1 in 2014.

A notable number of 30-day annual averages of discomfort were calculated in the cities of southern Siberia such as Abakan, Irkutsk, Kyzyl and Ulan-Ude (178–119 days). Among cities in the Far East, Khabarovsk has a rather high number of

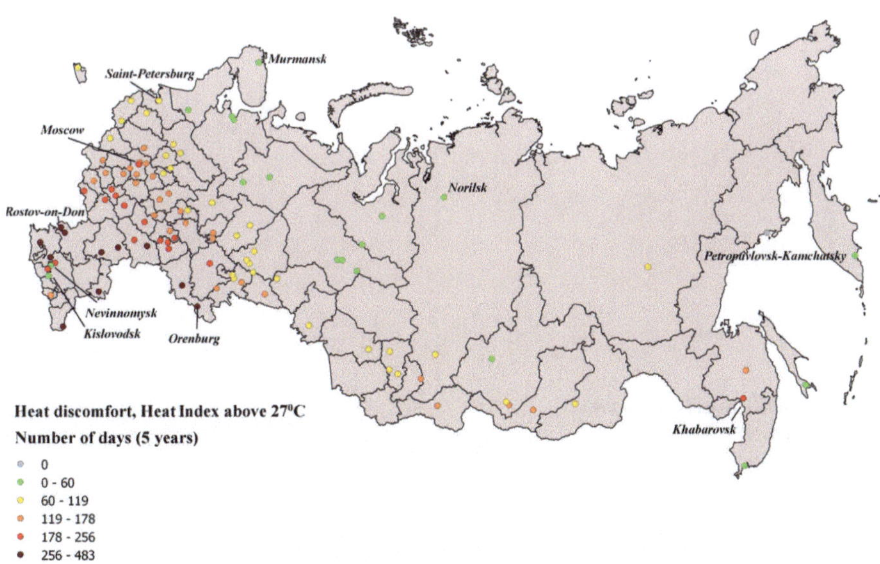

Fig. 2 Heat discomfort as described by the Heat Index, 2010–2014

days of discomfort, at 186 days. Over the period studied, only one day of discomfort was calculated in Petropavlovsk-Kamchatsky, eight in Murmansk and fifteen in Norilsk.

3.2 *Assessment Provided by the Humidex*

According to the HU, heat discomfort is most frequent for the six cities in southern Russia (Armavir, Astrakhan, Derbent, Krasnodar, Rostov-on-Don and Taganrog), with 496–324 days with a HU value above 30 °C (Fig. 3).

The group of cities with the second highest number of days of discomfort (324–216 days) includes Balakovo, Volgograd, Voronezh, Kamyshin, Orenburg, Stavropol, Saratov and Khabarovsk.

The group of cities with days of discomfort varying from 216 to 142 days primarily includes cities in European Russia.

The largest group of cities in the territory of Russia includes 41 cities with 142–86 days of discomfort. The least number of discomfort days was calculated in Murmansk, Norilsk (Russian Arctic) and Petropavlovsk-Kamchatsky (Far East).

The HI and HU results were compared, and these indices defined an equal number of days of discomfort for 50% of the cities analysed in this study (the difference in the number of days does not exceed 10 over five years). For 35% of the cities, the Heat Index stipulates a higher number of days of discomfort, and the difference may reach 80–85 days. This result is mainly characteristic of the cities in

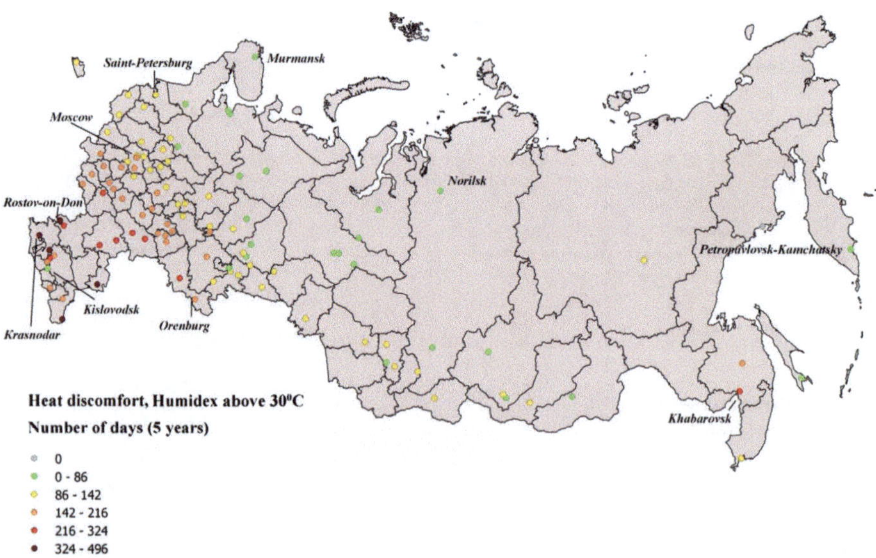

Fig. 3 Heat discomfort as described by the Humidex, 2010–2014

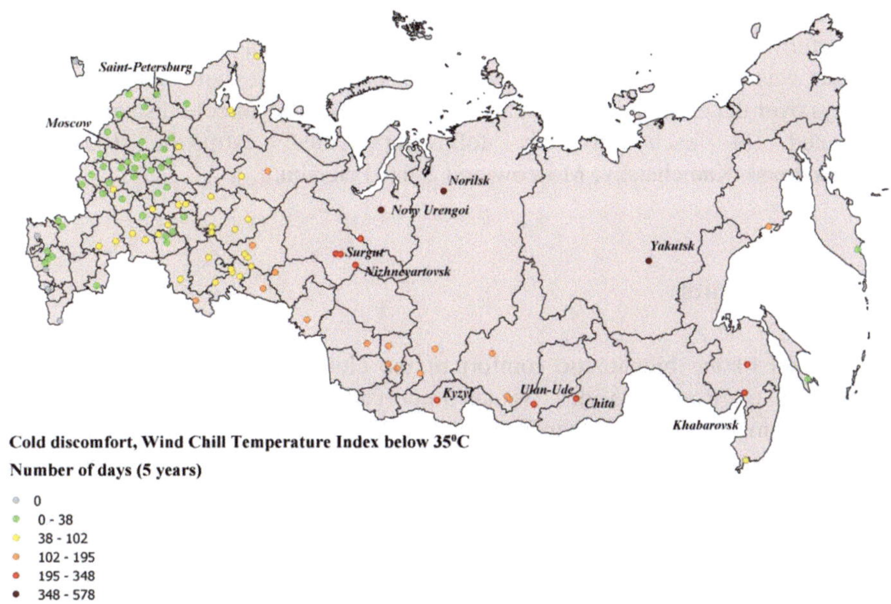

Fig. 4 Cold discomfort as described by the Wind Chill Temperature Index, 2010–2014

southern European Russia and Siberia. For 15% of the cities, the Heat Index calculates a lesser number of days of discomfort while the maximum difference is 58–66 days; these cities are mainly in Northwest Russia, the North Caucasus and the Far East.

Thus, any of the above indices may be used to assess the thermal discomfort in Central Russia. A more detailed analysis is required to choose the optimal index for hotter southern regions and more humid territories in the North Caucasus and the Far East.

3.3 Assessment Provided by the Wind Chill Temperature Index

The cold discomfort assessment using the WCT index clearly differentiates the cities by the number of days below the level of discomfort (Fig. 4). The cold discomfort (below 35 °C) is most vividly observed in Novy Urengoi, Norilsk, Noyabrsk and Yakutsk, where the number of days of thermal discomforts amounted to 578–348 days.

The cities with the second highest rated level of cold discomfort are located in western Siberia (Nefteyugansk, Nizhnevartovsk and Surgut), in southern Siberia (Kyzyl, Ulan-Ude, Chita), and the Far East (Komsomolsk-on-Amur, Khabarovsk), with 348–195 days of cold discomfort.

On average, over the 5 years studied, the cities in Siberia have 195–102 days of discomfort, and those in the Urals have 102–38 days. The lowest level of discomfort was calculated in the cities of European Russia, with a maximum of 38 days over the entire period of analysis. Only a few days of cold discomfort were calculated in each of the following cities: Armavir, Cherkessk, Petropavlovsk-Kamchatsky, Moscow and Saint Petersburg.

4 Conclusions

Assessment of the bioclimatic comfort of the cities, where the majority of the population of Russia resides, is an important task that can contribute to the development of climate change adaptation measures for the thermal comfort of the population. We developed software for calculation of bioclimatic indices as a tool to produce analytical data.

The overview of bioclimatic indices widely used in Russia highlighted their incompatibility with the global indices. This study is the first attempt to use global indices, namely the Heat Index, Humidex and Wind Chill Temperature, for evaluating thermal discomfort in Russia. The results obtained can be compared with the results for other cities in the world. These indices showed an adequate result for their application across Russia. A more detailed analysis is required to find the optimal index for hotter southern regions and more humid territories in the North Caucasus and the Far East.

Based on the validation of the newly developed software, we were the first to assess the day frequency of thermal discomfort in large cities in Russia during warm and cold seasons from 2010 to 2014. The preliminary analysis identified the most extreme cities, in terms of hot and cold climates. It is necessary to first develop adaptation measures to climate change for the cities that endure the extremes of thermal discomfort. For example, Derbent and Taganrog are identified as cities with predominantly unfavourable meteorological conditions, in terms of heat discomfort, with thermal discomfort for 75% of the days in July, as well as other cities in south Krasnodar (65%), Astrakhan (61%), Rostov-on-Don (60%), Armavir (43%). In terms of cold discomfort, the Arctic cities of Yakutsk (97% days of thermal discomfort in January), Norilsk (92%), Novy Urengoi (86%) and Noyabrsk (79%) are identified as cities with predominantly unfavourable meteorological conditions.

The obtained results will form the basis for further research on the categorization of Russian cities by level of comfort and the development of climate change adaptation measures.

Acknowledgements The research has been supported by the grant programme from the Russian Science Foundation (Project № 17-77-20070 "An initial assessment and projection of the bioclimatic comfort in Russian cities in XXI century against the context of climate change").

References

Akimov, L. M. (2012). Analysis of stiffness of climatic conditions of the Belgorod region in the winter, Vestnik of Voronezh State University. *Series: Geography, Geoecology, 2*, 76–80 (in Russian).

Andreev, S. S. (2009). Climatic comfort the southern Federal district and its zoning according to the calculated values of pathogenicity index and ratio-purification capacity of the atmosphere. *Meteorology and Hydrology, 8,* 100–105. (in Russian).

Andreev, S. S., & Popova, E. S. (2015). Estimation of climatic comfort coastal areas on the example of the city of Tuapse, Vestnik of Saint Petersburg University. *Series 7: Geology, Geography, 4*, 145–150 (in Russian).

Arkhipova, I. V., Lovetskaya, O. V., & Rotanova, I. N. (2005). Medico-geographical assessment of climatic comfort of the Altai territory. *Computational Technologies, 10,* 79–86. (in Russian).

Blazejczyk, K., Epstein, Y., Jendritzky, G., Staiger, H., & Tinz, B. (2012). Comparison of UTCI to selected thermal indices. *International Journal of Biometeorology, 56*(3), 515–535. https://doi.org/10.1007/s00484-011-0453-2.

Brauner, N., & Shacham, M. (1995). Meaningful wind chill indicators derived from heat transfer principles. *International Journal of Biometeorology, 39,* 46–52. https://doi.org/10.1007/BF01320893.

Budd, G. M. (2008). Wet-bulb globe temperature (WBGT)—Its history and its limitations. *Journal of Science and Medicine in Sport, 11*(1), 20–32. https://doi.org/10.1016/j.jsams.2007.07.003.

Costello, A., Abbas, M., Allen, A., Ball, S., Bell, S., Bellamy, R., et al. (2009). Managing the health effects of climate change. *Lancet, 373*(9676), 1693–1733. https://doi.org/10.1016/S0140-6736(09)60935-1.

Driscoll, D. M. (1992). Thermal comfort indexes. *Current Uses and Abuses, National Weather Digest, 17*(4), 33–38.

Dobrynina, I. V., Akimov, L. M., & Kurolap, S. A. (2013). Medico-geographical assessment of climatic comfort in Voronezh region, Vestnik of Voronezh State University, *1*, 120–128 (in Russian).

Erdyneeva, L. S. (2010). Estimation of influence of climatic factors on health of the population in Tyva Republic, Bulletin of V.P. *Astafiev Krasnoyarsk State Pedagogical University, 3*, 263–268 (in Russian).

Fanger, P. O. (1970). *Thermal comfort. Analysis and application in environment engineering* (p. 244). Copenhagen: Danish Technical Press.

Fedotov, V. I., Satula, K. S., & Nesterov, Y. A. (2001). Regional model comfort maps of natural environment, Vestnik of Voronezh State University. *Series: Geography, Geoecology, 1*, 3–15 (in Russian).

Frohlich, D., & Matzarakis, A. (2013). Modeling of changes in thermal bioclimate: Examples based on urban spaces in Freiburg, Germany. *Theoretical and Applied Climatology, 111,* 547–558.

Grigorieva, E. A. (2010). Integrated assessment of environmental and climate conditions in Jewish Autonomous region, Bulletin of RUFP. *Series "Ecology and Life Safety", 3*, 60–70 (in Russian).

Isaev, A. A. (2003). *Ecological climatology* (p. 472). Moscow: Nauchnyi Mir (in Russian).

Isaeva, M. V. (2009). Assessment of the bioclimatic conditions of the Udmurt Republic, Bulletin of Udmurt University. *Series: Biology. Earth Science, 6-2*, 124–130 (in Russian).

Isaeva, M. V., & Perevedentsev, Y. P. (2010). Peculiarities of bioclimatic conditions of the Volga Federal district. *Geographical Bulletin, 2,* 29–37. (in Russian).

Krasnoschekov, A. N., Kulagina, E. J., & Trifonova, T. A. (2013). Assessment of the bioclimatic comfort areas (exemplified by Central Federal district). *Problems of Regional Ecology, 3,* 46–50. (in Russian).

Landsberg, H. E. (1972). The assessment of human bioclimate, a limited review of physical parameters, World Meteorological Organization. Technical Note № 123. WMO, Vol. 331.

Latysheva, I. V., Lashenko, K. A., Potemkin, V. L., Potemkina, T. G., & Astaf'eva, N. V. (2014). Integral bioclimatological indicators in climate research of Irkutsk region for the period 1970–2010. *Biosphere, 6*(6), 265–274. (in Russian).

Litvinov, A. E., Bekukh, Z. A., & Kuznetsov, I. K. (2014). Analysis of bioclimatic potential of some of the resorts in the Western Caucasus. *Proceedings of the Samara Scientific Center of RAS, 16*(5), 286–290. (in Russian).

Masterson, J., & Richardson, F. A. (1979). *Humidex: A method of quantifying human discomfort due to excessive heat and humidity* (p. 45). Downsview, Ontario: Environment Canada.

McMichael, A., Woodruff, R., & Hales, S. (2006). Climate change and human health: Present and future risks. *Lancet, 367*(9513), 859–869. https://doi.org/10.1016/S0140-6736(06)68079-3.

Merkulov, I. P., Varfolomeev, A. F., Merkulova, S. V., et al. (2011). Mapping bioclimatic comfort on the basis of GIS-technologies (on the example of Privolzhsky Federal district), Cartography and Geodesy in the Modern World, Saransk: Ogarev Mordovia State University, pp. 42–52 (in Russian).

Parsons, K. C. (2003). *Human thermal environments: The effects of hot, moderate, and cold environments on human health, comfort and performance* (p. 538). New York, London: Taylor & Francis.

Perevedentsev, Y. P., Malkhazova, S. M., Aukhadeev, T. R., & Shantalinsky, K. M. (2016). Medical and demographic consequences of climate change and the assessment of comfort level of weather-climatic conditions in the Volga federal district. *Geography, Environment, Sustainability, 9*(4), 63–76. http://dx.doi.org/10/24057/2017-9388-2016-9-4-63-76.

Revich, B. A., Shaposhnikov, D. A., Podolny, M. A., Kharkova, T. L., & Kvasha, E. A. (2015). Heat wave in the southern cities of the European part of Russia as factor of premature mortality of the population. *Problems of Forecasting, 2,* 56–67. (in Russian).

Rothfusz, L. P. (1990). *The heat index equation*. NWS Southern Region Technical Attachment, SR/SSD 90–23, Fort Worth, Texas. https://www.weather.gov/media/ffc/ta_htindx.PDF. Last Accessed August 08, 2017.

Shartova, N. V., Krainov, V. N., & Malkhazova, S. M. (2015). Technology of integrated assessment of bioclimatic comfort and air pollution in urban areas. *Ecology and Industry of Russia, 1,* 24–29. (in Russian).

Varshanina, T. P., & Mitusov, D. V. (2006). Bioclimatic zoning of Adygea Republic, Bulletin of the Russian Academy of Sciences. *Geographical Series, 4,* 115–121 (in Russian).

Vinogradova, V. V. (2015). Climatic and bioclimatic conditions in the Murmansk region, Proceedings of the Russian Academy of Sciences. *Geographical Series, 6,* 90–99 (in Russian).

Natalia Shartova was a student in Geography (2005) and a Ph.D. candidate in Geography (2009) at the Lomonosov Moscow State University (Russia). She is currently a Senior Research Associate in the Department of Landscape Geochemistry and Soil Geography, Faculty of Geography, at the Lomonosov Moscow State University. Her main research interests are medical geography and human ecology, particularly in relation to the problems of urban ecology and urban population health. She is the author of 35 scientific publications.

Pavel Konstantinov is an Assistant Professor in the Department of Meteorology and Climatology, Faculty of Geography, at the Lomonosov Moscow State University (Russia). He is a permanent leader of the meteorological winter expedition from Moscow State University to the Arctic Zone (Cola Peninsula, Yamal Region, etc.). His group lead a study of the modern frequency of heat wave events in selected cities, which focused on the air quality, the complex climatology (spatial and temporal variability at the microscale) and the influence on the thermal comfort of inhabitants. He is a member of the International Association for Urban Climate (IAUC) and head of the Russian PEEX office (of the international PEEX research programme).

Integrating Biofertilizers with Conservation Agriculture Can Enhance Its Capacity to Mitigate Climate Change: Examples from Southern Africa

Isaiah I. C. Wakindiki, Mashapa E. Malobane and Adornis D. Nciizah

Abstract Climate change is now a palpable threat to sustainable agriculture in Africa. Whereas some advances have been made in the agricultural sector on climate change adaptation, the same cannot be said of mitigation. Climate-smart approaches have been experimented on and recommended, but stunning results are yet to be achieved. The objectives of this chapter were to examine the contribution of conservation agriculture (CA) to soil organic carbon (SOC) stock and how SOC management could enhance climate change mitigation under southern Africa conditions. Whereas application of CA increased SOC stock by up to 104% under experimental conditions, it was not sustainable in the field because the organic matter input was limited and unreliable. Increased use of inorganic fertilizers, especially nitrogen, has been recommended to boost biomass production. Unfortunately, inorganic fertilizer manufacture and application e.g. urea is associated with increased greenhouse gas (GHG) emission. Instead, this study proposes increased use of biofertilizers namely biological nitrogen fixation (BNF) bacteria and vesicular–arbuscular mycorrhizal (VAM) fungi. This review concluded that biofertilizers would sustainably boost SOC input without the danger of GHG emission and should be a compulsory aspect of CA.

I. I. C. Wakindiki (✉)
School of Agriculture, University of Venda, Private Bag X5050, Thohoyandou 0950, South Africa
e-mail: isaiah.wakindiki@univen.ac.za

I. I. C. Wakindiki · M. E. Malobane
Department of Agriculture and Animal Health, University of South Africa, Private Bag X6, Florida 1710, South Africa
e-mail: malobanem@arc.agric.za

M. E. Malobane · A. D. Nciizah
Agricultural Research Council, Institute for Soil Climate and Water, Private Bag X79, Pretoria 0083, South Africa
e-mail: nciizaha@arc.agric.za

© Springer International Publishing AG, part of Springer Nature 2019
W. Leal Filho and R. Leal-Arcas (eds.), *University Initiatives in Climate Change Mitigation and Adaptation*, https://doi.org/10.1007/978-3-319-89590-1_16

Keywords Greenhouse gas · Biological nitrogen fixation · Vesicular–arbuscular mycorrhizal fungi · Soil organic carbon · Biomass

1 Introduction

Climate change is causing significant rainfall and temperature variations around the world (Solh and van Ginkel 2014). Going forward, global warming will increase the intensity of extreme precipitation events, even in areas where mean rainfall decreases with longer periods between rainfall events (IPCC 2007). Such variations in rainfall patterns and intensity, coupled with rising temperature, affect agriculture adversely (Nciizah and Wakindiki 2014). Therefore, global warming will effectively erode the potential of smallholder rain-fed agriculture to meet the food requirements of the rural populace (Mupangwa et al. 2008).

The limited rainfall and drought in Africa are compounded by the additional momentous challenge of soil degradation (Tully et al. 2015). Soil degradation, especially low fertility, contributes to the chronic poor crop yield in Africa (Lal 1995). For example, cereals yield <1.5 ton ha^{-1} against a potential of >5 ton ha^{-1}, and legumes yield ~0.7 ton ha^{-1} against a potential of up to 3 ton ha^{-1} (Jager et al. 2001). The undesirable consequences of climate change, if unchecked by 2020, will reduce the paltry crop yields in southern Africa by 50% (IPCC 2007). Consequently, soil conservation and fertility restoration must be a priority effort (Barford et al. 2001; Lal 2001).

Calls for sustainable climate-smart strategies (Sithole et al. 2016; Thierfelder et al. 2016) are getting more urgent and frequent, mainly because about 80% of Africans are dependent on the low-yielding, rain-fed agriculture for their livelihood (UNDP 2006). One of the most popular climate-smart agriculture technologies that is being promoted in southern Africa is conservation agriculture (CA). Regrettably, its adoption remains low on smallholder farms in Africa (Corbeels et al. 2014) despite a sustained promotion for more than two decades (FAO 2010).

When CA is applied appropriately, it effectively captures most of the rainfall, thereby enhancing biomass production (Baudron et al. 2012). Due to the numerous other documented advantages of CA, it has heralded a feasible technology for climate change mitigation, especially in smallholder farms (FAO 2010). Whereas it is generally agreed that CA can significantly reduce soil degradation in the long term (Lal 1998), there is no consensus in the short term (Giller et al. 2009). Therefore, application of CA is limited mainly because it is generally difficult to sustain the required biomass input. Consequently, CA's ability to cushion smallholder farmers in southern Africa against the effects of climate change is doubtful.

Therefore, this review had two objectives: to examine the contribution of CA to SOC stock and how SOC management could enhance climate change mitigation by CA under southern Africa conditions.

2 Methodology

A systematic review was conducted to select research done on the application of CA in southern African. Data was collected in studies in the southern African region only. The selected studies compared CA with conventional practice. The search engines ScienceDirect and Google Scholar, with keywords "no-tillage", "conservation agriculture" and "southern Africa", as well as the local African journals in which results would most likely be published were used. A total of 23 publications met the criteria of being suitably focused on CA in southern African and were taken as a representative sample in this review. Only original research papers published in academic journals were selected. Conference proceedings, theses, fact sheets and project reports were excluded.

3 Results and Analysis

3.1 Conservation Agriculture Impact on Soil Organic Carbon Stock in Southern Africa

Table 1 shows the results of some CA experiments in southern Africa. Thierfelder et al. (2013b) observed a 12% increase in SOC stock in Zambia. Likewise, Thierfelder and Wall (2012) reported a 104% SOC increase in Zimbabwe. The increased SOC stock after implementing CA was attributed to site-specific management and conditions such as the level of soil fertility. For this reason, the general recommendation for CA application was criticized (Thierfelder et al. 2016).

Application of CA helps in climate change mitigation by reversing the deleterious effect of GHG emission (FAO 2010). Mchunu et al. (2011) found that the application of no tillage (NT) and residue retention of $\sim$10% reduced SOC loss by 52% compared to conventional tillage (CT) treatment (Busari et al. 2015; Ghosh et al. 2015). Thus, CA enriched the soil at minimal environmental impact. According to Gilbert (2012), enriching soil is the vital strategy to increase crop production in Africa.

Increased SOC stock under CA increases soil aggregate stability (Table 1). In some instances, aggregate stability under CA was up to 168% higher than in CT (Table 1). Increased aggregate stability enhances water infiltration (Nciizah and Wakindiki 2015). Water infiltration in CA was up to 331% higher than in CT (Table 1), thereby significantly increasing the soil profile water content. Because CA can improve soil water conservation (Table 1), it can safely be used as a climate change mitigating strategy (Thierfelder and Wall 2009).

Soil erosion, even at 10% residue retention, was low under CA (Table 1). In general, soil erosion decreased by $\geq$50% under CA compared to CT treatments. Conversely, tillage destroys soil structure and reduces infiltration rate, leading to an increase in run-off and soil erosion (Borie et al. 2006). Zhou et al. (2009) performed

Table 1 Conservation agriculture impact on soil organic carbon stock, aggregate stability and soil water

Location	SOC	Aggregate stability	Infiltration	Water conservation	Soil conservation	Reference
Zimbabwe	CA had 17.3–31.2% higher SOC than CT	CA increased aggregate stability				Nyamadzawo et al. (2009)
Zimbabwe	SOC increased by 46–104% under CA compared to CT	Aggregate stability was 123–168% higher under CA than in CT	CA had 38–65% higher infiltration than CT			Thierfelder and Wall (2012)
Zimbabwe	CA had 31% higher SOC than CT		CA increased water infiltration of up to 331% as compared to CT		Cumulative soil erosion was 32.5–36 t ha^{-1} less in CA compared to CT	Thierfelder et al. (2012)
Malawi, Mozambique, Zambia and Zimbabwe	CA increased SOC		CA increased infiltration by 70–238% compared to CT			Thierfelder et al. (2013a)
Zimbabwe				CA tillage practices increased soil moisture regimes and reduced run-off as compared to CT		Mupangwa et al. (2008)

(continued)

Table 1 (continued)

Location	SOC	Aggregate stability	Infiltration	Water conservation	Soil conservation	Reference
Zimbabwe				Run-off was higher under CT, 104 mm $year^{-1}$, compared to CA treatment, 34–40 mm $year^{-1}$	Soil losses were significantly higher under CT, 34 t ha^{-1} $year^{-1}$, compared to CA, 2 t ha^{-1} $year^{-1}$	Munodawafa and Zhou (2008)
Zimbabwe				CT practice had the highest run-off, 104 mm $year^{-1}$ as compared to CA tillage practices, 34–40 mm $year^{-1}$	Soil loss was higher in CT practice, 34 t ha^{-1}, than in CA tillage practice (mulch ripping and tied ridging), 1.7–2.2 t ha^{-1}	Munodawafa (2007)
Zimbabwe and Zambia			CA had 45–87% higher infiltration than CT	Generally, CA plots had highest soil moisture contents, and reduced run-off	CA (direct seeding, manually dug small basins, traction ripper) reduced soil loss as compared to CT during the study	Thierfelder and Wall (2009)
Zambia	CA increased SOC by 12% compared to CT					Thierfelder et al. (2013b)
South Africa					NT combined with residue return at less than 10% of the total harvested biomass reduced soil loss by 68% as compared to CT treatment	Mchunu et al. (2011)

(continued)

Table 1 (continued)

Location	SOC	Aggregate stability	Infiltration	Water conservation	Soil conservation	Reference
Tanzania				Soil water content was found to be higher in CA practices, 0.269 mm/mm, than in CT, 0.246 mm/mm		Mkoga et al. (2010)
Zimbabwe and Zambia			Infiltration rate was found to be three to five times higher in CA than in CT	CA practices (direct-seeded treatment and an animal-drawn, rip line-seeded) increased soil moisture content as compared to CT		Thierfelder and Wall (2010a)
Malawi			Infiltration rate was found to be high under CA practice, 85 mm h^{-1} compared to CT treatment, 32 mm h^{-1}	CA practice increased soil water content by an average of 22 mm as compared to CT	Sediment run-off was found to be higher in CT treatment, 698–1707 kg ha^{-1} than in CA treatments, 170–489 kg ha^{-1}	TerAvest et al. (2015)
Zamnia			Infiltration was higher in CA than under CT			Thierfelder and Wall (2010b)
Malawi			CA had 18–42% higher infiltration rate than CT			Ngwira et al. (2012)
Tanzania and Zambia				CA increased rainwater productivity compared to CT treatment. Thus, CA serves as water harvesting strategy		Rockström et al. (2009)

SOC Soil organic carbon, *CA* conservation agriculture, *NT* no tillage, *CT* conventional tillage a costs and benefits analysis and found that adoption of CA could increase the net benefit by up to \$300 ha^{-1} compared to the CT system after the cost of eroded soil is taken into account. This finding suggests that CA has greater environmental and economic benefits in areas with high water erosion potential than CT (Zhou et al. 2009).

3.2 Soil Organic Carbon Management to Enhance Climate Change Mitigation Through Conservation Agriculture

Decomposition of soil organic matter (SOM) is a temperature- and moisture-dependent process. Temperature and moisture are significantly influenced by climate change (Solh and van Ginkel 2014). Unlike in wetlands and peatlands, upland mineral soils accelerate decomposition process. If decomposition is fast enough, then it may lead to GHG emissions (Davidson and Janssens 2006) and need for increased application of inorganic fertilizers (Nyamangara et al. 2013; Vanlauwe et al. 2014) and little carbon sequestration. Nevertheless, CA promotes soil aggregate formation (Table 1), which in turn, provide physical protection to SOM and delay decomposition (Wakindiki and Njeru 2017). In such conditions, beneficial fungal hyphae proliferate (Davidson and Janssens 2006). Rhizosphere microbes provide vital root extensions that mine nutrients such as phosphorus and boost biomass production. Furthermore, nitrogen-fixing bacteria if maintained under CA can cut inorganic nitrogen requirements and associated GHG emission. Therefore, biofertilizers should be a compulsory aspect of CA.

3.3 Conservation Agriculture Impact on Crop Yields in Southern Africa

Table 2 shows some crop yields that were obtained under CA. Yields were found to be as high as 56% under CA compared to CT treatments. The increase in yields was attributed to the enhanced SOC stock and associated soil properties and processes (Table 1), and reduced evapotranspiration under CA treatments (Thierfelder and Wall 2012).

3.4 Challenges and Possible Solutions for Sustainable Application of Conservation Agriculture Beyond Field Trials

This section outlines some of the challenges faced by CA adoption beyond trial tests and possible solution under southern African climatic conditions.

Table 2 Impact of conservation agriculture on crop yields in southern Africa

Location	Crop yields	Reference
Zimbabwe	Crop yields were higher under CA by 35–56% than in CT treatment	Thierfelder and Wall (2012)
Zimbabwe	CA increased maize yield by 241–445 kg ha^{-1}	Mafongoya et al. (2016)
Malawi	Under CA, maize yields were 22.1–23.6% higher than in CT	Ngwira et al. (2013)
Malawi, Mozambique, Zambia and Zimbabwe	CA with residue retention increased crop yield, while CA without residue retention depressed yield by 50% compared to CT	Thierfelder et al. (2013c)
Malawi	CA increased crop yield compared to CT	Ngwira et al. (2014)
Mozambique	CA increased crop yield by 12–27% compared to CT	Thierfelder et al. (2016)
South Africa	CA increased maize yield compared to CT	Kutu (2012)
Zambia	Higher maize yield was observed under CA compared to CT	Thierfelder and Wall (2010b)
Malawi	CA increased maize yield by 28–51% compared to CT treatment	Ngwira et al. (2012)

CA Conservation agriculture, *CT* conventional tillage

The use of CA per se by smallholders in southern Africa is limited because it has not been possible to retain the crop residues as recommended (Giller et al. 2009; Valbuena et al. 2012), thereby denying the soil of the valuable SOC stock. The situation is worsened by frequent drought, run-off and soil erosion (Sanchez 2002; Giller et al. 2009). The vicious cycle leads to less SOM in subsequent cropping seasons.

CA requires adequate SOM input (Palm et al. 2013), for optimum crop yields and economic returns (Mupangwa et al. 2007). Approximately 4 t ha^{-1} of organic residues are vital to attain the desired results (Mupangwa et al. 2007). To achieve that amount of organic input, CA in southern Africa will require higher inputs, especially nitrogen fertilizer (Rusinamhodzi et al. 2011). The use of mineral fertilizer to increase biomass under CA in southern African region was also proposed by Nyamangara et al. (2013) and Vanlauwe et al. (2014).

The use of animal manure can serve as a low nutrients' input for smallholder farmers with livestock. Manure contains both macro- and micronutrients that are required by plants (Araji et al. 2001). The use of manure around the world is known to improve yields (Dordas et al. 2008; Martínez et al. 2017), give same yields as mineral fertilizer (Basso and Ritchie 2005) and improve soil fertility (Parham et al. 2002; Hati et al. 2008; Martínez et al. 2017). Liu et al. (2010) found manure treatment to have higher SOC and total nitrogen, 32.7 and 24.5%, than inorganic fertilizer 1.9 and 13.3%. Application of manure improves soil microbial communities (Gong et al. 2009) and reduces sediment run-off (Cabrera et al. 2009). Increase in macro-aggregates was also reported after application of manure (Mikha and Rice 2004; Wortmann and Shapiro 2007). However, besides the prohibitive cost of mineral fertilizers, its manufacture contributes to GHG emissions.

The use of biofertilizers namely nitrogen-fixing bacteria and vesicular–arbuscular mycorrhizal (VAM) fungi as a component of CA is not substantially explored. Biofertilizers can help to build up SOC stock and reduce GHG emission.

4 Conclusion

Detailed studies on the application of CA for climate change mitigation in southern Africa are few and scattered. Therefore, more studies are required in that direction to arrive at global conclusions. Nonetheless, increased decomposition of SOM in CA systems under southern Africa conditions tends to limit the capacity for reduced GHG emission, especially carbon dioxide. Incorporating biofertilizers into CA can enhance its capacity to mitigate climate change and should be a compulsory aspect of CA in future.

Acknowledgements We thank the Department of Science and Technology (DST) and the National Research Foundation (NRF) of South Africa for the doctoral research student bursary to Elvis M. Malobane under the supervision of Prof. Isaiah I. C. Wakindiki and Dr. Adornis D. Nciizah.

References

Araji, A. A., Abdo, Z. O., & Joyce, P. (2001). Efficient use of animal manure on cropland—Economic analysis. *Bioresource Technology, 79,* 179–191.

Barford, C. C., Wofsy, S. C., Goulden, M. L., Munger, J. W., Pyle, E. H., Urbanski, S. P., et al. (2001). Factors controlling long- and short-term sequestration of atmospheric CO_2 in a mid-latitude forest. *Science, 294,* 688–1691.

Basso, B., & Ritchie, J. T. (2005). Impact of compost, manure and inorganic fertilizer on nitrate leaching and yield for a 6-year maize–alfalfa rotation in Michigan. *Agriculture, Ecosystems & Environment, 108,* 329–341.

Baudrona, F., Tittonell, P., Corbeels, M., Letourmy, P., & Giller, K. E. (2012). Comparative performance of conservation agriculture and current smallholder farming practices in semi-arid Zimbabwe. *Field Crops Research, 132,* 117–128.

Borie, F., Rubio, R., Rouanet, J. L., Morales, A., Borie, G., & Rojas, C. (2006). Effects of tillage systems on soil characteristics, glomalin and mycorrhizal propagules in a Chilean Ultisol. *Soil and Tillage Research, 88,* 253–261.

Busari, M. A., Kukal, S. S., Kaur, A., Bhatt, R., & Dulazi, A. A. (2015). Conservation tillage impacts on soil, crop and the environment. *International Soil and Water Conservation Research, 3,* 119–129.

Cabrera, V. E., Stavast, L. J., Baker, T. T., Wood, M. K., Cram, D. S., Flynn, R. P., et al. (2009). Soil and runoff response to dairy manure application on New Mexico rangeland. *Agriculture, Ecosystems & Environment, 131,* 255–262.

Corbeels, M., de Graaff, J., Ndah, T. H., Penot, E., Baudron, F., & Naudin, K. (2014). Understanding the impact and adoption of conservation agriculture in Africa: A multi-scale analysis. *Agriculture, Ecosystems & Environment, 187,* 155–170.

Davidson, E. A., & Janssens, I. A. (2006). Temperature sensitivity od soil carbon decomposition and feedbacks to climate change. *Nature*, *440*, 165–173.

de Jager, A., Onduru, D., Van Wijk, M. S., Vllaming, J., & Gachini, G. N. (2001). Assessing sustainability of low-external input farm management systems with the nutrient monitoring approach: A case study in Kenya. *Agricultural Systems, 69,* 99–118.

Dordas, C. A., Lithourgidis, A. S., Matsi, T., & Barbayiannis, N. (2008). Application of liquid cattle manure and inorganic fertilizers affect dry matter, nitrogen accumulation, and partitioning in maize. *Nutrient Cycling in Agroecosystems, 80,* 283–296.

Food and Agricultural Organisation (FAO) of the United Nations. (2010). *Climate-Smart Agriculture Policies, Practices and Financing for Food Security, Adaptation and Mitigation*. http://www.fao.org/docrep/013/i1881e/i1881e00.pdf.

Ghosh, B. N., Dogra, P., Sharma, N. K., Bhattacharyya, R., & Mishra, P. K. (2015). Conservation agriculture impact for soil conservation in maize-wheat cropping system in the Indian sub-Himalayas. *International Soil and Water Conservation Research, 3,* 112–118.

Gilbert, N. (2012). Dirty poor. *Nature, 483,* 525–527.

Giller, K. E., Witter, E., Corbeels, M., & Tittonell, T. (2009). Conservation agriculture and smallholder farming in Africa: The heretics' view. *Field Crops Research, 114,* 23–34.

Gong, W., Yan, X., Wang, J., Hu, T., & Gong, Y. (2009). Long-term manure and fertilizer effects on soil organic matter fractions and microbes under a wheat-maize cropping system in northen China. *Geoderma*, *149*, 318–324.

Hati, K. M., Swarup, A., Mishra, B., Manna, M. C., Wanjari, R. H., Mandal, K. G., et al. (2008). Impact of long-term application of fertilizer, manure and lime under intensive cropping on physical properties and organic carbon content of an Alfisol. *Geoderma, 148,* 173–179.

IPCC. (2007). Climate change: Impacts, adaptation and vulnerability. In M. L. Parry, O. F. Canziani, J. P. Palutikof, P. J. van der Linden, & C. E. Hanson (Eds.), *Contribution of Working Group II to the Fourth Assessment Report of the Intergovernmental Panel on Climate Change*. Cambridge: Cambridge University Press.

Kutu, F. R. (2012). Effect of conservation agriculture management practices on maize productivity and selected soil quality indices under South Africa dryland conditions. *African Journal of Agricultural Research, 7,* 3839–3846.

Lal, R. (1995). Erosion-crop productivity relationships for soils of Africa. *Soil Science Society of America Journal, 59,* 661–667.

Lal, R. (1998). Soil erosion impact on agronomic productivity and environment quality. *Critical Reviews in Plant Sciences, 17,* 319–464.

Lal, R. (2001). World cropland soils as source or sink for atmospheric carbon. *Advances in Agronomy, 71,* 145–191.

Liu, E., Yan, C., Mei, X., He, W., Bing, S. H., Ding, L., et al. (2010). Long-term effect of chemical fertilizer, straw, and manure on soil chemical and biological properties in northwest China. *Geoderma, 158,* 173–180.

Mafongoya, P., Rusinamhodzi, L., Siziba, S., Thierfelder, C., Mvumi, B. M., Nhau, B., et al. (2016). Maize productivity and profitability in Conservation agriculture systems across agro-ecological regions in Zimbabwe: A review of knowledge and practice. *Agriculture, Ecosystems & Environment, 220,* 211–225.

Martínez, E., Domingo, F., Roselló, A., Serra, J., Boixadera, J., & Lloveras, J. (2017). The effects of dairy cattle manure and mineral N fertilizer on irrigated maize and soil N and organic C. *European Journal of Agronomy, 83,* 78–85.

Mchunu, C. N., Lorentz, S., Jewitt, G., Manson, A., & Chaplot, V. (2011). No-till impact on soil and soil organic carbon erosion under crop residue scarcity in Africa. *Soil Science Society of America Journal, 75,* 1502–1511.

Mikha, M. M., & Rice, C. W. (2004). Tillage and manure effects on soil and aggregate-associated carbon and nitrogen. *Soil Science Society of America Journal, 68,* 809–816.

Mkoga, Z. J., Tumbo, S. D., Kihupi, N., & Semoka, J. (2010). Extrapolating effects of conservation tillage on yield, soil moisture and dry spell mitigation using simulation modeling. *Physics and Chemistry of the Earth, 35,* 686–698.

Munodawafa, A. (2007). Assessing nutrient losses with soil erosion under different tillage systems and their implications on water quality. *Physics and Chemistry of the Earth, 32,* 1135–1140.

Munodawafa, A., & Zhou, N. (2008). Improving water utilization in maize production through conservation tillage systems in semi-arid Zimbabwe. *Physics and Chemistry of the Earth, 33,* 757–761.

Mupangwa, W., Twomlow, S., & Walker, S. (2008). The influence of conservation tillage methods on soil water regimes in semi-arid southern Zimbabwe. *Physics and Chemistry of the Earth, 33,* 762–767.

Mupangwa, W., Twomlow, S., Walker, S., & Hove, L. (2007). Effect of minimum tillage and mulch on maize (*Zea mays* L.) yield and water content of clayey and sandy soils. *Physics and Chemistry of the Earth, Parts A/B/C, 32,* 1127–1134.

Nciizah, A. D., & Wakindiki, I. I. C. (2014). Rainfall intensity effects on crusting and mode of seedling emergence in some quartz dominated South African soils. *Water SA, 40,* 587–595.

Nciizah, A. D., & Wakindiki, I. I. C. (2015). Soil sealing and crusting effects on infiltration rate: A critical review of shortfalls in prediction models and solutions. *Archives of Agronomy and Soil Science, 61,* 1211–1230.

Ngwira, A. R., Aune, J. B., & Thierfelder, C. (2014). DSSAT modelling of conservation agriculture maize response to climate change in Malawi. *Soil and Tillage Research, 143,* 85–94.

Ngwira, A. R., Thierfelder, C., Eash, N., & Lambert, D. M. (2013). Risk and maize-based cropping systems for smallholder Malawi farmers using conservation agriculture technologies. *Experimental Agriculture, 49,* 483–503.

Ngwira, A. R., Thierfelder, C., & Lambert, D. M. (2012). Lambert Conservation agriculture systems for Malawian smallholder farmers: Longterm effects on crop productivity, profitability and soil quality. *Renewable Agriculture and Food Systems.* https://doi.org/10.1017/S1742170512000257.

Nyamadzawo, G., Nyamangara, J., Nyamugafata, P., & Muzulu, A. (2009). Soil microbial biomass and mineralization of aggregate protected carbon in fallow-maize systems under conventional and no-tillage in Central Zimbabwe. *Soil and Tillage Research, 102,* 151–157.

Nyamangara, J., Masvaya, E. N., Tirivavi, R., & Nyengerai, K. (2013). Effect of hand-hoe based conservation agriculture on soil fertility and maize yield in selected smallholder areas in Zimbabwe. *Soil and Tillage Research, 126,* 19–25.

Palm, C., Blanco-Canqui, H., DeClerck, F., Gatere, L., & Grace, P. (2013). Conservation agriculture and ecosystems services: An overview. *Agriculture, Ecosystems & Environment, 187,* 87–105.

Parham, J. A., Deng, S. P., Raun, W. R., & Johnson, G. V. (2002). Long-term cattle manure application in soil I. Effect on soil phosphorus levels, microbial biomass C, and dehydrogenase and phosphatase activities. *Biology and Fertility of Soils, 35,* 328–337.

Rockström, J., Kaumbutho, P., Mwalley, J., Nzabi, A. W., Temesgen, M., Mawenya, L., et al. (2009). Conservation farming strategies in East and Southern Africa: Yields and rain water productivity from on-farm action research. *Soil and Tillage Research, 103,* 23–32.

Rusinamhodzi, L., Corbeels, M., van Wijk, M. T., Rufino, M. C., Nyamangara, J., & Giller, K. E. (2011). A meta-analysis of long-term effects of conservation agriculture on maize grain yield under rain-fed conditions. *Agronomy for Sustainable Development, 31,* 657–673.

Sanchez, P. (2002). Soil fertility and hunger in Africa. *Science, 295,* 2019–2020.

Sithole, N. J., Magwaza, L. S., & Mafongoya, P. L. (2016). Conservation agriculture and its impact on soil quality and maize yield: A South African perspective. *Soil and Tillage Research, 162,* 55–67.

Solh, M., & van Ginkel, M. (2014). Drought preparedness and drought mitigation in the developing world's drylands. *Weather and Climate Extremes, 3,* 62–66.

TerAvest, D., Carpenter-Boggs, L., Thiefielder. C., & Reganold, J. P. (2015). Crop |production and soil water management in conservation agriculture, no-till, and conventional tillage systems in Malawi. *Agriculture, Ecosystems & Environment, 212*, 285–296.

Thierfelder, C., & Wall, P. C. (2009). Effects of conservation agriculture techniques on infiltration and soil water content in Zambia and Zimbabwe. *Soil and Tillage Research, 105,* 217–227.

Thierfelder, C., & Wall, P. C. (2010a). Investigating conservation agriculture (CA) systems in Zambia and Zimbabwe to mitigate future effects of climate change. *Journal of Crop Improvement, 24,* 113–121.

Thierfelder, C., & Wall, P. C. (2010b). Rotation in conservation agriculture systems of Zambia: Effects on soil quality and water relations. *Experimental Agriculture, 46,* 309–325.

Thierfelder, C., & Wall, P. C. (2012). Effects of conservation agriculture on soil quality and productivity in contrasting agro-ecological environments of Zimbabwe. *Soil Use and Management, 28,* 209–220.

Thierfelder, C., Cheesman, S., & Rusinamhodzi, L. (2012). A comparative analysis of conservation agriculture systems: Benefits and challenges of rotations and intercropping in Zimbabwe. *Field Crops Research, 137,* 237–250.

Thierfelder, C., Cheesman, S., & Rusinamhodzi, L. (2013a). Benefits and challenges of crop rotations in maize-based conservation agriculture (CA) cropping systems of southern Africa. *International Journal of Agricultural Sustainability, 11,* 108–124.

Thierfelder, C., Matemba-Mutasa, R., Bundersonb, W. T., Mutenje, M., Nyagumboa, I., & Mupangwa, W. (2016). Evaluating manual conservation agriculture systems in southern Africa. *Agriculture, Ecosystems & Environment, 222,* 112–124.

Thierfelder, C., Mombeyarara, T., Mango, N., & Rusinamhodzi, L. (2013b). Integration of conservation agriculture in smallholder farming systems of southern Africa: Identification of key entry points. *International Journal of Agricultural Sustainability, 11,* 317–330.

Thierfelder, C., Mwila, M., & Rusinamhodzi, L. (2013c). Conservation agriculture in eastern and southern provinces of Zambia: Long-term effects on soil quality and maize productivity. *Soil and Tillage Research, 126,* 246–258.

Tully, K., Sullivan, C., Weil, R., & Sanchez, P. (2015). The state of soil degradation in Sub-Saharan Africa: Baselines, trajectories, and solutions. *Sustainability, 7,* 6523–6552.

UNDP. (2006). Human Development Report 2006. *Beyond scarcity: Power, poverty and the global water crisis*. UNDP, New York.

Valbuena, D., Erenstein, O., Homann-Kee Tui, S., Abdoulaye, T., Claessens, L., Duncan, A. J., et al. (2012). Conservation agriculture in mixed crop–livestock systems: Scoping crop residue trade-offs in Sub-Saharan Africa and South Asia. *Field Crops Research, 132,* 175–184.

Vanlauwe, B., Wendt, J., Giller, K. E., Corbeelsd, M., Gerard, B., & Nolte, C. (2014). A fourth principle is required to define conservation agriculture in sub-Saharan Africa: The appropriate use of fertilizer to enhance crop productivity. *Field Crops Research, 155,* 10–13.

Wakindiki, I. I. C., & Njeru, S. K. (2017). Tillage effects on aggregate-associated organic carbon in some quartz dominated South African soils. *South African Journal of Plant and Soil, 34,* 239–242.

Wortmann, C. S., & Shapiro, C. A. (2007). The effects of manure application on soil aggregation. *Nutrient Cycling in Agroecosystems*. https://doi.org/10.1007/s10705-007-9130-6.

Zhou, X., Al-Kaisi, M., & Helmers, M. J. (2009). Cost effectiveness of conservation practices in controlling water erosion in Iowa. *Soil and Tillage Research, 106,* 71–78.

Isaiah I. C. Wakindiki holds a B.Sc. in Agriculture Honours and M.Sc. in Soil Science from the University of Nairobi, Ph.D. in Soil Science from Egerton University and Post Graduate Diploma in Higher Education and Training from the University of Fort Hare. His other qualification is Advanced Certificate in Irrigation and Soil Management from Volcani Institute. He is a Research Professor of Soil Science at the University of Venda in South Africa. His research interests are in the soil physical properties, processes and conditions that enhance its resilience and sustainable use that promote human well-being and livelihoods.

Mashapa E. Malobane is currently a Ph.D. candidate in Agriculture (Soil Science) at the University of South Africa (UNISA). He holds M.Sc. in Agriculture (Soil Science) from the University of Pretoria where his work was on "Using the organic carbon fractions of the Van Soest method to determine compounds responsible for carbon and nitrogen mineralization from sludge amended soils". He also holds B.Sc. Agriculture (Agronomy) from the University of Limpopo. His research interests are in climate change, soil quality and sustainable biofuel feedstock production.

Adornis D. Nciizah holds a B.Sc. in Crop Production and Horticulture from the Midlands State University and M.Sc. and Ph.D. in Crop Science from the University of Fort Hare. He is currently a Researcher with the Agricultural Research Council–Soil, Climate and Water (ARC-SCW) in Pretoria, South Africa. His research interests are in soil surface structural conditions and properties related to soil quality and early crop growth and soil–climate interactions. His current research work focuses on climate-smart agriculture, sustainable bio-energy feedstock production, and micronutrient seed priming for micronutrient-deficient soils.

Exergy-Based Responsibility Allocation of Climate Change

Hossein Khajehpour, Yadollah Saboohi and George Tsatsaronis

Abstract Climate change as a serious environmental issue has been subjected to many management studies. Yet the issue of allocating global responsibilities to emission reduction in a reasonable way is an unconcluded challenge. Two main groups of responsibility allocation systems are production-based responsibilities (PBR) and consumption-based responsibilities (CBR). PBR takes territorial emissions within borders into account that is applied to global policy makings by the Intergovernmental Panel on Climate Change (IPCC). In contrast, CBR considers the emission transfers through international trade and allocates the responsibilities based on the goods and services being consumed by each country. Both two methods are discouraged by some logics. The PBR does not consider the effects of indirect emissions of imported materials/services. By contrast, in the CBR the consumer is the only one being charged for the emission costs, ignoring the profitability of the emissions for the producers, thus lowering incentives of exporters toward greener production (usually developing nations). Moreover, the calculation of the embodied emissions in the CBR method requires massive interindustry and intercountry trade data. This research aims at defining and applying a new less data-intensive method for the calculation of embodied carbon loads on streams on the one hand and sharing environmental responsibilities (SBRs) among producers and consumers on the other hand. The method is based on thermodynamic quantities to internalize the external environmental damages using the exergy concept. The main notion of the new method is that the traceable exergy destruction could be representative of environmental burdens. Therefore, shared responsibilities motivate net importers (i.e., developed nations) to buy from greener producers and simultaneously push net exporters (i.e., developing nations) toward greener production to reduce their responsibilities. In this study, the global model is simplified into three country categories: developed countries, BRIICS, and rest of

H. Khajehpour (✉) · Y. Saboohi
Energy Engineering Department, Sharif University of Technology, Tehran, Iran
e-mail: Khajehpour@energy.sharif.edu; Khajehpour@campus.tu-berlin.de

H. Khajehpour · G. Tsatsaronis
Energy Engineering and Environmental Protection, Energy Engineering Institute, Berlin Technical University, Berlin, Germany

© Springer International Publishing AG, part of Springer Nature 2019
W. Leal Filho and R. Leal-Arcas (eds.), *University Initiatives in Climate Change Mitigation and Adaptation*, https://doi.org/10.1007/978-3-319-89590-1_17

the world. The emission transfers are calculated using international trade statistics and are applied to the developed shared responsibility accounting model. Shared responsibilities lie between PBR and CBR. However, derived SBR recommends higher contribution of producers to reduce emissions.

Keywords Climate change · Responsibility · Exergy · Trade · Allocation

1 Introduction

Anthropocentric climate change is the consequence of ignoring the environmental capacity to bear greenhouse gas emissions resulting from human activities. Mitigation of the issue requires global action in order to cut the emission levels. In all integrated environmental action plans, after setting a cap on total emissions, the allocation of emission permits and responsibilities in act for reducing the emissions have always been major issues (Khajehpour et al. 2017). More specifically, many factors need to be considered while proposing a responsibility allocation method in the case of climate change: historical contributions to the accumulation of greenhouse gases in the atmosphere, current territorial emission rates, traded carbon load on imports, land-use management, transportation emissions of international trade (Peters et al. 2009), etc. It has been argued that an inadequate consideration of these factors has significantly impacted the participation and effectiveness of global climate agreements such as Kyoto (Peters and Hertwich 2008) and probably Paris.

Among the list of factors, fair allocation of the producer and consumer responsibilities to current global emission rates has been the subject of various studies. This is due to the increasing separation of where the production and consumption of goods and services occur in the world.

In this context, developed responsibility allocation techniques might be categorized into two main groups: production-based responsibilities (PBRs) and consumption-based responsibilities (CBRs). The PBR considers territorial emissions within the borders of each subsystem (i.e., countries), while the CBR recommends the use of embodied emissions in the provision chain of consumed products (carbon footprints). For any subsystem i with the territorial emission level Y_i, the values of PBR and CBR are defined as follows:

$$\mathrm{PBR} = Y_i \tag{1}$$

$$\begin{aligned} \mathrm{CBR} &= \mathrm{PBR} + \mathrm{EE}_{\mathrm{imports}} - \mathrm{EE}_{\mathrm{exports}} \\ &= E_{i,i} = Y_i + \sum_{j \neq i} E_{j,i} - \sum_{j \neq i} E_{i,j} \end{aligned} \tag{2}$$

In these formulations, the CBR of country i is the summation of national territorial emissions (Y_i) and embodied emissions on gross imports $(\mathrm{EE}_{\mathrm{imports}} = \sum_{j \neq i} E_{j,i})$

excluding embodied emissions of exported products ($EE_{exports} = \sum_{j \neq i} E_{i,j}$). Hence, CBR could be defined as the net embodied emissions on consumed products when they are produced either locally or imported from abroad. Figure 1 shows the temporal trend in PBR and CBR worldwide. In this figure, for the sake of simplicity, the world is categorized into three groups of countries: developed (as in Annex B excluding Russia), BRIICS (major developing countries: Brazil, Russia, India, Indonesia, China, and South Africa), and the rest of the world (RoW).

Both methods have been criticized for some reasons. The PBR has been disputed due to the ignorance of the indirect emission loads of imported materials/services. Numerous studies have investigated that the slight decline in emissions from developed countries (Annex B in the Kyoto protocol) since 1990 is the result of shifting the energy-intensive production to developing countries (non-Annex B countries), i.e., carbon leakage (Peters and Hertwich 2008). This shift has resulted in a rapid increase of emissions in these countries and therefore globally. CBR studies, (e.g., Caldeira and Davis 2011; Peters et al. 2011) show that accounting the emission loads on net imported consumptions in addition to the territorial emissions changes the view from a 2% decrease in PBR of the developed world to a more real analytic picture of 7% increase in CBR in the period of 1990–2008. This effect becomes more significant because of the rapid globalization trend (Fig. 1). More specifically, the tripled amount of global trade (Peters et al. 2009) has led to a 77% increase in carbon loads on trade in the same period (Peters et al. 2011). As a result, in 2004, 23% of the total fossil fuel-related emissions resulted from the production of goods which were consumed in a different country (Davis and Caldeira 2010). Therefore, the main drawback of the PBR method is ignoring the increasing effect of carbon leakage and its corresponding policy considerations. For more information about the extent of emission transfers on traded streams (see Davis and Caldeira 2010; Peters et al. 2011; Wiebe and Yamano 2016).

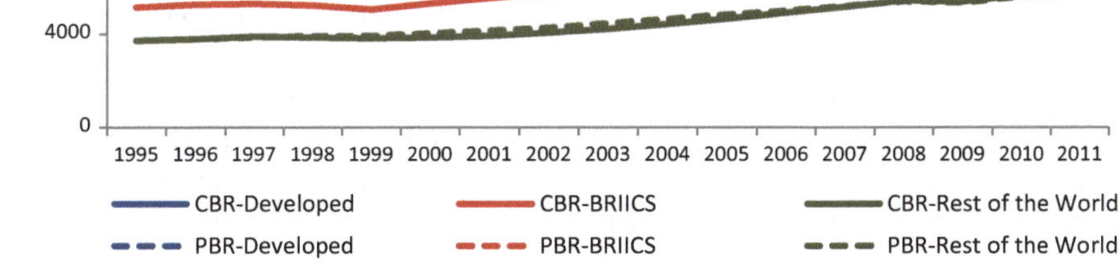

Fig. 1 Increasing difference between PBRs and CBRs, according to the information provided in (Wiebe and Yamano 2016)

On the other hand, the CBR has been also disputed by climate economists as it allocates full responsibilities to the consumers, ignoring the profitability of the emissions for the producers. Therefore, however, fairer from the causality point of view, CBR may decrease the incentives of net exporters (usually developing nations) for greener production. Also, calculation of the CBR requires massive data collection and manipulation to shape the multi-regional input–output (MRIO) tables. A MRIO table includes trade values among any economic sector of any country to any economic sector of the same and other countries. This costly and time-consuming task alongside the need for the sector- and country-specific emission factors could be also a source of the uncertainty in calculation of the CBRs.

The failure of both PBR and CBR methods in the mutually agreed holistic allocation of responsibilities together with other unnoticed factors (specifically historical contributions to the current global warming) has caused a failure in collective global actions against climate change. Therefore, considering the shortcomings of both PBR and CBR, shared responsibility accounting (SBR) methods have been recently evolved by some researchers in regional (Eder and Narodoslawsky 1999; Khajehpour et al. 2017) and global (Bastianoni et al. 2004; Lenzen et al. 2007; Rodrigues and Domingos 2008; Kander et al. 2015) analyses. It is expected that a SBR method will overcome the limitations of the PBR and CBR methods. Therefore, it should consider the emission loads on material/service transfer among units, and in the meantime, it should also be able to assign shares in a fair manner. A SBR method should also satisfy the three verification conditions of *sensitivity*, *monotonicity*, and *additivity* as defined by Kander et al. (2015). *Sensitivity* means that national emission inventories must be responsive to the changes in national policies, e.g., taxation on trade. Satisfying the condition of *monotonicity* avoids national emission reduction policies which increase global emissions. Finally, *additivity* means that the summation of national responsibilities must be equal to the global emissions (Kander et al. 2015). As it is discussed in detail in Kander et al. (2015) and Jakob and Marschinski (2013), neither PBR nor CBR is always verified based on the first two conditions (*sensitivity* and *monotonicity*). Hence, care must be taken when interpreting the outcomes of both CBR and PBR, especially when it comes to national and global policy making (see Jakob and Marschinski 2013 for more details). Rodrigues also discussed different implications of SBR results for environmental policy making (Rodrigues and Domingos 2008).

It is also worth mentioning that in addition to the sharing of responsibilities of upstream production emissions among downstream consumers, in some applications the identification of the producers' responsibilities to downstream consumptions matters as well (Lenzen et al. 2007; Rodrigues and Domingos 2008): For example, the producers of fossil fuel energy carriers have been benefitting from low production costs and relatively high prices exported products (which causes massive emissions outside). Davis et al. estimated that some 37% of global emissions result from fossil fuels traded internationally (Davis et al. 2011). However, this is not addressed in the applied literature and is not the subject matter of this study either. Furthermore, there are other challenges of global climate action plans especially in the identification of worldwide action priorities, the historical contributions to current greenhouse gas concentrations in the atmosphere, emissions in international transportations, effect of land-use managements as carbon sinks, and, finally, the design of appropriate policy tools which are not subject to this study either.

The sole focus of this study is the development and application of a new SBR method which also affords a new approach for the calculation of cumulative carbon loads on produced materials/services. It is shown that the new SBR method satisfies three conditionals of a proper allocation system. The new calculation approach also reduces the data requirements for calculation of the emission loads on the transported streams. This is achievable by implementing the exergy destructions as a metric for allocation of territorial emissions among different national economic activities.

In the rest of this paper, after an introduction to the methodological history of SBR, the method which is developed and used in this research work and its novelty over other SBR methods are presented. In Sect. 3, the modeling stage is explained. In this section, the procedure for application of the method, the defined system, and the data gathering and manipulation are introduced. Then, some main key metrics of the global production, trade, and emissivity of the productions in different subsystems are discussed. In Sect. 4, the results out of the application of the method to the modeled system are presented. This section is followed by a discussion on the results and their validity in Sect. 5. Finally, the key findings of the paper and the limitations of this study are concluded.

2 Methodology

Almost all SBR methods account responsibilities based on either methodological modifications into MRIO analysis (e.g., environmentally extended MRIO in Eder and Narodoslawsky 1999; Lenzen et al. 2007; Rodrigues and Domingos 2008) or processing the MRIO results as inputs to other accounting methods as in Bastianoni et al. (2004), Kander et al. (2015). However, there has been no consensus among scientific researchers about the best (fairest and most precise) way of sharing responsibilities. Furthermore, many of these proposed methods, seemingly due to the lack of access to required input data, have not yet been applied to the question of global climate change (i.e., Bastianoni et al. 2004; Lenzen et al. 2007; Rodrigues and Domingos 2008). Another common challenge of all these methods is also the high complexity of accounting concepts (especially the methods presented in, e.g., Lenzen et al. 2007; Rodrigues and Domingos 2008) which might be problematic with regard to the communication to the global community and climate policy makers.

This research aims at defining and applying a new less data-intensive method for the calculation of embodied carbon loads on streams, on the one hand, and sharing consumer and producer responsibilities concerning the climate change on the other hand. This is done by the so-called extended exergoenvironmental accounting method which has been recently defined and applied by Khajehpour et al. (2017) in an environmental analysis of complex energy systems.

The method is based on physical quantities to internalize the external environmental damages by using the exergy concept. The main notion of the method is that the traceable exergy destruction might be representative for environmental burdens. This argument is widely discussed in our previous work (Khajehpour et al. 2017) as environmental emissions are normally caused by exergy destructions, directly via exergy destructing conversions or indirectly via exergy provisions from upstream to payoff downstream destruction effects.

Aside from its sharing approach, the novelty of the method in estimation of carbon loads on streams avoids the necessity of extensive trade data and emission factor requirements which are needed for CBR studies which are based on the so-called environmentally extended MRIO analysis. Instead, in the proposed method in this research work, the emission loads of traded streams are calculated based on the allocation of national emissions according to exergy costs of different product streams. Hence, there is no need for the sector-to-sector trade data inside and among countries. For a typical analysis of n countries and k sectors, the MRIO-based analysis requires ($nk \times nk$) matrix of trade values, while the proposed method requires ($nk \times n$) dataset (one kth as of the MRIO-based methods). So that the use of this method in this research work reduces almost 97% of the data requirements.

The method considers both emissions within each subsystem and the carbon loads carried into the system through indirect emissions, e.g., imports. The total environmental load of each unit is therefore the summation of the environmental load input to the unit via feed streams plus the territorial emissions (or impacts) which must be divided among products/services being produced in the unit. This allocation is to be done proportionally to the exergy costs of the products. This idea is based on the notion that any environmental impact is due to irreversibilities in its production chain (which is traceable by exergy destructions in the system).

As of the exergoenvironmental method, in the presence of detailed data of the disaggregated subsystems, there would be no need for auxiliary equations. However, in order to avoid extensive data requirements (e.g., for the MRIO), the auxiliary rule of allocation based on exergy costs might be useful. Indeed, the more aggregated the model becomes, the less accuracy can be expected in the derived results (Khajehpour et al. 2017). This "error of aggregation" is also present in any input–output analysis and is generally accepted as a manageable source of uncertainty (Kander et al. 2015).

A general formulation of the proposed extended exergoenvironmental method with new representation from Khajehpour et al. (2017) is as below:

$$E_{i,\mathrm{out}} = E_{i,\mathrm{in}} + Y_i \tag{3}$$

Equation (3) presents the main balance equation for the calculation of the carbon loads on the streams. In this equation, the terms are defined as below:

$$E_{i,\mathrm{out}} = \sum_k E_{i,i,k} + \sum_k \sum_{j \neq i} E_{i,j,k}$$

$$E_{i,\mathrm{in}} = \sum_k \sum_{j \neq i} E_{j,i,k}$$

If we consider the term $E_{i,j,k}$ to be the result of carbon load intensity (e: carbon load per unit amount of the stream) and the quantity of stream (V: usually in monetary terms) ($E_{i,j,k} = e_{i,j,k} \times V_{i,j,k}$), then the auxiliary equation for allocating the responsibilities among jth outgoing stream from subsystem i is defined as:

$$\frac{E_{i,j,k}}{\mathrm{CE} \times C_{i,j,k}} = \frac{e_{i,j,k} \times V_{i,j,k}}{\mathrm{CE} \times C_{i,j,k}} = \frac{e_{i,j,k}}{\left(\mathrm{CE} \times C_{i,j,k}/V_{i,j,k}\right)} = \frac{e_{i,j,k}}{\mathrm{SCE} \times C_{i,j,k}} = \mathrm{cte}. \quad (4)$$

Equation (4) allocates the environmental loads $(E_{i,j,k})$ proportional to the cumulative exergy costs of the streams $(\mathrm{CE} \times C_{i,j,k})$. In other words, this equation allocates the environmental loads per unit monetary values of the streams $(e_{i,j,k})$ proportional to the specific cumulative exergy cost $(\mathrm{SCE} \times C_{i,j,k})$ (cumulative exergy cost per unit value $(\mathrm{CE} \times C_{i,j,k}/V_{i,j,k})$) of the streams. This method of allocation is an alternative to the MRIO analysis in which the emission factors and detailed intersector/-country trade data are being used to track the emissions.

As soon as the carbon loads on streams are calculated, shared responsibilities could be calculated. According to the idea of the embodied energy analysis (*emergy*) which was first introduced by Odum (1996), Bastianoni recommended the use of the normalized environmental loads on total inputs plus territorial emissions $(E_{i,\mathrm{in}} + Y_i)$ as the allocation measure of the responsibilities (Bastianoni et al. 2004): He named this approach "carbon emission added (CEA)." Table 1 shows the responsibility allocation method for a very simplified system:

In comparison with PBR and CBR definitions, the SBR approach presented above suggests the inclusion of imported environmental load on the subsystem responsibilities without excluding the exported environmental loads on products being consumed outside. This is to verify the *sensitivity* and *homogeneity*

Table 1 Comparison of the allocation approaches in a simplified example

Simplified model	Responsibilities	Prod.-based	Cons.-based	CEA
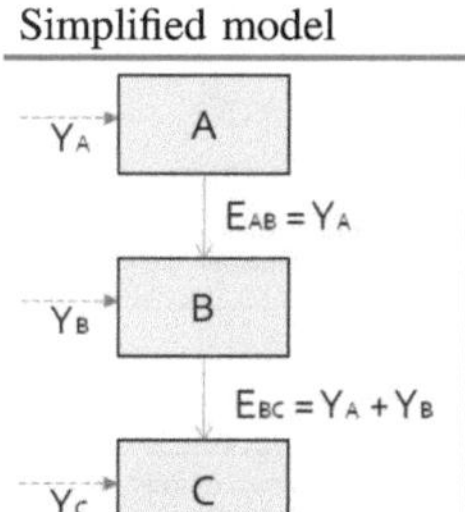	A	Y_A	0	Y_A*
	B	Y_B	0	$(Y_A + Y_B)$*
	C	Y_C	$Y_A + Y_B + Y_C$	$(Y_A + Y_B + Y_C)$*

*To be multiplied by the normalizing factor ((YA + YB + YC)/(3YA + 2YB + YC))

conditions of the allocation. However, in order to verify the *additivity* condition, the proposed SBR must be normalized with a normalizing factor of accumulated emission loads $((YA + YB + YC)/(3YA + 2YB + YC))$.

3 Modeling

3.1 Procedure

Figure 2 depicts the procedure for the application of a SBR method into the global climate change problem. Starting points of this procedure are the choices of the method of calculation of embodied emissions, the responsibility sharing method, and a definition of the system components and streams (products and service types).

After defining the method, system inputs, and boundaries, the input data to the model are gathered. The produced information out of the data manipulation stage must be verified and validated by comparing them with other studies. Then, the corresponding system of equations for the calculation of the emission loads is shaped. In our approach, the results out of the solution of this formulated system present the intensities of emission load per unit of streams $(e_{i,j,k})$. Using the absolute amounts of produced, consumed, and traded products, the emission loads on consumed and transferred streams are derived to be used in the responsibility allocation system. Based on the responsibility allocation system, other input data to the model might also be required [e.g., value added for the sharing method presented by Lenzen et al. (2007) or technical efficiency in Kander et al. (2015)].

3.2 System Definition

The aggregation level in choice of subsystem components and diversity of products/services is a trade-off among accuracy and data/time requirements (Khajehpour

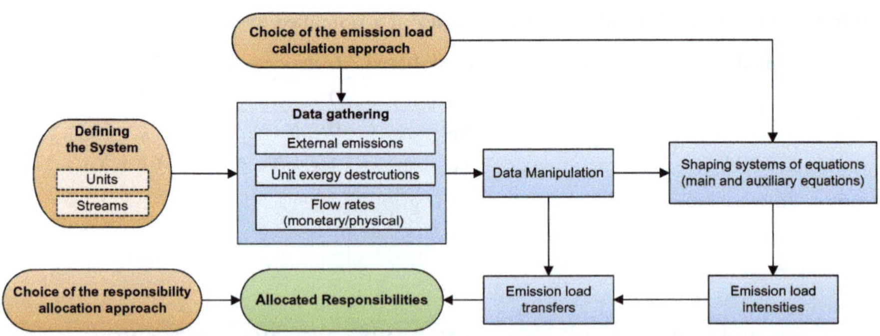

Fig. 2 Procedure for the application of the sharing method to the global climate change

et al. 2017). To address the main challenge of a fair allocation of global responsibilities to the climate change among global North and South, and for the sake of simplicity, the problem is formulated by categorizing the global emitters into three main groups of countries:

1. The developed countries [based on the definition of *developed* nations in the UN World Economic Situation Prospect (WESP) (United Nations 2017b)] which include the Annex B countries of the Kyoto protocol excluding Russia (Developed-D). These countries account for 59 and 66% of global trade in merchandise and service sectors correspondingly (WTO 2017).
2. The major developing countries: Brazil, Russia, India, Indonesia, China, and South Africa (BRIICS-B).
3. The rest of the world (RoW-R). This aggregation has always been used in CBR studies for countries of the rest of the world for which detailed input–output data in global context are either not available or, as in this study, are assumed as a global-averaged single economy for the sake of simplicity. It has been shown that this categorization in CBR studies is a good representation of the climatic effect on the rest of the world (Andrew et al. 2009).

Figure 3 shows the simplified model of the world and the streams being produced, consumed, and transferred worldwide. It should be noted that this categorization undermines the high potential effects of interregional or intercountry trade which could be as high as, e.g., 63% of the total trade in the EU (WTO 2017). However, this effect is only important when the allocation of responsibilities among the countries of any of these groups is to be investigated.

In this picture, the Y_i is representative for the production-based emissions within the territory of subsystem *i*, and E_{ij} is a vector of emission loads on streams produced in *i* and consumed in *j*. As any other emission load accounting study, global production is categorized into a reasonable number of material or service groups. In our study, the products are divided into 34 product groups in accordance

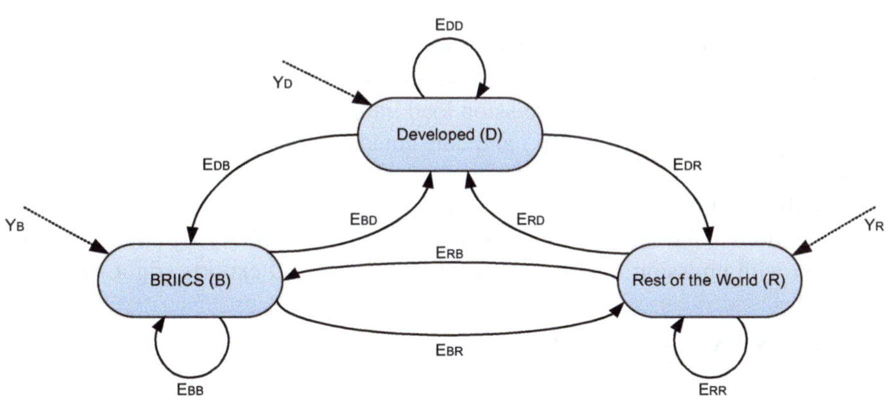

Fig. 3 Simplified model of global emission transfers

Table 2 List of product groups according to ISIC Rev. 4 implemented in this study

No.	Group name	No.	Group name
1	Agriculture, hunting, forestry, and fishing	18	Manufacturing nec; recycling
2	Mining and quarrying	19	Electricity, gas, and water supply
3	Food products, beverages, and tobacco	20	Construction
4	Textiles, textile products, leather, and footwear	21	Wholesale and retail trade; repairs
5	Wood and products of wood and cork	22	Hotels and restaurants
6	Pulp, paper, paper products, printing, and publishing	23	Transport and storage
7	Coke, refined petroleum products, and nuclear fuel	24	Post and telecommunications
8	Chemicals and chemical products	25	Financial intermediation
9	Rubber and plastics products	26	Real estate activities
10	Other nonmetallic mineral products	27	Renting of machinery and equipment
11	Basic metals	28	Computer and related activities
12	Fabricated metal products	29	R&D and other business activities
13	Machinery and equipment, nec	30	Public admin. and defense; compulsory social security
14	Computer, electronic and optical equipment	31	Education
15	Electrical machinery and apparatus, nec	32	Health and social work
16	Motor vehicles, trailers, and semitrailers	33	Other community, social and personal services
17	Other transport equipment	34	Private households with employed persons

with the 4th revision of the International Standard Industrial Classification (ISIC) standards (United Nations 2017a). A list of these product groups are presented in Table 2.

Prior to the data gathering and manipulation stage, the desired system of equations for solving the problem is to be formulated. For the simplified model in this study, 306 unknowns ($e_{ij,k}$: emission load intensities of products k (k: 1, …, 34) which are produced in land group i and consumed in land group j (i, j: D, B, R)) are be calculated.

According to Eq. (3), there are three emission load balancing equations. Each of these equations balances the emission loads on imports plus territorial emissions (left side) with emission loads on own consumptions and exports (right side):

$$Y_i + \sum_{j \neq i} \sum_{k=1}^{34} E_{ji,k} = \sum_{j} \sum_{k=1}^{34} E_{ij,k} \quad (i, j = D, B, R) \quad E_{ij,k} = e_{ij,k} \times V_{ij,k}$$

The remaining need for 303 equations is being met by two types of auxiliary equations. The first type is the use of distribution ratios of product type k among own consumption and export to other two subsystems proportional to their quantities. This type will provide 204 auxiliary equations:

$$e_{iD,k} = e_{iB,k} = e_{iR,k}$$

Furthermore, according to Eq. (4), the other type of equations accounts for the distribution of emissions among 34 product groups in each region proportional to the exergy destructions in the production chains of those emissions. This is the point of which the use of exergy destruction as in Eq. (4) will provide 99 more equations to fulfill the mathematical needs for finding the unknown emission loads:

$$\frac{E_{ij,k}}{E \times C_{ij,k}} = \frac{E_{ij,l}}{E \times C_{ij,l}} \quad (k, l = 1, 2, \ldots, 34)$$

3.3 Data Gathering and Manipulation

Based on the mentioned formulations, three groups of input data are required:

- Flow rates in physical or monetary values,
- Cumulated exergy destruction per unit of flow rate,
- Territorial production-based emissions.

Data for interregional flow rates could be either drawn from international reports of the global trade in monetary or physical units [e.g., UN International Trade Statistics Yearbook (ITSY) (UN-Comtrade 2017), Eurostat MFA (UEROSTAT 2017), World Trade Statistical Review 2017 (WTO 2017), and Global Trade Analysis Project (GTAP) database (Aguiar et al. 2016)], or processed data out of the MRIO tables with global coverage [e.g., WIOD 2016 (Timmer et al. 2015), EORA2013 (Lenzen et al. 2013), EXIOBASE2 (Wood et al. 2015), OECD ICIO2016 (Wiebe and Yamano 2016)].

Yet, physical databases with complete global coverage are not well-shaped and the available ones also do not include trade in services. On the other hand, the development of comprehensive MRIO tables is a very time- and data-consuming tasks, and these tables consider of the input and outputs of any sector of any country to other sectors of the same or other countries (for more information about the developed MRIOs, see Lenzen et al. 2013). To the best of authors' knowledge, one of the most updated MRIOs with a global coverage is for 2011 in OECD ICIO 2016 database (OECD 2016).

As the purpose of this study is to present a new method for calculating the emission loads and also the validation of the derived emission loads with a reference study matters, the trade data provided in the MRIO table developed by OECD (2016) are used. Accordingly, an aggregation of the detailed MRIO tables to the

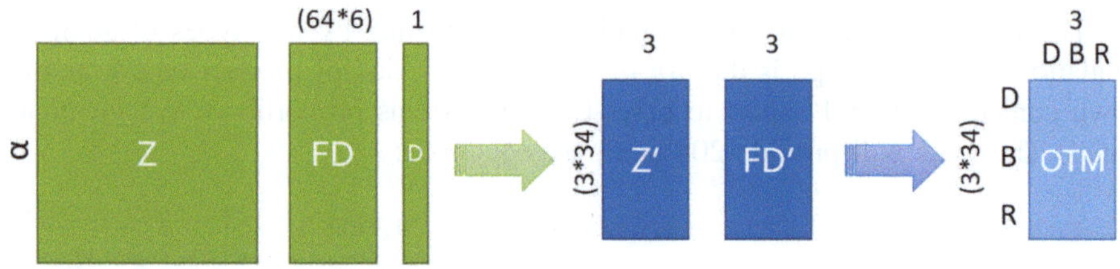

Fig. 4 Data manipulation stages to aggregate the MRIO data into desired form of this study

specifications of the method and the defined system in this study is necessary. This is because the OECD ICIO2016 is a categorized, updated, and comprehensive data base which provides values of the trade quantities among 63 + 1 countries (considering 63 countries and 1 as the rest of the world) and 34 product groups in year 2011. The carbon accounting study of Wiebe and Yamano (2016) could be also considered as a reference study for the comparison of the derived SBR and CBR values.

Furthermore, in order to fulfill the data requirements for Eq. (4), the cumulative exergy destructions per unit of monetary values are assumed from the exergetic input–output analysis of the Italian economy conducted by Rocco (2016). In his comprehensive research work, Rocco has conducted an exergy-extended input–output analysis at the national level. Furthermore, product categorization based on the ISIC fits his results $\left(\text{SCE} \times C_{i,j,k} \text{ in } \left(\frac{\text{kgoe}}{1000\$}\right)\right)$ as proper inputs to the modeling task of this study. It is worth mentioning that the values of exergy destruction intensities which are assumed from the Rocco study are to share the emissions among k products of any of the single subsystems. Hence, differences in energy intensities of the regional economies do not contradict with this assumption. However, in the presence of more accurate exergy destruction data for each region, a higher consistency of the derived results would be possible.

The data manipulation stage for this study is to aggregate the stream transfers among sectors and countries into sectors and regions. Figure 4 depicts the aggregation process implemented in this study:

Finally, the territorial emission inventory must be available for any subsystem. These data are also taken from the OECD report on production-based emissions of countries (Wiebe and Yamano 2016). The assumed PBRs are then summed up for three defined subsystems in this study.

3.4 Overview of the Input Data

A brief review of key indices out of the processed input data is presented in Table 3.

Table 3 Key indices of emissions and productions in three country groups

	Import/Export	Share of imported consumption (%)	Share of exported production (%)	Shares in global production (%)	Production-based emissions (%)	Emission intensity of productions (kg CO_2/1000$)
Developed	1.11	5.3	4.8	56.4	36.3	139.8
BRIICS	0.93	8.1	8.7	26.1	41.7	347.7
RoW	0.95	14.6	15.3	17.6	22.0	271.4

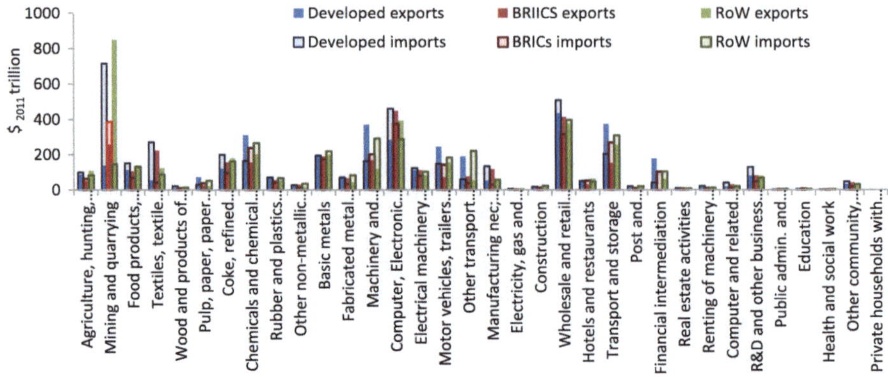

Fig. 5 Net export of emission-intensive products from developing to developed countries

According to the information provided in Table 3, developed countries are generally net consumers. Hence, although a large share of the global production (in monetary value) is being produced in these countries, 95% of their own production is consumed domestically (column 2 of Table 3). However, the production in developed countries has much less environmental impact than the one of emerging economies of BRIICS so far (last column in Table 3). The country group of the RoW is very trade-oriented in a way that a very large amount of their production is being consumed outside and vice versa, and a high portion of their consumption is also imported from other countries. A main message of this information is the focus of energy-intensive productions in the major developing countries which yield in high emission intensities and corresponding high shares in global emissions. This outcome is in agreement with the comparative results in previous CBR studies such as Davis and Caldeira (2010), Peters et al. (2011), Kander et al. (2015). More specifically, Fig. 5 shows this fact in more detail by tracking the emission transfers in sectors:

Table 4 presents data of net values of exports and the dominant providing country group in the world for each product category. According to the information presented in Table 4, the RoW group of countries is the provider of raw materials and agricultural products to the two other groups with an increasing trend [70% increase from 2006 to 2017 (WTO 2017)], while BRIICS is the main provider of labor-, energy-, and carbon-intensive products. Finally, the developed group is the dominant provider of high-technology products as well as most services to the world. These facts give insight about the carbon intensity of economic activities presented in Table 3.

Furthermore, the track of the exergy destruction of the productions (of exports) and consumptions (of imports) confirms the outsourcing of energy-intensive industries from global North to South. Figure 6 (left) depicts a comparison among exergy destruction per unit monetary value of productions in different country groups.

This yields a meaningful difference in shares of monetary values and exergy destructions in global trade, as depicted in Fig. 6 (right).

Table 4 Net values of exports and dominant provider of each product category ($2011 trillion)

Product category	Developed	BRIICS	RoW	Main provider	Exergy destructions (kgoe/100$)
Agriculture, hunting, forestry, and fishing	−11.97	−12.19	24.16	RoW	12.27
Mining and quarrying	−577.37	−127.88	705.2	RoW	129.57
Food products, beverages, and tobacco	−38.71	37.21	1.51	BRIICS	12.61
Textiles, textile products, leather and footwear	−216.67	181.21	35.46	BRIICS	10.6
Wood and products of wood and cork	−9.86	11.44	−1.58	BRIICS	11.56
Pulp, paper, paper products, printing, publishing	42.16	−10.58	−31.6	Developed	13
Coke, refined petroleum products, nuclear fuel	−79.16	62.50	16.66	BRIICS	210.62
Chemicals and chemical products	146.01	−82.22	−63.8	Developed	22.67
Rubber and plastics products	−1.94	13.94	−12.00	BRIICS	16.71
Other nonmetallic mineral products	2.51	17.13	−19.64	BRIICS	33.49
Basic metals	7.00	14.52	−21.52	BRIICS	21.51
Fabricated metal products	8.62	31.93	−40.55	BRIICS	21.51
Machinery and equipment	207.38	−33.60	−173.8	Developed	13.24
Computer, electronic and optical equipment	−175.66	71.29	104.4	RoW	12.29
Electrical machinery and apparatus	−12.85	33.80	−20.95	BRIICS	12.29
Motor vehicles, trailers and semi-trailers	99.02	−70.67	−28.34	Developed	13.67
Other transport equipment	129.51	37.26	−166.8	Developed	13.67
Manufacturing; recycling	−80.13	86.68	−6.55	BRIICS	11.95
Electricity, gas, and water supply	−2.28	−0.23	2.51	RoW	106.79
Construction	3.25	3.07	−6.32	Developed	11.12
Wholesale and retail trade; repairs	−74.17	96.04	−21.87	BRIICS	9.24
Hotels and restaurants	3.37	−19.38	16.01	RoW	8.83
Transport and storage	169.53	−116.56	−52.97	Developed	16.9

(continued)

Table 4 (continued)

Product category	Developed	BRIICS	RoW	Main provider	Exergy destructions (kgoe/100$)
Post and telecommunications	−4.69	6.18	−1.49	BRIICS	6.56
Financial intermediation	136.77	−95.66	−41.10	Developed	2.77
Real estate activities	−0.47	−0.80	1.27	RoW	1.21
Renting of machinery and equipment	−5.73	−0.88	6.61	RoW	6.15
Computer and related activities	−22.70	20.13	2.57	BRIICS	6.15
R&D and other business activities	−48.51	35.69	12.82	BRIICS	6.15
Public admin. and defense; social security	6.88	−2.80	−4.08	Developed	5.4
Education	10.21	−5.56	−4.65	Developed	2.16
Health and social work	1.49	0.16	−1.65	Developed	5.79
Other community, social and personal services	−14.75	23.12	−8.37	BRIICS	6.86
Private households with employed persons	0.00	0.00	0.00		6.86
Total	−403.92	204.27	199.6		

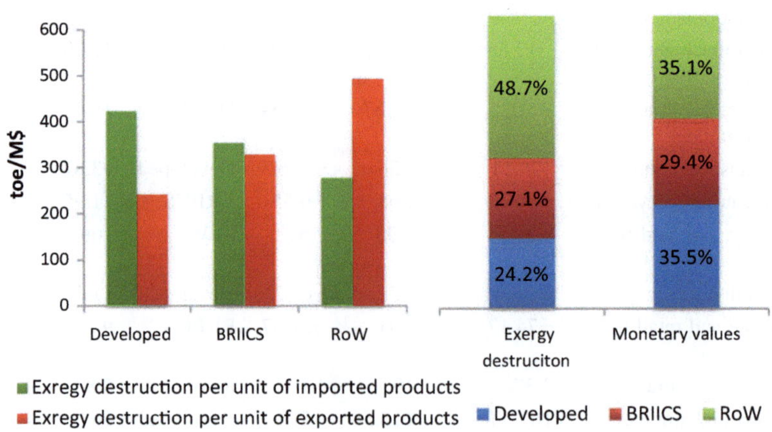

Fig. 6 Energy intensity of imported and exported products (left) and comparison of exergy destruction and monetary values of global trade from different country groups (right)

4 Results

The outputs of the solution to the developed system of equations are in terms of environmental loads per unit (exergy) of the streams. These values, multiplied by the quantity (exergy contents) of streams, result in the environmental loads of each stream.

Figure 7 shows the calculated emission loads on streams as the result out of the formulated system in Fig. 3.

A deeper view of the emission transfers by different product categories shows the extent of emissions resulting from energy-intensive products, specifically mining products and fossil fuels in developing countries (Fig. 8).

Similar to the results of tracking global exergy destructions in Fig. 6, tracking the emission loads on productions and consumptions in Fig. 8 coincides with the

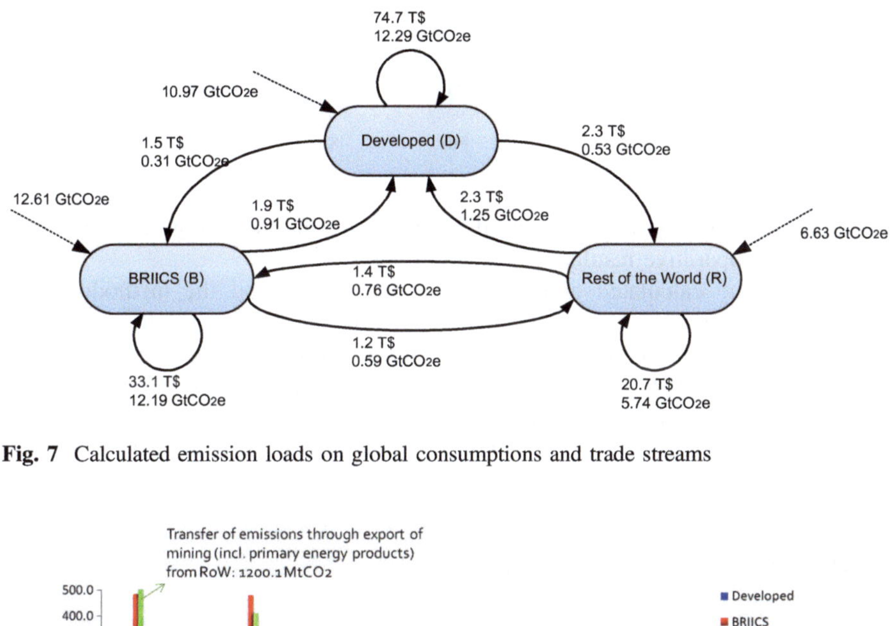

Fig. 7 Calculated emission loads on global consumptions and trade streams

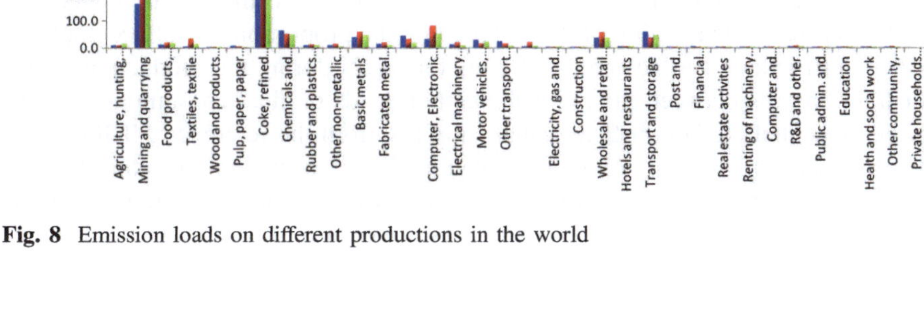

Fig. 8 Emission loads on different productions in the world

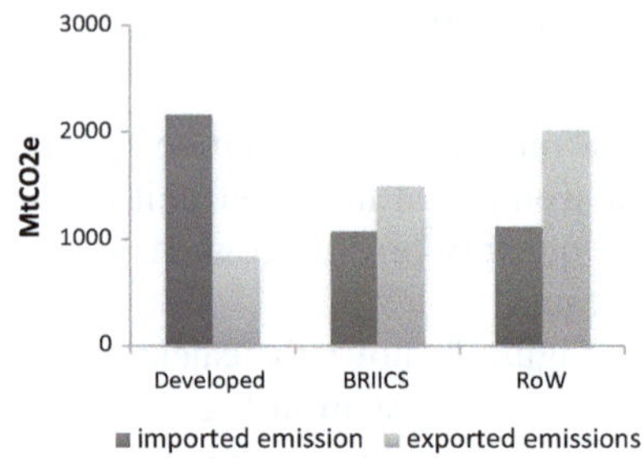

Fig. 9 Comparison of the imported and exported carbon emissions in the world

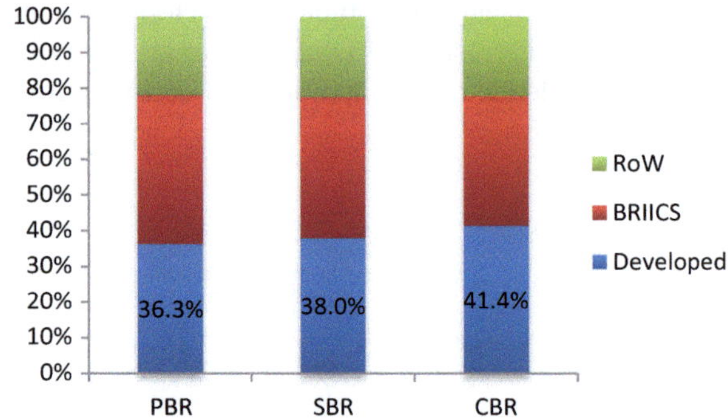

Fig. 10 SBR responsibilities lie between PBR and CBR

outsource of emission-intensive production to developing countries. Figure 9 depicts the comparative results of emission transfers.

Based on the calculated emission loads on streams and the methodological explanation presented in this study, the CEA approach is used to allocate shared responsibilities among country groups. As any other sharing method, the exergy destruction-based SBRs are to allocate responsibilities of emissions among producers and consumers of the final products at once. Hence, it is expected that the derived exergy destruction SBR lies between PBR and CBR (Fig. 10).

5 Discussions

The main innovation of this research study is the presentation of a new calculation method for the estimation of carbon loads on produced streams that are traded and finally consumed. The notion of this is based on the proportionality of accumulated exergy destructions and accumulated emission loads during the production chain. Small differences between the estimated carbon loads on final consumption, based on the presented method and the values of CBR from the reference MRIO study, show an acceptable validity of the presented method, especially if one considers the error of aggregation in this simplified analysis. However, it is important to check this appropriateness of exergy destruction as a measure of environmental load, at

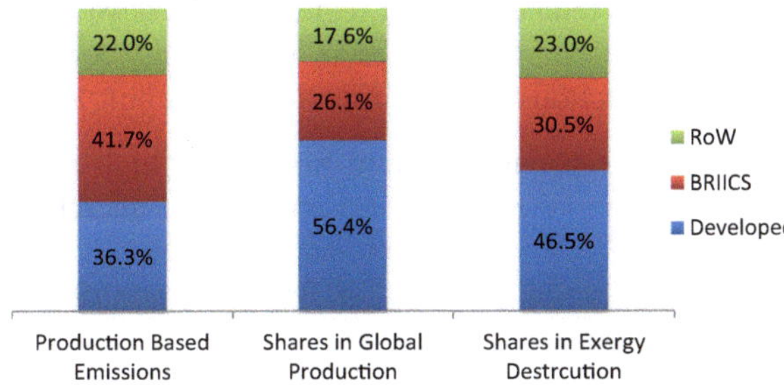

Fig. 11 Comparison of the proportionality of monetary value of productions and exergy destructions with territorial emissions

least in comparison with the monetary value added during the production chain (Fig. 11). The main role of exergy destruction is to reduce the massive data requirements of trade details among sectors of countries and corresponding sectoral emission factors to countrywide aggregated values. This reduces the data requirements enormously; in systems with 34 industrial and service sectors, this amount of data requirements would be reduced by 97%. This would be more important in cases in which the sectoral imports are not being measured or recorded in the context of MRIO analysis; this lack of data exists for many developing countries. Therefore, based on the presented method, general global trade data would be sufficient for tracking emission loads. Nevertheless, in comparison with the MRIO method, this method assumes the proportionality of exergy destruction and carbon emissions which is a source of uncertainty (error of aggregation). At least in comparison with other allocation approaches for environmental emissions (e.g., market values in ISO 2006), exergy destructions are physically more reasonable for accounting the contributions. Figure 11 provides a comparison among the proportionality of exergy destructions and monetary added values of productions with territorial emissions.

In this study, the world is assumed to be a system with three components in which each of them is performing economic activities and attributing to exergy destructions and carbon emissions. According to the data provided in Fig. 11, the symmetry of emissions to exergy destructions is much higher than the monetary added values of productions. This is also true in comparison among representability of the emission loads by exergy destructions rather than exergy contents of products. For more discussions on this, see Khajehpour et al. (2017). However, it should be noted that in this approach for SBR, the exergy destruction has been used only for the allocation of emission loads among k type of products of each country group. The shares of responsibilities are then affected by the amount of trades ($V_{i,j,k}$) and territorial emissions (Y_i).

Furthermore, this research proposes the rationality of allocation responsibilities among countries based on the CEA approach, which is originated in embodied energy as well as exergoenvironmental analyses. High emission intensities in BRIICS countries and very low emission intensities in the developed group show a clear shift in global carbon-intensive industries into emerging economies which seek a competitive advantage in their economic activities. This competitive advantage is relaxed carbon regulations in developing countries, which in contrast to restrictive carbon policies in developing countries motivates carbon leakage. Developed nations are net importers of carbon loads as a result of two factors:

- High rate of consumption: Since their high rate of consumption exceeds their production, they are net importers of merchandise and, to a lesser extent, services.
- Higher carbon intensity of imported products in comparison with a cleaner production of their exports.

Hence, developed countries are the most carbon-importing countries followed by BRIICS. RoW countries are in total the least carbon-importing countries as they are exporters of high carbon-intensive products and importers of cleaner products.

The input data analysis also show that RoW countries are more dependent on international trade since higher shares of their productions (consumptions) are being exported to (imported from) the other two groups of countries. But when it comes to absolute values of traded products, BRIICS is the most exporting group and the developed group is the one that imports the most. The RoW countries are the main providers of raw materials as well as energy carriers and agricultural products to the world, especially since all OPEC countries are present in this group. In contrast, developed countries are providing more science-based and high-tech products to the world.

The derived SBR values lie between the ones of PBR and CBR. However, in comparison with the averaged responsibilities among producers and consumers, the exergy destruction-based responsibilities are rather more oriented toward the PBR than CBR.

This bias might be understood based on the methodological definition of share responsibilities in the CEA. According to Eqs. (1) and (2), the averaged responsibilities account for production emissions plus gross imported emissions. However, the averaged responsibilities are the production emissions plus half of the net imported emissions.

$$\text{Averaged Responsibilities}_i \equiv \frac{\text{PBR}_i + \text{CBR}_i}{2} = Y_i + \frac{\sum_{j\neq i} E_{ji} - \sum_{j\neq i} E_{ij}}{2}$$
$$= Y_i + \sum_{j\neq i} E_{ji} - \frac{\sum_{j\neq i} E_{ji} + \sum_{j\neq i} E_{ij}}{2}$$

$$=> \text{Averaged Responsibilities}_i \equiv \text{SBR}_{\text{CEA},i} - \frac{\sum_{j\neq i} E_{ji} + \sum_{j\neq i} E_{ij}}{2} \quad (5)$$

According to Eq. (5), the SBR might be lower or higher than the averaged responsibilities, and this depends on the trade condition of the countries. In this context, net importer countries will have more SBR than the average of the production-based and consumption-based responsibilities ($\text{SBR}_{\text{CEA},i}$ > Averaged Responsibilities$_i$). Hence, the developing world as the main consumer of global

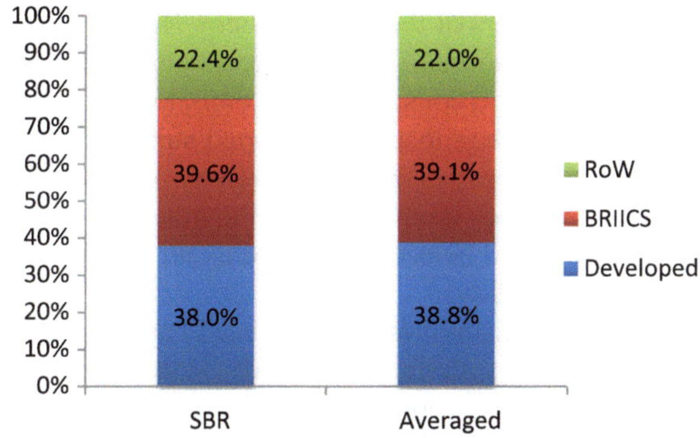

Fig. 12 Relatively higher shares of responsibilities for producers in SBR in comparison with the averaged responsibilities

products in the world will have higher SBR than in the condition in which the responsibilities have been shared by averaging PBR and CBR (Fig. 12).

6 Conclusions

The necessity for global cooperative actions against climate change has been realized by all countries in the world. This fact has led to intergovernmental agreements in different conferences of parties in the past 20 years. However, there have been many practical failures in the application of the agreed action plans. Even in the Paris agreement, which supposed to be the state of the art for a global agreement for action on a common target, it has not been well clarified how much the countries worldwide should contribute concerning the funding of mitigation and adaptation actions. A lack of realistic allocation knowledge might have resulted in disagreements in the past international negotiations. Therefore, sharing approaches are developed to create allocated responsibilities in a more reasonable way in order to shape a consensus among countries for the participation in integrated actions against climate change. From the production and consumption point of view, this fairness might be achieved by considering the utility of emissions for different nations (e.g., monetary value added from production, imports, and exports) or by equally considering the physical emission loads on both productions and consumptions.

The approach which is used in this study is a trial for the allocation of responsibilities by considering both the profitability of territorial production emissions for emitters and the utilities in the consumption of imported products for consumers. Furthermore, this approach, in trade-off between data and precision, provides possibility of easier available estimations of the emission loads on global trades.

Similar to the synergetic benefits of the global economic system, the sum of the contributions to emissions of all system elements are more than the total emissions within the borders of the whole system. Hence, subsystem responsibilities are to be calculated by normalizing their total contributions to the emissions. Since the allocation method considers direct emissions within the borders of the components and indirect carbon loads on imported streams, the results motivate the consumers toward the imports of cleaner products, on the one hand, and encourage the producers for cleaner productions on the other hand. Therefore, it is expected that the derived SBR will be better accepted by all donors as they consider all aspects of the creation of liability.

These responsibilities are calculated based on a thermodynamic-based approach in which exergy destructions have been used as measures of carbon loads on streams. In this concept, tracking of exergy destructions, together with the monetary values of trades, has been used as a proxy for tracking of environmental loads for national allocations. This has led to a huge (97%) reduction in data requirements for the calculation of emission loads on products of each subsystem.

The derived results out of this study show that the major developing countries (BRIICS) have two times more emission intensities in comparison with the developed group. This might be because of the shift of energy-intensive (and consequently the carbon-intensive) industries into countries with more relaxed climate regulations (non-Annex B countries) and higher long-term economic growth rates (BRIICS). As a result, the emissions embodied on exports from the RoW and BRIICS countries to the developed countries occur. The consideration of the carbon trades and consumption liabilities result in some 5.1% shift of responsibilities from global South to the global North. Moreover, in a more comprehensive view to the liabilities, the SBR analyses take also the profitability of emissions in emitting bodies into account. This is done by considering both carbon loads on trades (indirect) and territorial (direct) emissions. As a result, the SBR recommends 3.4% higher responsibilities in comparison with the CBR and 1.4% less responsibilities in comparison with PBR of the developing countries. The holistic view to the benefits in the SBR reduces the gap between the two different paradigms of responsibility allocations of the CBR and the PBR based on a scientific rationale and recommends fairer expectations from the global involvements in the integrated climate actions.

It is worth mentioning here that the derived results out of responsibility allocation studies do not recommend emission permits or appropriate policies for mitigation and adaptation actions. Instead, these responsibilities will identify the role of each country (or country groups) in funding prioritized action plans worldwide. The scientific base for rational allocation of the expected contributions of all parties for funding globally distributed climate action projects, which is presented in this study, will enhance meeting climate targets.

Lower precision in assessing the emission loads on streams, in payment for reducing data requirements, is the main critique of the method. Use of country-specific SCE $\times$ $C_{i,j,k}$ will also increase the accuracy of the national allocation of emission loads. This analysis has been done for the year 2011. This is due to the limited availability of trade data for more recent years and also for the sake of comparability of the results with the developed CBR studies.

The comparative analysis of the outcomes of the application of other SBR methods as well as disaggregation of the model for accounting responsibilities of individual countries and unions is recommended for further studies. Moreover, further application of the method for assessing the temporal distribution of the SBRs will also help in the investigation of the climate equity which is related to the historical contributions to the climate change.

Acknowledgements This study has been done with the support of the Alexander von Humboldt Foundation/Stiftung, the Climate Protection Program.

References

Aguiar, A., Narayanan, B., & McDougall, R. (2016). An overview of the GTAP 9 data base. Journal of Global Economic Analysis, 1(1), 28. https://doi.org/10.21642/jgea.010103af.

Andrew, R., Peters, G. P., & Lennox, J. (2009). Approximation and regional aggregation in multi-regional input–output analysis for national carbon footprint accounting. *Economic Systems Research, 21*(3), 311–335. https://doi.org/10.1080/09535310903541751.

Bastianoni, S., Pulselli, F. M., & Tiezzi, E. (2004). The problem of assigning responsibility for greenhouse gas emissions. *Ecological Economics, 49*(3), 253–257. https://doi.org/10.1016/j.ecolecon.2004.01.018.

Caldeira, K., & Davis, S. J. (2011). Accounting for carbon dioxide emissions: A matter of time. *Proceedings of the National Academy of Sciences of the United States of America, 108*(21), 8533–8534. https://doi.org/10.1073/pnas.1106517108.

Davis, S. J., & Caldeira, K. (2010). Consumption-based accounting of CO_2 emissions. *Proceedings of the National Academy of Sciences of the United States of America, 107*(12), 5687–5692. https://doi.org/10.1073/pnas.0906974107.

Davis, S. J., Peters, G. P., & Caldeira, K. (2011). The supply chain of CO_2 emissions. *Proceedings of the National Academy of Sciences of the United States of America, 108*(45), 18554–18559. https://doi.org/10.1073/pnas.1107409108.

Eder, P., & Narodoslawsky, M. (1999). What environmental pressures are a region's industries responsible for? A method of analysis with descriptive indices and input–output models. *Ecological Economics, 29*(3), 359–374. https://doi.org/10.1016/S0921-8009(98)00092-5.

ISO. (2006). ISO 14044:2006 Environmental management—Life cycle assessment—Requirements and guidelines. Geneva, Switzerland, International Standard Organization: ISO/TC 207/SC 5 Life cycle assessment. 13.020.60 Product life-cycles: 46.

Jakob, M., & Marschinski, R. (2013). Interpreting trade-related CO_2 emission transfers. *Nature Climate Change, 3*(1), 19–23. https://doi.org/10.1038/nclimate1630.

Kander, A., Jiborn, M., Moran, D. D., & Wiedmann, T. O. (2015). National greenhouse-gas accounting for effective climate policy on international trade. Nature Climate Change, 5(5), 431–435. http://dx.doi.org/10.1038/nclimate2555.

Khajehpour, H., Saboohi, Y., & Tsatsaronis, G. (2017). Environmental responsibility accounting in complex energy systems. Journal of Cleaner Production, 166(Supplement C), 998–1009. https://doi.org/10.1016/j.jclepro.2017.08.013.

Lenzen, M., Moran, D., Kanemoto, K., & Geschke, A. (2013). Building Eora: A global multi-region input–output database at high country and sector resolution. Economic Systems Research, 25(1), 20–49. https://doi.org/10.1080/09535314.2013.769938.

Lenzen, M., Murray, J., Sack, F., & Wiedmann, T. (2007). Shared producer and consumer responsibility—Theory and practice. Ecological Economics, 61(1), 27–42. https://doi.org/10.1016/j.ecolecon.2006.05.018.

Odum, H. T. (1996). *Environmental accounting: Emergy and environmental decision making*. New York: Wiley.
OECD. (2016). OECD inter-country input-output (ICIO) tables (2016 ed.). Paris, France, Organisation for Economic Co-operation and Development (OECD). Retrieved October 20, 2017, from http://www.oecd.org/sti/ind/inter-country-input-output-tables.htm.
Peters, G. P., & Hertwich, E. G. (2008). CO_2 Embodied in international trade with implications for global climate policy. *Environmental Science and Technology, 42*(5), 1401–1407. https://doi.org/10.1021/es072023k.
Peters, G. P., Marland, G., Hertwich, E. G., Saikku, L., Rautiainen, A., & Kauppi, P. E. (2009). Trade, transport, and sinks extend the carbon dioxide responsibility of countries: An editorial essay. Climatic Change, 97(3), 379. https://doi.org/10.1007/s10584-009-9606-2.
Peters, G. P., Minx, J. C., Weber, C. L., & Edenhofer, O. (2011). Growth in emission transfers via international trade from 1990 to 2008. Proceedings of the National Academy of Sciences of the United States of America, 108(21), 8903–8908. https://doi.org/10.1073/pnas.1006388108.
Rocco, M. V. (2016). Chapter 6: Case studies: Applications of the Exergy based Input-Output analysis. In Primary exergy cost of goods and services: An input–output approach (pp. 101–137). Cham: Springer International Publishing. https://doi.org/10.1007/978-3-319-43656-2_6.
Rodrigues, J., & Domingos, T. (2008). Consumer and producer environmental responsibility: Comparing two approaches. *Ecological Economics, 66*(2), 533–546. https://doi.org/10.1016/j.ecolecon.2007.12.010.
Timmer, M. P., Dietzenbacher, E., Los, B., Stehrer, R., & de Vries, G. J. (2015). An illustrated user guide to the world input–output database: The case of global automotive production. Review of International Economics, 23(3), 575–605. http://dx.doi.org/10.1111/roie.12178.
UEROSTAT. (2017). Material flow accounts—Flows in raw material equivalents. Luxembourg, Eurostat - Statistical Office of the European Union. Retrieved October 23, 2017, from http://ec.europa.eu/eurostat/web/products-datasets/-/env_ac_mfa.
UN-Comtrade. (2017). 2016 International trade statistics yearbook: Volume I—Trade by Country. New York, USA, United Nations Statistics Division/Department of Economic and Social Affairs. http://dx.doi.org/10.18356/78d35343-en.
United Nations. (2017a). International standard industrial classification of all economic activities, Rev. 4. United Nations Statistics Division. Retrieved October 22, 2017, from https://unstats.un.org/unsd/cr/registry/regcst.asp?Cl=27.
United Nations. (2017b). World economic situation and prospects 2017. New York, United Nations publication Sales No. E.17.II.C.2. https://www.un.org/development/desa/dpad/wp-content/uploads/sites/45/publication/2017wesp_full_en.pdf.
Wiebe, K., & Yamano, N. (2016). Estimating CO_2 emissions embodied in final demand and trade using the OECD ICIO 2015. Paris: OECD Publishing. http://dx.doi.org/10.1787/5jlrcm216xkl-en.
Wood, R., Stadler, K., Bulavskaya, T., Lutter, S., Giljum, S., de Koning, A., et al. (2015). Global sustainability accounting: Developing EXIOBASE for multi-regional footprint analysis. Sustainability, 7(1). http://dx.doi.org/10.3390/su7010138.
WTO. (2017). World Trade Statistical Review 2017, Economic Research and Statistics Division and International Trade Statistics Section, World Trade Organization. https://www.wto.org/english/res_e/statis_e/wts2017_e/wts17_toc_e.htm.

Hossein Khajehpour is a Ph.D. candidate of Energy Systems Engineering at the Sharif University of Technology, Tehran, Iran. He is also a Guest Researcher at the Chair for Energy Engineering and Environmental Protection of the Energy Engineering Institute of the Technical University, Berlin. He is currently a Fellow of the Climate Protection Program of the Alexander von Humboldt Foundation. His research work is focused on the responsibility allocation of climate change. He has experience in many projects in the field of environmental management of energy systems at the Sharif Energy Research Institute, Tehran, Iran.

Yadollah Saboohi is a Professor at the Energy Engineering Department of the Sharif University of Technology. He is also the Founder and Director of the Sharif Energy Research Institute. He has supervised dozens of Ph.D. theses, has led a variety of national and international projects, and has many publications to his credit.

George Tsatsaronis is a Professor and Director of the Chair for Energy Engineering and Environmental Protection of the Energy Engineering Institute of Berlin Technical University. He has supervised dozens of Ph.D. theses, has led a variety of national and international projects, and has many publications to his credit.

Geo-Web Services and New Exchange Formats to Develop Future Services Supporting Climate Change Adaptation Measures

Martin Scheinert, Hardy Pundt and Andrea Heilmann

Abstract Climate change and a rising number of extreme weather events increase the demand for regional climate adaptation concepts. Consequently, the need for supporting the development of such concepts, for example through the usage of digital climate services, is growing. In particular, the adequate use of geospatial data and Web services could accompany actors in the development and primarily in the implementation of climate adaptation measures. The paper describes the potential of current geo-Web services and innovative data exchange formats for future climate services. The hypothesis is to determine whether the use of innovative exchange formats and interoperable geo-Web services can contribute to an increase of quality and an expanded usage of Web services during the definition of regional climate adaptation measures. Aiming at the verification, or falsification, of this hypothesis, various investigations were and are being carried out. They are embedded in a research project, focusing on climate adaptation measures in small river catchments. Initial results suggest that the hypothesis can be confirmed. It will be tested based on the development of innovative climate services aimed at producing maps that inform users about erosion vulnerabilities in two exemplary catchments. Resulting findings will be presented, and conclusions are drawn.

Keywords Climate change adaptation · Climate services · Geo-Web services GeoPackage

M. Scheinert (✉) · H. Pundt · A. Heilmann
Department for Automation and Computer Science, Harz University of Applied Sciences, Friedrichstr. 57-59, 38855 Wernigerode, Germany
e-mail: mscheinert@hs-harz.de

H. Pundt
e-mail: hpundt@hs-harz.de

A. Heilmann
e-mail: aheilmann@hs-harz.de

© Springer International Publishing AG, part of Springer Nature 2019
W. Leal Filho and R. Leal-Arcas (eds.), *University Initiatives in Climate Change Mitigation and Adaptation*, https://doi.org/10.1007/978-3-319-89590-1_18

1 Introduction

Climate change, climate variability and adaptation are the major challenges for societies worldwide. A cross-sector communication and detailed, up-to-date information on climate change play a crucial role in the development of global, national and regional adaptation strategies. Climate services will support the creation, explanation, application and exchange of climate-related knowledge which is useful for all participating actors. Such services support decision-making processes that aim at mitigating potential risks. The services contribute furthermore to a climate-friendly policy. Due to the current relevance as well as the previously described application scenarios, climate services could be established worldwide (Cortekar et al. 2014). The offerings of the services vary from consultancy to the creation of graphics, maps and reports, up to the provision of all kinds of relevant data. However, due to the multi-layer offerings and the variety of providers, ranging from research facilities through local authorities to private companies, there is currently no standard definition of the term "climate service" (Hewitt et al. 2012), came up, for example, with a combination of different definitions and describing the term as follows:

> A climate service can be considered as the provision of climate information in such a way as to assist decision-making. The service needs to be based on scientifically credible information and expertise, have appropriate engagement from users and providers, have an effective access mechanism and meet the users' needs. Effective climate services facilitate climate-smart decisions that reduce the impact of climate-related hazards and increase the benefits from benign climate conditions.

While the establishment of the United Kingdom Climate Impact Programme (UKCIP) was preceded by the development and application of climate services at the European level in 1997, the establishment of a national service facility in Germany took place in 2009 with the founding of the Climate Service Center Germany (GERICS).

The hypothesis of this paper is based on the assumption that climate services that are grounded on new and more suitable interoperable formats can lead to an improved quality and therefore an expanded usage by the involved actors. Chapter two (Current Suppliers of Climate Services) describes an up-to-date overview of the current providers of climate services. Chapter three (Political and Technical Developments in the Area of Climate Services) presents relevant political and technical development steps in the context of climate services. They form the basis for the services described in Chapter four (Standardized Geo-Sensor Networks and New Data Concepts as a Chance for Qualitatively Better Future Climate Services), aimed at the standardization and linkage of different technological approaches. Based on this, Chapter five (A Case Study: Application of GeoPackage Within Climate Services) describes various new approaches and methods for establishing climate service offerings in the municipal sector. They have to be tested and evaluated critically, which is part of the aforementioned research project.

2 Current Suppliers of Climate Services

The portfolio of climate services and its suppliers in Germany is foremost oriented at the increased demand of a wide range of users. Cortekar et al. (2014) conducted an analysis of the providers and users of climate services in Germany. One result of their study is an overview of locations of different German climate service providers. They were collected within the context of the Climate Knowledge Hub. This Internet-based platform includes an interactive map, which enables providers of climate services from member countries of the European Joint Programming Initiative (JPI) to register. In such a way, it serves as an overview of the providers of climate services in Europe. With its help, users and the interested public can access contact information as well as hints on the capabilities of the services. An integrated filter function enables users to perform the search either spatially, according to the application sector, core competency, thematically or chronologically.

In addition to this overview, another Web portal provides similar information, the "Klimanavigator" (Klimanavigator). However, this portal bundles the access to climate knowledge and its providers in Germany, but it is based on a different approach. Categorized by geographic location and currently 18 thematic priorities, information on research facilities and networks can be obtained together with detailed information on their fields of activity, in particular climate services. The search results are displayed in an interactive location overview similar to the Climate Knowledge Hub. The resulting map (Fig. 1) contains also vendors (they are not listed in the Climate Knowledge Hub) and allows the identification of spatially nearby cooperation partners that provide thematically suitable offers.

3 Political and Technical Developments in the Area of Climate Services

At the European political level, the Global Monitoring for Environment and Security (GMES) program, jointly set up in 1998 by the European Commission (EU) and the European Space Agency (ESA), deals with the construction of an infrastructure for earth observation. This program was renamed to "Copernicus" in December 2012. Within the framework of a conceptual draft (Uppala et al. 2011), the foundation stone for specific "Copernicus climate services" has been laid in June 2011. The aim of the Copernicus Climate Services is the establishment of various climate information services based on satellite earth observation and in situ (non-spatial) data. Copernicus services address six main thematic areas (atmosphere, marine, land, climate, emergency, security).

Fig. 1 Overview map of climate knowledge providers in Germany. *Source* Excerpt of an screenshot from the website www.klimanavigator.de

The services have reached different degrees of maturity. Some were already declared operational several years ago (in 2012 for the Land Monitoring Service and the Emergency Management Service – Mapping, in 2015 for the Atmosphere Monitoring Service and the Marine Environment Monitoring Service) while others were declared operational more recently (in 2016 for the Border Surveillance and Maritime Surveillance components of the Security service and in May 2017 for the Support to External Action component) or are still in their development phase (Climate Change Service).

(Copernicus)

From the perspective of data harmonization, the development of spatial data infrastructures (SDI) is a political goal that has a deep impact in the area of climate services (Vescovi et al. 2015). Within the scope of the Infrastructure for Spatial Information in Europe (INSPIRE) initiative, the cross-border use of geodata in Europe will be facilitated. The European Union therefore wants to support joint environmental policy decisions, an objective envisaged also in the Copernicus program, concerning the latter with a special focus on climate services.

From a technological point of view, an uncomplicated, and timely access to decision-relevant and scientifically sophisticated, location-related information for climate services is considered as indispensable (Cortekar et al. 2014). This should be achieved via INSPIRE on the administrational level. Technically, various important developments took place during the foregoing years. Standardized geo-Web services such as the Web Map Service (WMS), Web Coverage Service (WCS) and the Web Feature Service (WFS) provide technologies and protocols for a comprehensive global access to spatial data via the World Wide Web. They pave the way for an active communication and information exchange between service producers and different user groups. These standardized Geo-Web services originate from the Open Geospatial Consortium (OGC) and provide a frequently used form of data provision and retrieval. Figure 2 describes a typical processing chain aimed at the provision of geo- and climate data [illustration based on (Schöner 2010), modified].

4 Standardized Geo-Sensor Networks and New Data Concepts as a Chance for Qualitatively Better Future Climate Services

Modelling of local environmental processes, especially at the regional (including municipal and district) level, implies major challenges for planners, administrative authorities and science. Modelling geometry, and topology, as well as the automatic and integrated analysis of settlement, vegetation, terrain forms and other data are important requirements of actors involved in planning adaptation measures. GIS technologies provide such functionalities. For instance, virtual 3D city models, e.g. based on the CityGML standard (Gröger et al. 2012), close the gap and offer the possibility to carry out extensive spatial analyses. Concrete examples are noise mapping in cities or heat demand analysis (Eicker et al. 2012). Such analyses are carried out under explicit consideration of climate variables. Solar potential analysis in conjunction with data on sunshine duration as well as the regional air temperature for a 30-year time period is another example.

The use of technological monitoring tools to describe the status quo of the environment has undergone a steady change in the last decades. Previously, isolated measuring stations, often based on cabled setups, were necessary. Nowadays, a large number of miniaturized sensors integrated into smartphones, cars, residential

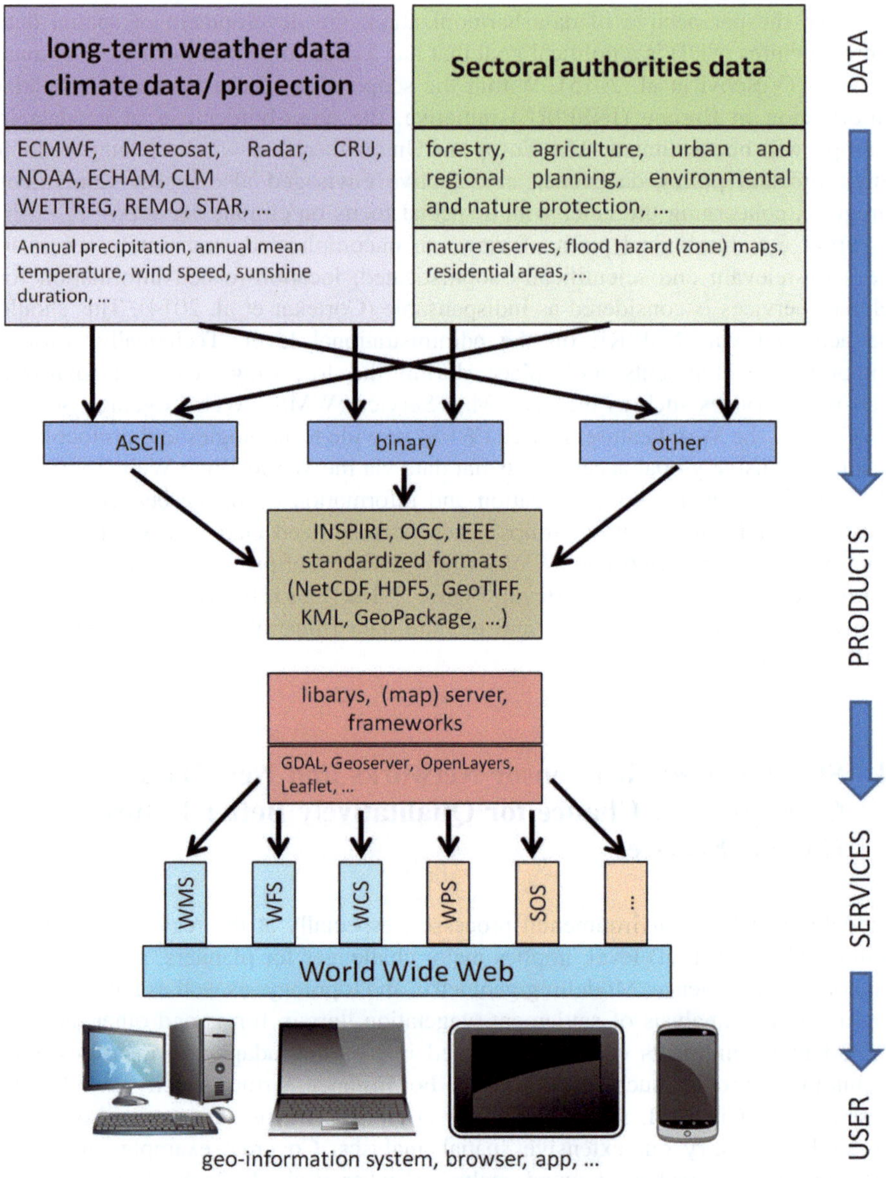

Fig. 2 Schematic representation of a processing chain of geo-/climate data

buildings as well as in a wide variety of infrastructure systems are used. A typical field of application of classical sensor networks is, among others, the detection of water levels and the collection of values describing the flow of running waters. They are needed to collect data to be integrated into flood warning systems. Adaptations of such sensor networks are also installed to support weather forecast systems (Meissen and Röthlein 2010). By using the interoperable approaches of the Sensor Web Enablement (SWE) initiative (Botts et al. 2007), attempts were carried out to achieve standardization at the level of data and services. The SWE initiative brought significant progress concerning the integration of measurement data from different geo-sensor networks. Standards such as the Sensor Observation Service (SOS), Sensor Planning Service (SPS) or the Sensor Model Language (SensorML) were created and are successfully established meanwhile, e.g. in disaster management (Babitski et al. 2009).

Originally used comprehensively in urban regions, sensor networks are increasingly used in rural areas to identify and collect variables that concern climate protection and adaptation. An example is the determination of fresh air corridors or air-conditioning ventilation axes. Concerned investigations could be carried out by regional, and local, administrations, if the related climate services would be available via the WWW. This is also true concerning the analysis and assessment of flooding risks. For this purpose, digital terrain models would help to identify the vulnerabilities for safety-relevant and highly sensitive infrastructures such as hospitals, data centres or power plants, as well as settlements in general. Climate services to support the investigation of rigole systems, i.e. drainage systems for rainwater infiltration, could be offered to regional and municipal users that are dealing with new systems to retain critical surface run-offs (Reinstorf and Ohme 2014), an important issue with regard to climate change.

According to (Resch 2013), sensors installed in smartphones and cars could also provide valuable real-time data. Such crowd-sourcing concepts would lead to more comprehensive datasets supporting strategies to handle urban mobility, e.g. dynamic traffic management systems. An integration into sensor networks that detect carbon dioxide could represent a contribution to the reduction of greenhouse gas emissions.

4.1 GeoPackage as a Format to Improve Interoperability of Mobile Applications

With the global spread of smartphones and their permanently increasing performance, including the integration of a wide range of sensors, a change happened concerning the collection and management of mobile geodata. A new data format standard (Daisey 2015), called GeoPackage, was realized. GeoPackage is an open, standardized, self-contained, platform-independent, portable and self-describing file format based on SQLite. The standard describes a set of conventions for storing a number of content types, such as vector features, tile matrix sets (pyramids) of imagery and raster maps at various scales, schema, metadata and a number of extension data (Simonis et al. 2016). For example, Rashidan and Musliman (2015), Adams and Suykens (2013) as well as Masó et al. (2013) describe the enormous potential of this data exchange format. Normally, traditional mobile GIS applications always require individual data formats for the data storage of raster and vector formats, which in turn leads to the limitation of storage space and generally to the reduction of the devices performance. This is avoidable through the use of the GeoPackage format. As described by (Kalberer 2013), GeoPackage could represent "*the shapefile of the future*" and replace a long-overdue standard for geospatial data.

The use of such a data format for climate services could be a suitable choice due to its already existing implementation in open source software, such as the GeoServer, the GDAL and Geotools program libraries as well as the free desktop geographic information system QGIS. The worldwide mostly applied ArcGIS geographic information system from Esri, which is often used in municipal authorities, also supports this format. Since prepared data and information on climate change and depictions of vulnerable areas and adaptation measures are often offered in application-specific and competing file formats, the use of a universal exchange format such as GeoPackage seems more than sensible. This could in particular increase the networking of different climate service providers and sustainably improve their offers.

5 A Case Study: Application of GeoPackage Within Climate Services

To support the evaluation of the aforementioned hypothesis, some research was invested into the application of current geo-Web services and exchange formats for climate services. These concern the selection, support or evaluation of climate adaptation measures for the municipal sector. The need for such research resulted from a series of projects carried out in close cooperation with various actors in regional and local administrations, and organisations. Their feedback (wishes, suggestions and expectations concerning new Web services) and their hints on current limitations and obstacles were driving forces to investigate opportunities to overcome current deficits.

Within the framework of a pilot project ("Klimpass", 2012–2013), carried out in the administrative district of Mansfeld-Südharz, Germany, a local climate change adaptation strategy has been developed and finalized in 2012. The aim of the follow-up project ("Klimpass-Aktiv", 2013–2016) was to conduct and evaluate the development and implementation of various adaptation measures to climate change. During the project, an interactive online communication platform was created. It includes a comprehensive online mapping system that is based on GIS functionalities. In order to support collaboration between different stakeholders, a cross-sector approach was pursued. This means that not only actors from different sectors in practice, but also practice and science are enabled to cooperate. The aim of the cross-sector approach was to support integrated, comprehensive decision-making under consideration of possibly all relevant data from different sectors. A basic system was implemented during "Klimpass-Aktiv", but meanwhile, there is a significant need for innovative extensions. To achieve these, supporting online services is envisaged. The follow-up project ("BebeR", 2017–2019) extends the idea of online participation. Not only administrative actors and scientists, but also citizens should participate in action planning and decision-making. The goal of "BebeR" is the adaptation of existing, and the development of new methodological approaches to reduce soil erosion in river catchments in view of future climate change induced, extreme weather events.

Approaches from Citizen Science are foreseen to be integrated to improve cooperation between administrational employees and citizens, as well as scientists, on the one hand, and to enhance the cross-sectors, multidisciplinary database, on the other. The latter goal, however, is only achievable if new formats like GeoPackage are applied.

5.1 GIS-Related Methods for Improving Local Climate Adaptation Measures—The Current Situation

In this section, an overview is given on methods that aim at achieving an improvement concerning the integration and exchange of data sources from different sectors. Multidisciplinary information sharing, however, is needed for the creation, assessment and evaluation of sustainable climate change adaptation measures. Examples of data sources are official, administrative data, scientific data and proofed data provided by citizens. Data are foremost related to space, and some might be non-spatial. Up to now, many cases exist which show that partly large datasets remained unused due to obstacles concerning access, usability or different, not sharable data formats. Furthermore, it has been identified in previous projects that there is often a lack of knowledge about existing data and expertise how to integrate these into "own" decisions or applications. These deficits can be summarized under the problem of data heterogeneity. An answer is the approach suggested here, open geoinformation standards and the development of services based on such standards, requiring as little expertise concerning technology as possible from the user's perspective.

To follow a systematic approach, in-depth investigations regarding the use of the previously mentioned spatial data standards will be carried out, focusing on their usability for re-engineered or new (future) climate services.

As a first step, the current state of knowledge among actors in the BebeR project concerning geodata standards was recorded and evaluated in a survey. A total of 15 mostly administrative institutions, with various sub-divisions, were included in the survey. Table 1 gives an overview of the results.

Approximately 60% of the actors have a basic understanding or a good/advanced knowledge related to geographic information systems. The most common geographic information system seems to be ArcGIS (7), followed by QGIS (4) and SAGAGIS (2). Other geographical information systems, such as "gvSIG", seem to

Table 1 Survey on knowledge about geographic information systems (GIS)

Desktop-GIS	No knowledge	Basic knowledge	Good knowledge	Advanced knowledge
ArcGIS	5	2	4	1
SAGAGIS	10	2		
QGIS	8	2	2	
gvSIG	12			
Other (name)		Geomedia (1)		

be unknown. Only "Geomedia" was mentioned as a supplement. Additionally, the knowledge about OGC Web Services and vector, raster or hybrid data formats was queried. The results (number of entries) are shown in Table 2.

The knowledge and application of typical "OGC Web Services", in particular INSPIRE-compliant Web services, was queried as well. Among the project actors, only the Web Map Service and Web Feature Service are known. Two actors and thus only 17% stated that they use these Web services in their daily work. Established Web services which, for example, ensure access to various sensor

Table 2 Survey knowledge geo-Web services, geodata formats

Web services/Formats	No knowledge	Existing knowledge	Used/ Operated	
OGC Web Services				
Web Map Service (WMS)	9	3	2	
Web Feature Service (WFS)	10	2		
Web Coverage Service (WCS)	12			
Web Processing Service (WPS)	12			
Sensor Observation Service (SOS)	12			
Sensor Planning Service (SPS)	12			
Other (please name)				
Vector formats				
Esri Shape Files (.shp,.shx,.dbf)	3	9	7	
Keyhole Markup Language (.kml,.kmz)	11	1		
Geography Markup Language (.gml)	12			
GeoJSON (.geojson)	11	1		
Other (please name)				
Raster formats				
(Geo-)Tiff (.geotiff,.tiff)	8	4	3	
Esri Grid (.grid)	9	3	3	
ASCII-Grid (.xyz,.asc)	9	3	2	
Other (please name)		.fltu (1)		
Hybrid-formats				
GeoPackage (.gpkg)	11	1		
Other (please name)				

networks, seem to be unknown and therefore not in use yet. An opposing image can be observed with regard to the use of vector data formats. Seventy-five per cent of the respondents stated that they have a basic knowledge about specific vector data formats. Seven use them in their daily work. Only the "Esri Shape File" format seems to be relevant. Other vector formats do not appear. However, a wider user experience seems to be present when using raster data formats. At least around 33% of the actors use raster formats. However, innovative, interoperable data standards like the hybrid file format "GeoPackage" are still completely unknown.

5.2 *Conceptual Design for a Vulnerability Map System Based on New Standardized Data Formats and Web Services*

As part of the "Klimpass-Aktiv" project, mentioned in Chapter five (A Case Study: Application of GeoPackage Within Climate Services), vulnerability assessments were initially carried out on a qualitative basis. This was mainly done through the integration of different data layers. In such a way, it was possible to detect vulnerabilities in different sectors. The resulting maps enable users to get insight into areas that are threatened by harmful phenomena, e.g. flooding, erosion or the spread of invasive plants. An exemplary map of vulnerable areas threatened by flooding and soil erosion in case of heavy rainfall is presented in Fig. 3.

The BebeR project aims at a more in-depth analysis, focusing on the vulnerability concerning soil erosion, thus applying quantitative methods. This will be achieved by using the Revised Universal Soil Loss Equation (RUSLE). The equation is an evolution of the Universal Soil Loss Equation (USLE) (see NSERL 2017). Under consideration of the GeoPackgae format, a GIS service for planners and authorities will be developed for this purpose. One goal is to generate a version that is easy to handle, thus requiring low computer expertise. The findings of the survey described in Chapter five (A Case Study: Application of GeoPackage Within Climate Services) can help to achieve this. A second aim is the integration of the resulting GIS service and the extension of the online mapping system.

Summarizing the various conceptual and practical sub-goals mentioned before, the research should result in a system that contains services that are based on open and innovative geodata standards. It will be created following the schematic structure shown in Fig. 4. However, if the conceptual design and the following implementation will be assessed positively, this would serve as a verification of the hypothesis mentioned at the beginning.

The data integration process should be achieved by the adjustment and extension of the existing system that must be compliant to INSPIRE data interfaces.

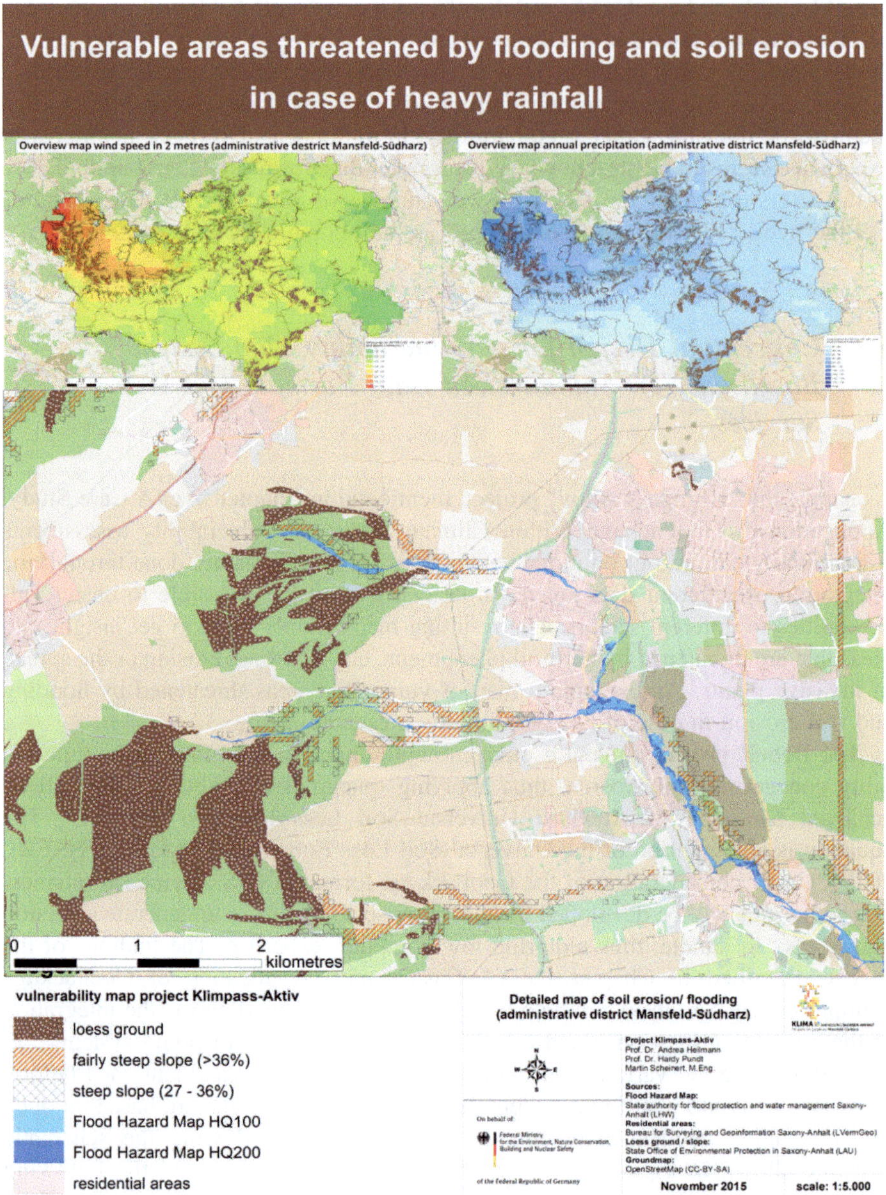

Fig. 3 Vulnerable areas threatened by flooding and soil erosion in case of heavy rainfall

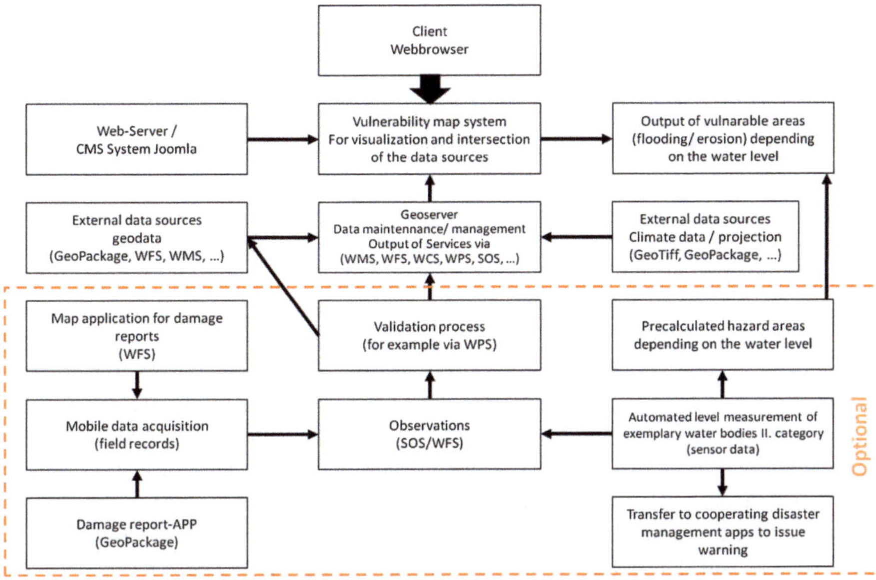

Fig. 4 Schematic structure of the planned system that integrates climate services to achieve a comprehensive user support during the cross-sector decision-making process about climate change adaptation measures

Due to the focus on small river catchments, specific requirements concerning water level measurements must be considered. Since constant level measurements in Germany are only carried out for larger streams and rivers, data on small running waters are lacking. This is also true for the exemplary rivers that serve as investigation objects within the BebeR project. An extension of the previously described online mapping systems could be envisaged by means of a test setup aiming at real-time data collection at these rivers. For this purpose, a cost-effective data collection system (sensor observation service) could be installed and tested. It would make sense to link the level measurements with existing disaster management warning applications, such as the apps NINA or KATWARN (see Reuter et al. 2017; Klafft and Reinhardt 2016), which are widely used. The development of a damage reporting application, which has an upload function of georeferenced images, would result in a quite comprehensive support of users. Precalculated hazards or flooding surfaces, however, could be represented in the mapping system. In such a way, the significance of open standards and the benefits of climate services for all actors during the development of climate change adaptation measures would become visible. Their contribution to more effective and qualitatively better decision support could be proofed.

6 Conclusion

Climate services have undergone a rapid change during the foregoing ten years. While in the past, the focus of the institutions and initiatives was on global or nation-wide considerations, nowadays the need for climate services on a regional or local scale has been identified. The aim of future climate services is to respond to the increasing demand from stakeholders who ask for current and usable information as well as online solutions that support an adequate assessment of climate change impacts, vulnerabilities and risks. However, the support during the evaluation of opportunities, their uncertainties and probabilities, as well as the assessment of different options for actions, is highly needed to achieve decisions that satisfy as many stakeholders, and possibly citizens, as possible (ERA4CS 2016).

The implementation of various new guidelines (for example INSPIRE) and standards (OGC Web Services, GeoPackage) helps to develop services as they were described above. In particular, smaller institutions and administrations often suffer a lack of employees with appropriate skills to use and integrate multidisciplinary data sources within their daily work. Innovative and open online services could help to close the gap and parallel support qualitatively better decisions.

Future research and a stakeholder-based test of the system (see Fig. 4) will give insights whether it fulfils the requirements and helps to achieve the envisaged goals. A broader participation of relevant actors, including citizens, and the usage of innovative data formats and mapping opportunities seem to be important steps towards deeply discussed, well-grounded and sustainable climate adaptation measures.

References

Adams, B., & Suykens, F. (2013). Astute: Increased situational awareness through proactive decision support and adaptive mapcentric user interfaces. In *Proceedings of the 2013 European Intelligence and Security Informatics Conference (EISIC)*, Uppsala, Sweden, August 12–14, 2013, pp. 289–293. http://ieeexplore.ieee.org/stamp/stamp.jsp?arnumber=6657185.

Babitski, G., Bergweiler, S., Hoffmann, J. Schön, D., Stasch, C., & Walkowski, A. C. (2009). Ontology-based integration of sensor web services in disaster management. In K. Janowicz, M. Raubal, & S. Levashkin (Hrsg.), *GeoSpatial semantics*. Third International Conference, GeoS 2009, Mexico (pp. 103–121). Berlin, Heidelberg: Springer.

Botts, M., Percivall, G., Reed, C., & Davidson, J. (Hrsg.) (2007). OGC sensor web enablement: Overview and high level architecture. *OGC Doc. No. OGC 07-165, Version 3*. portal.opengeospatial.org/files/?artifact_id=25562. Last Accessed July 12, 2017.

Climate Knowledge Hub—A network of climate service providers. http://www.climate-knowledge-hub.org/#/. Last Accessed July 13, 2017.

Copernicus—A European Union Programme aimed at developing European information services based on satellite Earth Observation and in situ (non-space) data. www.copernicus.eu/main/services. Last Accessed July 14, 2017.

Cortekar, J., Máñez, M., & Zölch, T. (2014). Klimadienstleistungen in Deutschland - Eine Analyse der Anbieter und Anwender. *CSC Report 16*, Climate Service Center, Germany.

Daisey, P. (2015). OGC® GeoPackage Encoding Standard with corrigendum. *OGC Doc No. OGC 12-128r12*. http://www.opengis.net/doc/IS/geopackage/1.0. Last Accessed July 12, 2017.

Eicker, U., Nouvel, R., Schulte, C., Schumacher, J., & Coors, V. (2012). 3D-Stadtmodelle für die Wärmebedarfsberechnung. In *Fourth German-Austrian IBPSA Conference—BauSim 2012*, Berlin.

ERA4CS—European Research Area for Climate Services. (2016). ERA4CS Joint Call on Researching and Advancing Climate Services Development by (A) Advanced co-development with users, (B) Institutional integration, Joint Programming Initiative "Connecting Climate Knowledge for Europe" (JPI Climate). http://www.jpi-climate.eu/media/default.aspx/emma/org/10869130/ERA4CS_joint+call_04march.pdf.

Gröger, G., Kolbe, T. H., Nagel, C., & Häfele, K.-H. (Hg.). (2012). OGC City geography markup language (CityGML) encoding standard. *OGC Doc. No. OGC 12-019*.

Hewitt, C., Mason, S., & Walland, D. (2012). The global framework for climate services. *Nature Climate Change, 2*(2012), 831–832. https://doi.org/10.1038/nclimate1745.

Kalberer, P. (2013). GeoPackage, das Shapefile der Zukunft. *FOSSGIS Konferenz 2013*. Rapperswil, Schweiz.

Klafft, M., & Reinhardt, N. (2016). Information and interaction needs of vulnerable groups with regard to disaster alert apps. Gesellschaft für Informatik e.V. 2016. In B. Weyers & A. Dittmar (Eds.), Mensch und Computer 2016 – Workshopbeiträge, September 4–7, 2016, Aachen. http://dx.doi.org/10.18420/muc2016-ws01-0003.

Klimanavigator - Der Wegweiser zum Klimawissen in Deutschland. www.klimanavigator.de. Last Accessed July 13, 2017.

Masó, J., Diaz, P., Riverola, A., Díaz, D., & Pons, X. (2013). Exchanging the Status between Clients of Geospatial Web Services and GIS applications using Atom. In *Proceedings of the International Multi Conference of Engineers and Computer Scientists* (Vol. 1, pp. 385–390). https://www.researchgate.net/publication/289780174_Exchanging_the_Status_between_Clients_of_Geospatial_Web_Services_and_GIS_applications_using_Atom.

Meissen, U., & Röthlein, B. (2010). SAFE - Ein künftiges Extremwetterwarnsystem im Praxistest, Schadenprisma. http://www.schadenprisma.de/pdf/sp_2010_1_2.pdf.

NSERL. (2017). (National Soil Erosion Laboratory, Purdue University Indiana), Revised Universal Soil Loss Equation, Version 2 (RUSLE2). http://fargo.nserl.purdue.edu/rusle2_dataweb/RUSLE2_Index.htm.

Rashidan, M. H., & Musliman, I. A. (2015). GeoPackage as future ubiquitous GIS data format: A review. *Jurnal Teknologi, 73*(5), 47–53. https://doi.org/10.11113/jt.v73.4319.

Reinstorf, F., & Ohme, J. (2014). Hydrologische Bewertung des Mulden-Rigolen-Systems im Bebauungsgebiet Poetengang der Stadt Sangerhausen, Projektbericht, Hochschule Magdeburg-Standal, Fachbereich Wasser- und Kreislaufwirtschaft.

Resch, B. (2013). Standardisierte Geosensornetzwerke für Umweltbeobachtungen in naher Echtzeit. In A. Koch, R. Bill, & A. Donaubauer (Hg.), *Geoinforationssysteme 2013*, Beiträge zum 18. Münchner Fortbildungsseminar (pp. 260–264). Berlin: Wichmann Verlag.

Reuter, C., Kaufhold, M., Leopold, I., & Knipp, H. (2017). KATWARN, NINA, OR FEMA? MULTI-METHOD STUDY ON DISTRIBUTION, USE, AND PUBLIC VIEWS ON CRISIS APPS. In *Proceedings of the 25th European Conference on Information Systems (ECIS)*, Guimarães, Portugal, June 5–10, 2017 (pp. 2187–2201). ISBN 978-989-20-7655-3. Research Papers. http://aisel.aisnet.org/ecis2017_rp/139.

Schöner, W. (2010). Basics of climatological and meteorological observations for GIS applications. In P. Carrega (Ed.), *Geographical information and climatology 2010*. Wiley-ISTE: Hoboken, NY.

Simonis, I., De Lathouwer, B., & Taylor, T. (2016). Sensor web enablement (SEW) for citizen science. In *2016 IEEE International Geoscience and Remote Sensing Symposium (IGARSS)*, Beijing, 2016, pp. 3618–3620.

Uppala, S., Broad, A., Delécluse, P., Köchy, M., Strahlendorff, M., & Doherty, M. (2011). GMES Climate service: Towards a European knowledge base in support of mitigation and adaptation. European Commission Enterprise and Industry Directorate-General. Draft Document DG ENTR/H.4/CC.

Vescovi, F. D., Lankester, T., Coleman, E., & Ottavianelli, G. (2015). Harmonisation initiatives of Copernicus data quality control. *The International Archives of the Photogrammetry, Remote Sensing and Spatial Information Sciences, Volume XL-7/W3*, 2015 36th International Symposium on Remote Sensing of Environment, May 11–15, 2015, Berlin, Germany.

Interdisciplinary Research on Energy Efficiency Standards and Climate Change Mitigation: Methods, Results, and Communication

Carl J. Dalhammar and Jessika Luth Richter

Abstract Energy efficiency regulations are one of the important policy packages employed for climate mitigation. Recent evaluations from the USA and the European Union (EU) indicate that the setting of minimum mandatory energy performance standards (MEPS) for different products—such as electric motors, TVs, and dishwashers—are among the best-performing climate policies and that these regulations are also cost-effective. In the EU, MEPS are adopted under the Ecodesign Directive. Despite the positive evaluations, MEPS have also come under criticism from media and some researchers, questioning their unintended effects and their cost-effectiveness. This chapter reflects the research of a five-year, interdisciplinary research program (2013–2018) which researched the effects and potential of the Ecodesign Directive. Multiple methods were employed in the project, which have resulted in various insights. The aim of this chapter is to discuss the main experiences with this interdisciplinary research project, including the variety of research methods, the ways the results have been communicated to policymakers and other actors outside academia, and the need for future research to better understand the interactions between resource efficiency and climate mitigation in the context of product regulation. Two issues of high importance in the climate change mitigation research arena are highlighted. The first issue concerns the need for research that looks at interactions between energy efficiency and resource efficiency in the case of MEPS, and the need to develop new methods and approaches. Secondly, in the case of the "policy mix" for effective climate mitigation, there is a need to develop new theories, and also to consider the political feasibility of proposed policies.

Keywords Ecodesign · Energy efficiency · MEPS · Climate policy
Ecodesign Directive

C. J. Dalhammar (✉) · J. L. Richter
International Institute for Industrial Environmental Economics (IIIEE),
Lund University, Stora Algatan 2, 221 00 Lund, Sweden
e-mail: carl.dalhammar@iiiee.lu.se

© Springer International Publishing AG, part of Springer Nature 2019
W. Leal Filho and R. Leal-Arcas (eds.), *University Initiatives in Climate Change Mitigation and Adaptation*, https://doi.org/10.1007/978-3-319-89590-1_19

1 Introduction: MEPS and the Ecodesign Directive

Generally, there are three main categories of policies for inducing climate change mitigation (Hood 2013):

1. *A price on carbon*, which induces economy-wide mitigation efforts, primarily (at least in theory) among actors that have the lowest marginal abatement costs. A price can be set through different policies such as a carbon tax or a trading scheme;
2. *Renewable technology policies*, which provide support for renewable technologies which are not yet competitive on the market. These policies are required to make new technologies competitive, increasing the scale of technological uptake, to reduce their cost[1];
3. *Energy efficiency policies*, which can unlock cost-effective carbon mitigation investments that are not otherwise realized due to barriers such as lack of information, split incentives, information asymmetries, and limited rationality among users.

In this third category of policies, minimum energy performance standards (MEPS) are used to set mandatory energy efficiency requirements on buildings, vehicles, and products. In this chapter, we will concentrate on MEPS for products and exclude buildings and vehicles.

While MEPS may not be a well-known acronym, MEPS for products appear to be the best-performing European Union (EU) climate policy. MEPS work in combination with mandatory energy labeling; MEPS remove the worst performers from the EU Internal Market, whereas energy labeling enables consumers to differentiate among compliant products on the market based on energy efficiency performance. In the EU, estimated savings from the first MEPS regulations are provided in Table 1 (Bertoldi and Atanasiu 2009). These savings are calculated until 2020 and they accumulate over the years as old products are substituted for new, more efficient, products.

In the study, electric motors stood out as the product group with the highest savings, as electric motors account for 40–50% of global electricity use. It is expected that the electric motor MEPS can save about 5% of the current EU electricity use. Later regulations addressed additional product groups with large potential, such as boilers, ventilation, and lighting products. In a large evaluation from 2015 (VHK 2015), these product groups were considered (altogether product groups representing approximately 50% of EU energy use are currently addressed by MEPS/labeling), and it was estimated that all EU MEPS and labels would lead to (compared to a business-as-usual, BAU, scenario):

[1]There are a vast number of such policies, aimed at various stages of technology development and diffusion, which are both general and technology specific in nature. They include feed-in tariffs, green certificates, consumer subsidies, funding of R&D, and technology demonstration projects.

Table 1 Expected savings under the first 12 implementing measures adopted under the Ecodesign Directive in combination with energy labeling

Adopted regulations	Estimated savings (yearly by 2020) (TWh)
Standby and off mode losses, electric, and electronic equipment	35
Simple set top boxes	9
Domestic lighting	39
Tertiary sector lighting (office and street)	38
External power supplies	9
Televisions	43
Electric motors	135
Circulators	23
Domestic refrigeration	8
Domestic dishwashers	2
Domestic washing machines	1.5
Fans	34
= 376 TWh = 14% of the electricity consumption of the EU in 2009	

Source European Commission. Based on Bertoldi and Atanasiu (2009)

- By 2020, net savings representing approximately 9% of the current EU energy consumption and 7% of the carbon emissions. The average saving for the product groups addressed by policies is 18%;
- By 2030, net savings growing to 15% of current EU energy consumption and 11% of carbon emission totals.

Further, the consumers' monetary savings, as a percentage of the current GDP of the EU, were estimated to be close to 1% in 2020 and 2.3% in 2030. Additional benefits included €55 billion extra revenues for industries (e.g., from lower energy bills), reduced pollution, savings of water and paper, and job creation. The report claims that the numbers presented were quite conservative, as potential new MEPS/labels not yet enacted were not part of the study.

In the USA, a study from 2003 (Meyers et al. 2003) estimated that US savings from adopted MEPS would reduce residential primary energy consumption and carbon dioxide (CO_2) emissions by 8–9% in 2020. Based on data, between 1987 and 2000, the standards saved US consumers an $17 billion USD and the total combined net present value of consumer benefit amount to nearly $80 billion USD by 2015, and to grow to $130 billion USD by 2030.

Thus, such reviews show MEPS have an impressive track record. Furthermore, researchers like Noah Sachs have argued that MEPS for products and fuel efficiency standards have been the main drivers for energy efficiency improvements in the USA so far: "*Although information disclosure, financial incentives, and other softer alternatives to regulation play a vital role in reducing energy demand, these should be viewed as complements to efficiency regulation, rather than replacements*." (Sachs 2012, p. 1633).

Despite these achievements, energy efficiency policies like MEPS have received less attention and research than renewable energy and carbon pricing policies, although this is now changing at the EU level. Further, the MEPS have been under attack from media in the USA and Europe (Sachs 2012; Barford and Dalhammar 2015), and some economists are also concerned that they can distort the functioning of carbon markets, are less cost-effective than carbon pricing and may increase costs, and can face political obstacles (Böhringer et al. 2016; IETA 2015; see also Aldy and Stavins 2012 and Van Buskirk et al. 2014).

This chapter details the research themes, approaches, findings, and communication of a five-year research project on the impact and potential impacts of the Ecodesign Directive. The project started in January 2013 and concludes in February 2018. The project name is "Promoting ecodesign for the advancement of innovation, competitiveness, and environmental improvements." The project is funded by the Swedish Energy Agency (Energimyndigheten). The project originated at a time when MEPS for energy using products through the Ecodesign Directive had been in force for approximately 5 years, and a number of interesting research topics had started to emerge. In the project application, four overarching research themes were outlined; see Table 2. The overall aim was to research potential of the Directive in a dynamic context. This included how more stringent MEPS could be set, innovation effects of the Directive, and how the Directive could interact with other climate and energy policies. At this time (around 2012), the current topic of discussion was whether the Directive and MEPS could also address resource efficiency, an issue which has renewed interest with the emergence of the Circular Economy concept in the last couple of years. This topic was also included in the research project.

Thus, the program has several different objectives, which required an interdisciplinary approach. The interdisciplinary nature of the project, and realization of the multiple objectives, in turn required the use of various methods, including use of both qualitative and quantitative research methods. While most of research has made use of qualitative methods employed in social sciences, modeling (e.g., life-cycle assessment (LCA)) and more quantitative approaches have also been employed in the project. It should be noted that unlike many other research projects, a specific aim of the project has been to produce policy-relevant results, and to communicate these results to policymakers and other researchers.

The aim of this chapter is to discuss the main experiences with this interdisciplinary research project, including the variety of research methods, the ways the results have been communicated to policymakers and other actors outside academia, and the need for future research to better understand the interactions between resource efficiency and climate mitigation in the context of product regulation.

Section 2 presents the main results and publications (completed and planned) and also outlines the methods employed. Further, the need for new theories and approaches in the project is discussed. Section 3 accounts for the main communication events in relation to academia and policymakers. In Sect. 4, we discuss some reflections on the processes and learnings we have made in the project.

Table 2 Themes in the research project

1. Energy efficiency and energy sufficiency of systems and appliances (a) How can the Directive trigger even more stringent energy efficiency standards? (b) Can it be used to drive "sufficiency" type of standards, where technological innovations are required if larger appliances are to be allowed on the market? (c) Can the standards focus more on systems—as opposed to single products—and what is the environmental potential if such an approach is widely used? What are the barriers?
2. Resources and material use (a) Can the Directive be an effective tool to trigger environmental improvements related to resource use and material choice; and can it be a driver for the Circular Economy objectives, e.g., by setting standards for recycled content, phasing out hazardous substances, and through material restrictions? (b) Can the Directive aim in the pursuit of new business models with high potential for resource savings, based on selling functions instead of products? (c) What are the potential clashes with WTO law?
3. Innovation and competitiveness (a) How can the standards set under the Directive and related standards be designed in order to boost the competitiveness of EU industries? (b) What are the most important policy issues to consider? (c) How should legal standards interact with industry-driven standardization?
4. The policy mix and the potential for CO_2 reductions in relation to other instruments (a) How should the Directive interact with other relevant policy instruments, including green procurement, technology procurement, eco-labels, and energy labels? (b) How well does the Directive function as a tool for CO_2 reductions in comparison with the EU-ETS and carbon taxes when evaluated through criteria including environmental effectiveness, cost-efficiency, and political feasibility?

2 Publications and Methods

This section describes the completed publications and methods in the project, followed by an analysis of the progression of methods, approaches, and theories employed.

An overview of the completed and soon to be completed publications is provided in Table 3 (the full reference list of publications is provided in Annex). Approximately 25 publications are expected by completion of the project. The existing publications include publications by the project researchers, but also a few publications co-authored with students, policymakers, and people in companies involved in market analysis of relevant products subjected to MEPS.

Most methods were grounded in social science, but toward the end of the project more methods associated with economics, engineering and natural science were employed. Among the different methods, and their objectives, we find:

- Semi-structured, anonymous interviews with people in industry to inquire the effects of regulations and their attitudes toward proposed regulations, and drivers and barriers on the market;
- Media content analysis of UK media, to see how they portrayed MEPS, and whether there seem to be any correlation with BREXIT arguments;

- Market data and costs of products, to inquire the relationship between regulations and costs;
- The use of LCA software and life-cycle cost (LCC) modeling, to inquire under what conditions it makes sense to prolong the lifetime of products, and the factors that are of relevance in determining this.[2]
- Mapping of theoretical and "real" interplays between MEPS and other relevant schemes such as the EU emission trading scheme (EU-ETS), now and in the near future.

Regarding methods, it can be seen that several initial publications were based on qualitative methods, which include legal analysis, policy analysis, and semi-structured interviews. This is due to the fact that much focus in the research was devoted to understand the policies and their interactions and the studied effects. In the case where this understanding could not be obtained from the existing literature, interviews with industry were employed to get a better understanding of (1) how the industry had been affected by existing policies, (2) what industry expected from future policies, and (3) input from industry on the "appropriate" policy mix.

As understanding of the topic evolved, so did the understanding of the most interesting research gaps. Again, reflecting a common reality of interdisciplinary research, often these gaps were best researched with different methods than those employed in the beginning of the project, and beyond the expertise of any single researcher. The inclusion of some of the methods in the latter part of the project necessitated the hiring of new researchers in Lund University and that cooperation with researchers and corporations outside the university have been initiated. In one case, an experienced researcher in another university was hired ("paid per hour") in order to assist the research. The methods progressed during the project and came to include semi-structured interviews, media analysis, statistical methods, modeling, and life-cycle assessment (LCA) software.

With the emergence of the Circular Economy (European Commission 2015) concept, there has been increasing interest in how MEPS can address not only energy efficiency, but also resource efficiency. The potential was analyzed through a large report (Dalhammar et al. 2014) and through interviews with industry. One potential option identified is to prolong the lifetime of products to save resources. This can be done, e.g., by forcing manufacturers to provide evidence that appliances —or components in appliances that often break down—will last for a specific period and provide evidence for the durability. While this makes sense for many product groups, there are some caveats, such as:

1. If consumers switch products not because they do not function, but because consumers want novelty (or better-performing products), it does not make sense

[2]Whether it makes ecological sense to prolong the lifetime of a product depends on inter alia the life-cycle impacts associated with the product, the rate of (technical) change, and user behavior (for instance: if people want novelty and will replace the products anyway, making it more durable may lead to waste resources); these factors are however not static but change over time.

Table 3 Publications and methods

Title	Type of publication/Year	Main methods used	Comments
Industry attitudes towards ecodesign standards for improved resource efficiency	Journal article, 2016	Semi-structured interviews with people in industry	Provided input form industries on effects of potential future regulations
Regulating planned obsolescence: a review of legal approaches to increase product durability and reparability in Europe	Journal article, 2016	Legal analysis, literature reviews	Outlined EU and national policies
The application of 'life cycle thinking' in European environmental law: theory and Practice	Journal article, 2015	Legal analysis, literature reviews	
The Setting of Progressive Energy Efficiency Performance Standards for Products through the Ecodesign Directive	Journal article, 2015	Legal analysis, literature reviews	
Promoting energy and resource efficiency through the Ecodesign Directive	Journal article, 2014	Legal analysis, literature reviews	
Rethinking the Ecodesign Policy Mix in Europe	Book chapter, 2017	Policy analysis	
Addressing resource efficiency through the Ecodesign Directive: a review of opportunities and barriers	Report, published by the Nordic Council of Ministers, 2014	Literature reviews, assessments of national and European policies, legal analysis, semi-structured interviews	The report was written by researchers at Lund and Aalborg universities. It was developed in a separate project but material from the studied project was used in the report
Addressing resource efficiency through the Ecodesign Directive—A review of opportunities and barriers	Policy brief, published by the Nordic Council of Ministers, 2014	Summary of report	Policy brief based on the report with the same name
Accounting for durability in least life cycle cost methods	Peer reviewed conference paper, 2017	Modelling, statistical analysis, market data analysis, literature reviews	The paper was written together with people from a company doing market analysis and one policy-maker from a government agency

(continued)

Table 3 (continued)

Title	Type of publication/Year	Main methods used	Comments
When prices don't steer—mimicking ambitious carbon pricing with energy performance standards	Peer reviewed conference paper, 2017	Modelling, statistical analysis, market data analysis, literature reviews	The paper was written together with people from a company doing market analysis
The Top Runner and the Ecodesign Directive: what can we learn from current policies for product energy efficiency and how can we apply it in the construction sector?	Peer reviewed conference paper, 2013	Policy analysis, literature reviews	The paper was written for a workshop with policymakers
Regulating complex products: The policy mix for product and building energy efficiency	Conference paper, 2015	Policy analysis, legal analysis, literature reviews	
Policies promoting ecodesign for energy and resource efficiency in Europe: experiences, barriers and future options	Conference paper, 2014	Policy analysis, legal analysis	
The setting of ecodesign standards to promote improved waste recycling	Conference paper, 2014	Policy analysis, legal analysis	
Addressing resource efficiency through the Ecodesign Directive: a review of opportunities and barriers	Conference paper, 2013	Policy analysis	
Potential for rare earth element resource efficiency improvements in permanent magnet motors through and extension of the electric motor product group regulation under the Ecodesign Directive	Conference paper, 2013	Semi-structured interviews, policy analysis, technical analysis	
Climate change and energy efficiency standards: a study of the UK media backlash to EU Ecodesign standards for vacuum cleaners	Working paper, 2015	Media analysis; metadata and content analysis	Written by a master student and one project researcher

(continued)

Table 3 (continued)

Title	Type of publication/Year	Main methods used	Comments
Options for lifetime labeling: design, scope and consumer interfaces	Paper for introducing a workshop in a conference, 2017	Policy analysis	Made to prepare workshop participants for a workshop, were we want input from designers and researchers
Considering Optimal Lifetimes for LEDs: a mixed approach and policy implications	Conference paper, 2017	Life cycle costs modeling and life cycle assessment software	Written with LCA lighting expert to introduce more dynamic modelling with LCAs for LEDs
Energy efficiency regulations, market and behavioral failures, and standardization	Book chapter, forthcoming	Literature reviews	Applies new theories to try to understand why MEPS work better in practice than standard neoclassical theory would suggest
Interaction between carbon pricing and energy efficiency: A Case study on interactions between the EU ETS and the Ecodesign Directive	Student master thesis (expected Sept 2017)	Policy evaluation, statistical methods	The aim is to outline policy interactions in theory and practice
A model for estimating the optimal lifetime for LEDs	Journal article (2018)	Life cycle cost and impact modeling and life cycle assessment software	

to prolong the lifetime; it may actually lead to wasted resources due to "over-engineering."

2. For products that are rapidly evolving (e.g., since the product group is rather new and under technological development), it may not be a good idea to make it very durable: If the product is durable, it means it cannot be replaced by a more energy-efficient product until a later date. This means that the environmental gain from a resource perspective (if the product lasts longer, it takes longer before it is replaced, which saves resources) may be offset due to losses in energy efficiency (by keeping an old energy-inefficient product in use longer). This problem may be very relevant for product groups undergoing rapid energy efficiency improvements. In previous years, this has included fridges, freezers, and more lately TVs (due an increased use of LED technology). Currently, this may be relevant for lighting products as these are undergoing rapid improvements in energy efficiency. The problem is less relevant for mature products, where the potential future energy efficiency improvements are small (e.g., for vacuum cleaners).

The issues above represented significant research gaps in how resource efficiency should best be included in MEPS and were researched in several ways, such as:

1. Interviews with industry to get their views on potential MEPS related to resources. This provided insight into some complex issues. For instance, consumers may want novelty in the case of consumer electronics (CE), and switch model often, but if manufacturers would be legally mandated to make more durable CE, they may be interested in changing their business model. For instance, they may make CE that are upgradable, and therefore in use longer.
2. Through life-cycle costing and LCA software modeling, it is possible to model scenarios that can highlight potential trade-offs and influential factors when it comes to the interactions between resource efficiency/durability and energy use/ CO_2 emissions. It also allows taking regional variations—e.g., differences in energy mix—into account when designing an "optimal policy." This research not only added such sensitivity analyses to LCA and LCC modeling, but also added dynamics such as technological innovations (e.g., increasing energy efficiency of products), which had only been considered in a limited way by previous research approaches. Initial modeling has indicated optimal durability can be modeled, and while it is context specific, generally as energy mixes decarbonize and products mature, durability is increasingly desirable. Furthermore, initial LCC modeling also demonstrated a method that could be easily employed by policymakers to assess the theoretical optimal durability on the market as a means of more easily incorporating and monitoring this in MEPS.
3. Mandatory MEPS are not the only option to incentivize durability. Consumer law (e.g., consumer guarantees) and public procurement (where risks and durability can be included in the contract, or where the supplier can retain

ownership of the product), as well as mandatory consumer information (i.e., the manufacturer must state on the product how long it will last) are other options. These have been explored in the project and will be further explored through future workshops.

In addition, while public acceptability of MEPS has been argued as an advantage of this policy approach, there were media reports during this research project that contested this argument. This topic that had not received much attention in academia in order to better understand this development, media analysis was utilized in order to gain insights into: (1) how media framed the MEPS (positive vs. negative), and (2) whether this framing was associated with the political orientation of newspapers.

Regarding theories, many publications relied on traditional "policy evaluation" theory and methods (cf. Mickwitz 2006), particular the research topics reviewing individual policies and policy mixes. Policy evaluations in this research generally considered criteria of environmental effectiveness, political feasibility as well as criteria based on more economic theory including cost-effectiveness and transaction costs. One interesting issue with MEPS is that they appear to be good not only for triggering product energy efficiency improvements, but also appear to be associated with decreased product costs (for consumers) and improved product quality over time (cf. Taylor et al. 2015; Brucal and Roberts 2017). This is at odds with the opinion of many neoclassical economic scholars, which tend to associate regulation with higher product prices and reduced innovation activities among corporations. In the chapter "Energy efficiency regulations, market and behavioral failures, and standardization," we have therefore looked at various theoretical contributions that could explain this phenomena. Concepts like "transaction costs," "network effects," and "behavioral failures" can help explain why command-and-control regulation may indeed lead to win-win-win outcomes and be preferred to economic instruments in some cases.

An additional area of the research currently being explored concerns the question of what are the "best-"performing climate mitigation policies, with an aim of enhancing climate policy packages. Some of the initial findings are:

1. For various reasons, we need a "policy mix"; therefore, one should be a bit skeptical about research that recommends the use of one single policy.
2. Much research, especially neoclassical economics, originates from an "ideal" situation, and assumes that the best policy can be implemented. In reality, many policies cannot be implemented, and in any case, they will not work as well as projected. One benefit of MEPS compared to economic instruments (taxes and charges) is that they are politically feasible to implement.
3. MEPS seem to work much better in practice than would be expected from theory (especially neoclassical economic theory). They can both cut energy use and save money for consumers, but also provide incentives for product design among manufacturers beyond what is legally required and induce innovation. This implies that some current approaches and theories need to be

Table 4 Outreach activities

Event	Stakeholders	Aim	Comments
Speech at the workshop "Eco-inovação e compras públicas em mercados da fileira habitat - Boas práticas e instrumentos de política" Lisbon, Dec. 2013	The workshop was arranged by the ECOPOL project. Attendants came from several sectors, but a majority was policymakers from Portugal	Presented how MEPS were regulated in the EU and Japan, and the coordination between the Ecodesign Directive and other policies	One project researcher was invited as speaker
Speech at the Workshop "Ecodesign as a Tool for Resource Efficiency and Circular Economy" Brussels, June 2014	The WS was arranged by the Nordic Council of Ministers Speakers came for inter alia the DG Energy, DG Env, the European Parliament, national ministries, corporations, academia, and NGOs	Knowledge exchange and initiating discussions on how to move forward with the Ecodesign Directive	
Speech on 'research about the Ecodesign Directive', at network meeting for Swedish authorities Stockholm, Nov. 2015	8 Swedish authorities	Inform authorities working with the Ecodesign Directive and associated legislation about ongoing research/ planned research	One project researcher was invited as speaker
Arranging workshop on 'New policy approaches for Eco-design' at the Global Cleaner Production and Sustainable Consumption Conference Sitges, Nov. 2015	Workshop arranged together with researchers at Aalborg University and the Joint Research Centre (The European Commission's science and knowledge service)	Attract more presentations related to the Ecodesign Directive; especially policy-relevant research and presentations by practitioners Initiate discussions among researchers and practitioners	

(continued)

Table 4 (continued)

Event	Stakeholders	Aim	Comments
Arranging the workshop “Options for lifetime labeling: design, scope and consumer interfaces” at the PLATE conference Delft Nov. 2017	Participants are expected to be designers in academia and industry, as well as some other stakeholders	The main aim is to discuss and get good input on how mandatory Ecodesign standards on products lifetimes can be designed, and the barriers and drivers; and how the potential for regulation varies between products groups	
Arranging a workshop with the preliminary title “The Ecodesign Directive in a changing policy climate- challenges and opportunities!” Brussels, Jan. 2018	The WS will be arranged with other actors, which may include the Swedish Energy Agency and ECEEE (www.eceee.org)	The main aim is to present the main research findings in the project and to initiate discussion of importance for the future, such as: the interaction btw the Ecodesign Directive and other standards; how to regulate product lifetime	It is expected that there will be three presentations from the project, and six additional presentations from other academics, NGOs and policymakers

complemented with new theories and insights. For instance, one chapter in the project looks at transaction costs and some insights from behavioral economics, as these may explain some of the positive outcomes of MEPS.

3 Communication and the "Third Task"

The research has been presented at several academic conferences, and also at conferences with practitioners from industry and governments. In this chapter, however, we focus on discussing communicative activities that are aimed specifically aimed at policymakers, industry, and NGOs. The project has aimed at providing concrete policy recommendations, and as discussed previously, there has been some papers co-authored with policymakers and corporations.

Table 4 provides an outline of completed and planned activities, the cooperating partners, and the main aim of the events.

The workshops above include both events where project researchers have been invited to speak and events where projects researchers have been involved in the planning. The two final events on the list are initiated by the project researchers.

The planned event in January 2018 may be considered as the final activity in the research project. The final event will bring up the most complex issues involved in the research project, such as the interaction between the Ecodesign Directive and other policies, particularly climate policies. It will also provide a link to the developments in the Circular Economy, where the Ecodesign Directive is considered to be a key policy. Lastly, it is a stepping stone toward the future as it will discuss the key opportunities and barriers that lie ahead for the Ecodesign Directive and the need for future research.

4 Conclusions

In this short chapter, we have only been able to outline some main contours of the research project. The uniqueness of the project lies in its interdisciplinary nature and its objective to come up with relevant policy recommendations, as well as the intent to communicate them to a broader audience. In the project, we have learned that being truly interdisciplinary requires hiring of new resources and takes time, as some of the activities and methods employed are time-intensive. The project was initially planned as a four-year project, but an additional year was required to conduct the more advanced modeling and statistical analysis, and an outside researcher was hired to provide support. Further, some research could not have been done without the support of master students.

There are a few key findings from the project that are very relevant for the broader discussion on climate mitigation. Firstly, although MEPS are considered to be the one of the best performing, if not *the* best performing, climate policy in the EU, it attracts rather limited research. This seems to be about to change, but the fact remains that renewable technologies and carbon markets seem to be more popular, whereas energy efficiency policies seem to be considered somewhat boring. Secondly, there is a need to develop new methodologies to explore the complexities, and potential trade-offs, between MEPS for increased lifetime/product durability (related to Circular Economy objectives) and MEPS for improved energy efficiency (related to climate mitigation objectives). This requires the innovative use of existing tools and methods, in combination. If we are to reach the objectives of both the Circular Economy and the climate change, we need to identify and model trade-offs, and the importance of different assumptions. Thirdly, there is a need to reflect upon some "orthodoxies" often expressed in policy debates. The results indicate—in line with other research—that command-and-control regulation may indeed be a better option than economic instruments in some cases. Likewise, the oft-expressed idea that product regulation will increase costs for producers and consumers may be wrong in some cases; this calls for new theories and perspectives that can complement those applied in neoclassical economic research.

The learnings from the project will be used in future projects and programs. The project researchers are involved in a new large Swedish research program[3] related to the Circular Economy, with participants from industry and municipalities, and some of the research activities will be continued in this program.

The communicative activities in the program, such as workshops with practitioners, increased the relevance of the project and added value to the researchers. However, it is also time consuming, as it requires planning and coordination. This project demonstrated that while effective communication of policy-relevant research can influence policy, the contribution of single projects is limited. Based on our discussions with policymakers, the research has helped in bringing policy forward and opens up the new discussions among policymakers in Brussels, but it is the fact that several other research projects have done similar research and provided similar recommendations that seems to have been influential. Thus, when there is a critical mass of researchers communicating similar findings, policymakers will listen.

Acknowledgements This research was funded by the Swedish Energy Agency (Energimyndigheten) through the project "Promoting ecodesign for the advancement of innovation, competitiveness, and environmental improvements" (grant no. 36936-1).

[3]Mistra-Rees, Web site: mistrarees.se.

Annex: Published Project Publications

Barford A, Dalhammar C (2015) Climate change and energy efficiency standards: a study of the UK media backlash to EU Ecodesign standards for vacuum cleaners. Working paper, IIIEE, Lund University.

Dalhammar C (2014) Promoting energy and resource efficiency through the Ecodesign Directive. Scand Studies in Law Vol. 65.

Dalhammar C (2014) The setting of ecodesign standards to promote improved waste re-cycling. Paper presented at the 25th Annual Conference of the Japan Society of Material Cycles and Waste Management (JSMCWM), Hiroshima, Sep 2014.

Dalhammar C (2014) Policies promoting ecodesign for energy and resource efficiency in Europe: experiences, barriers and future options. Paper presented at EcoBalance, Tsukuba, Oct 2014.

Dalhammar C (2015) The Setting of Progressive Energy Efficiency Performance Standards for Products through the Ecodesign Directive. Nordic Env Law Journal 2015(1):21–41.

Dalhammar C (2015) The application of 'life cycle thinking' in European environmental law: theory and Practice. Journal for European Env & Planning Law 12: 97–127.

Dalhammar C (2016) Industry attitudes towards ecodesign standards for improved resource efficiency. Journal of Cleaner Production 123(1):155–166.

Dalhammar C (2017) Rethinking the Ecodesign Policy Mix in Europe. In: Sustainability Through Innovation in Product Life Cycle Design. M. Matsumoto, K. Masui, S. Fukushige and S. Kondoh (Eds.) Singapore, Springer Singapore: 539–549.

Dalhammar C, Kiss B, Machacek E (2013) The Top Runner and the Ecodesign Directive: what can we learn from current policies for product energy efficiency and how can we apply it in the construction sector? Presented at ECOPOL—Workshop Eco-innovation and public procurement markets in the habitat value chain- Good practices and policy instruments. Dec 5th, 2013, Lisbon, Portugal.

Dalhammar C, Kiss B (2015) Regulating complex products: The policy mix for product and building energy efficiency. Paper presented at Global Cleaner Production & Sustainable Consumption Conference, Sitges, Nov 2015.

Dalhammar C, Machacek E (2013) Addressing resource efficiency through the Ecodesign Directive: a review of opportunities and barriers. In proceeding from LCM2013, the 6th International Conference on Life Cycle Management in Gothenburg Aug 2013.

Dalhammar C, Machacek E, Bundgaard A, Overgaard Zacho K, Remmen, A (2014) Addressing resource efficiency through the Ecodesign Directive: a review of opportunities and barriers. TemaNord Report 2014:511. Copenhagen: Nordic Council of Ministers.

Dalhammar C, Machacek E, Bundgaard A, Overgaard Zacho K, Remmen, A (2014). Addressing resource efficiency through the Ecodesign Directive—A review

of opportunities and barriers. Policy Brief: Nordic Working Papers, Nordic Council of Ministers Copenhagen.
Luth Richter J, Dalhammar C, van Buskirk R., Bennich P (2017) Accounting for durability in least life cycle cost methods. Proceedings from ECEEE2017, France.
Machacek E, Dalhammar C (2013) Potential for rare earth element resource efficiency improvements in permanent magnet motors through and extension of the electric motor product group regulation under the Ecodesign Directive. In proceedings from LCM2013, the 6th International Conference on Life Cycle Management, Gothenburg Aug 2013.
Maitre E, Dalhammar C (2016). Regulating planned obsolescence: a review of legal approaches to increase product durability and reparability in Europe. Review of European, Comparative & International Environmental Law (RECIEL) 25(3): 378–394.
Sonnenschein J, Luth Richter J, Dalhammar C, van Buskirk R (2017) When prices don't steer - mimicking ambitious carbon pricing with energy performance standards. Proceedings from ECEEE2017, France.
Tasaki T, Koichi S, Nakajima K, Dalhammar C (2014) Application of Life Cycle Assessment and Economic Rationality: A Cost Survey. Paper presented at EcoBalance, Tsukuba, Oct 2014.

References

Aldy, J. E., & Stavins, R. N. (2012). The promise and problems of pricing carbon theory and experience. *The Journal of Environment & Development*, *21*, 152–180.

Barford, A., & Dalhammar, C. (2015). Climate change and energy efficiency standards: A study of the UK media backlash to EU Ecodesign standards for vacuum cleaners. Working paper, IIIEE, Lund University.

Bertoldi, P., & Atanasiu, B. (2009). *Electricity consumption and efficiency trends in European Union*. Report: Joint Research Centre.

Brucal, A., & Roberts, M. (2017). Do energy efficiency standards hurt consumers? Evidence from household appliance sales. Working Paper No. 266. Grantham Research Institute on Climate Change and the Environment.

Böhringer, C., Keller, A., Bartolamedi, M., & Seyffart, A. R. (2016). Good things do not always come in threes: On the excess cost of overlapping regulation in EU climate policy. *Energy Policy, 94,* 502–508.

Dalhammar, C., Machacek, E., Bundgaard, A., Overgaard, Z. K., & Remmen, A. (2014). *Addressing resource efficiency through the Ecodesign Directive: A review of opportunities and barriers*. TemaNord Report 2014:511. Copenhagen: Nordic Council of Ministers.

European Commission. (2015). Closing the loop—An EU action plan for the Circular Economy. COM (2015) 614 final. European Commission, Brussels.

Hood, C. (2013). *Managing interactions between carbon pricing and existing energy policies. Guidance for Policymakers*. IEA Insight Series, IEA/OECD: Paris.

IETA. (2015). Overlapping Policies with the EU ETS. IETA, July 2015. Available at: http://www.ieta.org/resources/EU/IETA_overlapping_policies_with_the_EU_ETA.pdf. Accessed July 29, 2017.

Meyers, S., McMahon, J. E., McNeil, M., & Liu, X. (2003). Impacts of US federal energy efficiency standards for residential appliances. *Energy, 28,* 755–767.
Mickwitz, P. (2006). *Environmental policy evaluation: Concepts and practice.* Helsinki: The Finnish Society of Sciences and Letters.
Sachs, N. (2012). Can we regulate our way to energy efficiency? Product standards as climate policy. *Vanderbilt Law Review, 65,* 1631–1678.
Taylor, M., Spurlock, C. A., & Yang, H.-C. (2015). Confronting regulatory cost and quality expectations: An exploration of technical change in minimum efficiency performance standards. Discussion paper, Resources for the Future.
Van Buskirk, R., Kantner, C. L. S., Gerke, B. F., & Chu, S. (2014). A retrospective investigation of energy efficiency standards: Policies may have accelerated long term declines in appliance costs. *Environmental Research Letters, 9*(2014) 114010, 11 pp.
VHK. (2015). *Ecodesign impact accounting Final—Status May 2015.* Report to the European Commission. VHK, Delft.

Carl J. Dalhammar is Associate Professor at the International Institute for Industrial Environmental Economics (IIIEE) at Lund University. His research interests include law and innovation, government–industry interactions, technology transfer, energy law, and product-oriented environmental law and policy. In his research, he has analyzed several environmental laws and policies, including the Ecodesign Directive, EU energy efficiency laws, government procurement programs, eco-labeling schemes, energy labeling, environmental licensing, and public funding for cleantech. He teaches environmental law and politics in several European and Swedish master and undergraduate programs.

Jessika Luth Richter is a research associate at the International Institute for Industrial Environmental Economics (IIIEE) at Lund University. Her research focuses on policies for a resource efficiency and low-carbon economy, including ecodesign policies, extended producer responsibility, green procurement, and life-cycle assessments. In particular, she has focussed on the case of energy-efficient lighting and how questions related to energy efficiency, lifetimes, design, disposal, collection, labels, procurement, and value in closing material loops can be addressed in a synergistic way through a policy mix. She has also researched environmental economic instruments such as emissions trading schemes and green stimulus packages. She has been awarded her licentiate degree in engineering and is working toward completion of her PhD from Lund University. She is also involved in teaching Masters' courses and massive open online courses at the IIIEE.

Counting the Hot Air: Greenhouse Gas (GHG) Accounting Principles for National GHG Emission Inventories (NEIs)

Sooksiri Chamsuk

> *Truths are not relative. What is relative are opinions about truths.*
>
> Nicolas Gomez Davila, 2001.

Abstract It cannot be denied that a robust national emission inventory (NEI) serves as a powerful policy tool to formulate effective mitigation policies. The literature agrees that the official principle of the UNFCCC to construct an NEI is not powerful enough to capture causes of national emissions and has agued about alternative principles to replace the existing principle. Nonetheless, since each principle offers different emission profile, and all profiles could supplement each other in policymaking process therefore, a gap remains in which conditions additional principles could be implemented to supplement the existing official principle. Therefore, a gap remains in which conditions additional principles could be implemented to supplement the existing official principle. This paper fills this gap by examining conditions in which additional principle(s) can be adopted as secondary NEIs. Sitting in a realm of policy research with Thailand as a case study, and borrowing lens of a theory of institutions defined in New Institutional Economics, the paper offers perspectives from policymaking in a developing country, which is sparse in the literature landscape. The paper's results inform policymakers that they should prepare secondary NEIs for their own benefits in effectively governing climate change mitigation. These extensions could be achieved by collective advocacy from climate change research communities acting as an agent for change. Lastly, the paper concludes that none of currently used principles is without a flaw, and therefore it is crucial that the climate research communities continue to research for a best possible principle to count the hot air.

Keywords Climate change mitigation · National emission inventory
Policy research · Consumption-based GHG accounting

S. Chamsuk (✉)
The Institute for Policy Research, University of Bath, 10 West, Room 3.56,
Bath BA2 7AY, UK
e-mail: s.chamsuk@bath.ac.uk

© Springer International Publishing AG, part of Springer Nature 2019

W. Leal Filho and R. Leal-Arcas (eds.), *University Initiatives in Climate Change Mitigation and Adaptation*, https://doi.org/10.1007/978-3-319-89590-1_20

1 Introduction

Mountainous goals for climate change mitigation efforts that have been agreed upon include achieving a 2 °C limit for global temperature rises from the pre-industrial level. National greenhouse gas emissions lie at the core of countries when formulating their climate change mitigation policies. The national emissions are accounted for mainly by calculations and not direct measurements. Hence, the problem arises as to how these emissions should be calculated, and this is far from straightforward. A few methodological adjustments in the calculation process could alter the emission results, and, hence, climate change mitigation policy shifts accordingly.

Scholars agree that the methodology by which parties to the United Nations Framework on Climate Change Convention (UNFCCC) have been obligated to follow since 1995[1] is grounded on territorial-based GHG emissions (TBA) and accounted for emissions generated at the point of production (Afionis et al. 2016) and is incapable of effectively reflecting a country's cause of emissions and allocating responsibility to reduce the emissions accordingly. In this connection, the consumption-based GHG accounting-NEI (CBA) could serve as a more effective tool to allocate responsibility for parties than the TPA and production-based GHG accounting (PBA), even though the current scholarship is divided about the advantages of having the CBA over the TBA and PBA. In addition, differences in the calculation methodology and emissions between three accounting principles to construct NEI, namely TBA, PBA, and CBA, are apparent to some, in particular countries in the European Union. Because they are required to report their emissions to Eurostat using PBA-NEI following the EU's National Accounting Matrices including Environmental Accounting (NAMEA) (Peters and Hertwich 2008). On the contrary, others, in particular Thailand, are still left with vague perceptions about those differences as well as the benefits of each principle.

In this paper, the author highlights differences between the three principles[2] as well as explores conditions under which countries adopt additional accounting

[1]The 1995 IPCC Guidelines for National Greenhouse Gas Inventory which contains three volumes marks the very first guideline adopted and used by the parties. The volumes were presented at the UNFCCC's Conference of Parities (COP) 1 in Berlin, Germany, and published in 1995. Subsequently in 1996, at the IPCC's Twelfth session held in Mexico City, 11–13 September, additions and revisions were accepted by the IPCC. The additions and revisions were included and published as the Revised 1996 IPCC Guidelines replacing the 1995 Guidelines. The Revised 1996 Guidelines were subsided by the 2006 IPCC Guidelines, which are the latest volumes used by the parties (IPCC 2016). In 2013, supplements to the 2006 IPCC Guidelines were introduced on the wetlands sub-sector, and good practice arising from the Kyoto Protocol.

[2]The paper discusses the accounting principles for NEI construction only. There are other accounting principles discussed in the literature not included in this review due to their early development, such as technology-adjusted CBA (TCBA). In addition, those principles are designed mainly to serve as tools to allocate emissions' responsibility at the convention including the historical responsibility and income-based principles which are not discussed in this review either.

principles to be a secondary NEI. Based on the policy research's principle and framework, the paper aims to advocate for benefits a country would gain when adopting additional principles as its secondary NEI. In this paper, through the lens of institutions and institutional change in the New Institutional Economics (NIE) the author is enabled to reveal what hinders change of the institutions in question—adoption of additional accounting principles—and what could be done to succeed in implementing the change. This paper supplements the discussion on CBA by proposing that policy entrepreneurs advocating for change at a country level are needed. Apart from discussion from the global perspective, this paper offers views through the lens of a middle income, and manufacturing-led economy like Thailand under which it stands a challenge of achieving annual economic growth and at the same time, meeting its commitment of 20%[3] GHG reduction emission stated in the country's national determined contribution (NDC) to the Paris Agreement in 2015. It is therefore hoped that this exploration will illustrate a set of issues that has been underappreciated in the climate change mitigation debate.

2 Policy Research Framework

2.1 Rationale

This paper is conducted as a part of a project formulation to support policymakers in Thailand and Southeast Asian countries on the carbon intensity database and reporting under the Sustainable Development Goals framework. According to the United Nations 2030 Agenda for Sustainable Development adopted by the United Nations General Assembly in September 2015[4] (or Sustainable Development Goals —SDG in short), member states will be required to report their carbon intensity in a unit of CO_{2e} emission per sectoral value added, and the material footprint of nation per GDP. The carbon intensity unit is in line with the PBA-NEI not the IPCC's TBA-NEI. In addition, the material footprint is directly linked with the CBA. Some countries, in particular Thailand, are not aware of these connections.

Policy research differs from an academic research study that policy research focuses on answering questions relevant to policymakers and uses malleable variables (Majchrzak 1984). These characteristics of the policy research constitute the research framework of this paper.

[3]From Thailand's national determined contribution submitted to UNFCCC, 2015.

[4]The country's carbon intensity is included in Sustainable Development Goal (SDG) 9, *Build resilient infrastructure, promote inclusive and sustainable industrialization and foster innovation*, Target 9.4 to adopt sustainable production and low carbon technologies for industrial processes, from https://sustainabledevelopment.un.org/sdg9 and Target 12.2.2 (and Target 8.4.2) to achieve the sustainable management and efficient use of natural resources, from https://sustainabledevelopment.un.org/sdg12.

This paper employs a key method, focused synthesis, among other methods recommended for policy research by Majchrzak (1984). Apart from the reviewing of literature and written materials important to policy in question, the focused synthesis is comprised of information acquired from various forms including discussions with experts and stakeholders, unpublished documents, congressional hearings, anecdotal stories, and the personal past experience of the researchers (Majchrzak 1984).

In this connection, this paper uses a mix of two techniques combined with the author's own experience. The two techniques include the reviewing of selective literature, written and unpublished materials on the three GHG accounting principles, and discussions with experts and stakeholders. Consequently, the paper can overcome the limitation on the availability of literature and written materials concerning governing of the NEI in a country level.

2.2 Methods

In total, the author reviewed 20 papers, policy documents and reports, four book chapters and the four latest national GHG inventory reports—the UK, the USA, Norway, and Thailand.

The paper's findings benefit from two levels of face-to-face discussion, one at an overarching level of UNFCCC, and the other focuses on a case of Thailand. The discussions were conducted with an international expert attending the 44th IPCC session in Bangkok[5], and three UNFCCC's colleagues in a meeting in November 2016. During June 2016, structured interviews were conducted with four interviewees who have been involved in constructing NEIs and formulating climate change mitigation-related policies. The aim of the discussions and interviews was to answer following questions:

1. Is it possible for the CBA to be introduced to parties of the UNFCCC for adoption in near future?
2. What are the main barriers for Thailand to establish NEIs based on PBA and CBA as secondary NEI?
3. What conditions make it possible for Thailand to establish PBA and CBA-based NEIs as secondary NEI?

Therefore, the paper employs the adoption of an additional accounting principle as the dependent variable. Explanatory variables consist of external factors and internal factors. The external factors include (i) relevant decisions on NEI made by the Conference of Parties (COP) to UNFCCC; (ii) IPCC guidelines for NEI; (iii) decision-making actors on NEI in Thailand; and (iv) mandates of the NEI governing agency in Thailand. The internal factors focusing on endogenous aspects

[5]17–20 October 2016.

of NEI governing agency in Thailand encompass (i) human resource capacity, (ii) technical knowledge, and (iii) availability of data to process to PBA and CBA.

2.3 Definitions and Theories

New Institutional Economics (NIE) offers a set of tools to analyse how institutions affect human behaviour and social outcomes (Purkus 2016). This segment intends to focus on key definitions and theories relevant to the paper.

(a) **Institutions**

Institutions have been researched in many disciplines with no universally agreed definition of institutions. Even within the same discipline such as NIE scholars define the term differently.

This paper then adopts North's definitions as follows (North 1995): *Institution* refers to the laws, rules, customs, and norms that govern social behaviour. *Organization* refers to groups of actors that operate under formal or informal rules and norms. Hence, institutions are the rules of the game; organizations are the teams that play it. Institutions may be created—such as a country's constitution—or they may evolve over time—such as the common law (North 1990). Organizations include political bodies (political parties, a regulatory agency), economic bodies (firms, trade unions, family farms, cooperatives), social bodies (churches, clubs), and educational bodies (schools, universities). In other words, these are groups of individuals with shared objectives and bounded by legal rules (schools) or non-legal rules (farmer groups).

Consequently, institutions can be classified into formal, informal, and enforcement practices. *Formal institutions* encompass written constitutions and laws. *Informal institutions* refer to norms and codes of conduct. *Enforcement practices* set the terms for when and how institutions are, or are not, applied (North 1995).

(b) **Institutional change**

There is no grand theory of institutional change (Alston 2008). Most literature on institutional change focuses on the evolution of formal rules such as laws, regulations, and political orders. Institutional change analysis thus centres on the strategies used by those actors involving alliances, conflicts, bargaining and lobbying (North 1990; Brousseaua et al. 2011).

Interestingly, according to North (1990) a crucial element of institutional change is the actor, either individuals or organizations, acting as agents driving for institutional change in order to favour their own interests. Mahoney and Thelen (2010) remark otherwise. They argue that institutional change can occur without a result of efforts from various parties' intention to overcome problems, where no actor intentionally strives for the change.

Mahoney and Thelen (2010) state that the *power-distributional implications of institutions* trigger *change*. Hence, institutional change occurs endogenously. They further explain that patterns of institutional change are driven by the characteristics of the political context and the characteristics of the institution in question. Both characteristics can directly shape the types of institutional change or they influence the types of institutional change through strategies used by various types of dominant-change agents.

(c) **Typologies of Institutional changes**

Four modes of intuitional change are drift, conversion, layering, and displacement (Mahoney and Thelen 2010). Layering means adding new elements to an existing institution without intending to dismantle it. Conversion occurs when an existing institution changes function within the existing institutional framework to promote new goals. Drift occurs when goals of certain institutions or rules change without changing the institutional structure. Lastly, the displacement of an institution means to remove an existing institution and replace it with a new one.

(d) **Applying to the NEI's context**

Political context concerned in this paper is characterized as a regime, referring to the UNFCCC. The UNFCCC is the climate regime. Regimes refer to sets of principles, rules, norms, and procedures and exist to serve the preferences of one or more major states. Regime's rules often change when the preferences of one or more major states change or when power balances among those states alter (Lane and Montgomery 2014). Hence, the UNFCCC is the climate regime setting the rules of the game, including the principles and rules for the IPCC guideline direction. These rules may change when the power balance among the party states change.

Hence, before analysing how change to the NEI can occur, this paper begins by examining characteristics of and differences between each principle.

3 NEI's Accounting Principles

3.1 Calculation and National Emissions Differences

Calculation methods differences between PBA and TBA

Although the IPCC guidelines are formed based on the TBA, the literature agrees that the TBA may induce "carbon leakage" from Annex I countries and place the emissions in non-Annex I countries (Kander et al. 2015; Afionis et al. 2016; Storm and Mir 2016). In addition, there are a number of essential aspects that this principle overlooks, and these have triggered debate within the research literature.

Firstly, the definition of the territorial boundary used in this principle within the IPCC guidelines follows the International Energy Agency's energy statistics system, which limits the inclusion of emissions associated with economic activities occurring outside of a national territory. This differs from the system boundary of economic accounting (SNA)—a system used to record and draw up a country's economic profile including GDP and sectoral value added. The SNA includes economic values stemming from any economic activity by a nation even if it's outside of its territorial borders (SNA, United Nations 1993 cited in Peters 2008). Hence, this territorial-based principle does not allow for the inclusion of emissions arising from international territories including international trade and bunker fuels used for international aviation and shipping. Interestingly, some countries including the UK, the USA, and Norway do report the emissions from international bunker fuels in their NEIs submitted to the UNFCCC. The UNFCCC, however, does not allocate these emissions; instead, they are recorded as memo items (Barret et al. 2013).

The discrepancy between the system boundary of SNA and GHG emissions accounting may produce incompatible results when calculating a country's carbon intensity or carbon emissions (equivalent) divided by GDP. A country needs to consider this calculation discrepancy when performing such a calculation for future reporting under the SDG Framework.

Secondly, the principle might involve a high degree of uncertainty due to process variability in space and time, coupled with the dependency on appropriate country-level models, which may not always be available (Bastianoni et al. 2004). Rypdal and Winiwarter (2001) investigated the NEIs from five industrialized countries and found the degree of uncertainty to range from ±5 to 20%. This makes it difficult for countries to accurately plan and cost out their climate change mitigation policy, especially the associated investment costs for low carbon technologies.

The PBA holds the following key advantages over the TBA. Firstly, by aligning the GHG emissions with the SNA's definition, the PBA accounts for the GHG emissions stemming from both the domestic and foreign activities of a country depending on where the value is added for the country (Peters and Hertwich 2008). A PBA-NEI and the GDP accounting are comparable and compatible in calculating and updating the country's record, particularly the carbon intensity. Regarding this connection, the second benefit is that use of this principle can allocate emissions from international bunker fuels according to resident institutional units. Subsequently, the issue of unallocated emissions from the international bunker fuels could be overcome. Thirdly, the reporting system based on this principle has been established and systematized. Most EU countries are obliged to report their NEIs to Eurostat using this principle under the EU's NAMEA consulting the mentioned manual to convert the UNFCCC's NEI to the production-based NEI (Peters and Hertwich 2008). The steps to convert the production-based NEI are additional to the UNFCCC's NEI steps and do not contradict existing IPCC guidelines. Therefore, countries could perform the conversion without further complications. Lastly, because of the fact that differences in the national emissions from the two principles can be significant (Peters et al. 2016), it would be beneficial for countries to remain informed of their own differences.

Although the PBA resolves some issues, the PBA does not address two other issues—allocation of responsibility and emissions embedded in internationally traded goods. Therefore, both issues have led to the rise of a strand of literature supporting the CBA.

The CBA accounts for emissions at the point of consumption totalling all the emissions that occur over the course of production and distribution right through to the end consumers of goods and services (Afionis et al. 2016). Originally, the principle was conceptualized to overcome the issue of the allocation of emissions embedded in internationally traded goods. Hence, the principle has emerged over the past decade as an alternative accounting principle to PBA and TBA due to its ability to capture the embedded emissions in international trade.

In principle, there are two ways to construct a CBA-NEI from a PBA-NEI using Input/Output (I/O) Analysis: by considering total bilateral trade between regions (emissions embodied in bilateral trade (EEBT)), and trade to final consumption with segregated intermediate consumption (multi-regional input–output analysis (MRIO)).

Although Peters (2008) asserts EEBT to be a more suitable method for constructing an NEI due to its greater transparency, recently Afionis et al. (2016) note that MRIO is the preferred concept to construct a CBA-NEI. Its greatest advantage is that by being based on the input–output models, MRIO is consistent with the SNA (Wiedmann 2009). Most countries have the IO tables and models established to monitor their manufacturing and trading performances and to observe economic interactions between subsectors. Therefore, this consistency enables countries to extend the PBA-NEI to MRIO-based CBA, providing that the countries have the former in place. In addition, it would make national emissions accounted for using all three principles (territorial-, production-, and consumption-based) compatible and comparable.

Differences in national emissions between TBA, PBA, and CBA

Differences in the calculation of national emissions based on PBA compared to TBA can be fairly significant, depending on the economic profile. Indeed, they can even shift the country's perceptions towards its national emissions. Take, for example, for Denmark—a country whose economy depends largely on the international shipping industry—the emissions arising from economical resident units operating aboard—according to PBA—comprised mainly the emissions from international shipping and aviation. During 2000–2013, these emissions continued to rise from around 20 million tonne of carbon dioxide equivalent ($MtCO_{2e}$) to almost 40 $MtCO_{2e}$ in 2013; however, the country reported to the UNFCCC—based on TBA—that the emissions had been reduced from almost 60 $MtCO_{2e}$ in 2000 to a mere 40 $MtCO_{2e}$ in 2013 (Peters et al. 2016).

In other words, with the PBA calculation the emissions had in fact doubled in 2013, while the TBA calculation ignored half of the amount. It can be concluded that from these calculated differences between the two principles, Denmark had not reduced the emissions and the reported reduction during the period had benefited from carbon leakage—emissions consumed elsewhere or abroad (Peters et al. 2016).

The UK's case demonstrates a similar finding. The national emissions calculated using the PBA provided a different outlook to the figures from the TBA. On the one hand, by using the TBA-NEI submitted to the UNFCCC, the UK fulfilled its commitment to the Kyoto Protocol by demonstrating an annual decline of 1.4% in national emissions from 1990 to 2009 according to a report from the UK's Department of Environment, Food, and Rural Affairs (DEFRA 2012 cited in Barret et al. 2013). On the other hand, the very same report also revealed that the difference in the UK's total emissions between the two principles had increased by 75% during the very same period (Barret et al. 2013). Therefore, the UK's effort to reduce the emissions worked well to some extent, and at the same time, emissions that were consumed abroad also may have contributed to the continual emission reduction of the UK.

The CBA-NEI results in different national GHG emissions (compared with the PBA-NEI) mainly due to the allocation of the embedded emissions in international trade. The differences depend largely on a country's economic structure as either an export-led or import-led economy. Following their multi-dimension comparison of GHG emissions from major economies using PBA and CBA, Fan et al. (2016) noted that from 1995 to 2009 PBA emissions from China, India, Korea, Russia, and Taiwan were larger than the corresponding CBA emissions, whereas PBA emissions from each developed country (Japan, Italy, Germany, Spain, France, the UK, and the USA) were lower than their CBA emissions. These contrasting figures indicate that export-led economies emit more carbon emissions in their own territory in response to demand for the production of goods (and services) to be consumed outside of their borders. On the flipside, consumption-led economies enjoy imported goods produced outside of their country, and hence the emissions associated with the production are emitted and counted at the producer countries. It is important to note that in the study the definition of PBA was the same as that for TBA (Fan et al. 2016). Nonetheless, in the author's opinion, even if the study employed the differing definitions for PBA and TBA the difference between the emissions of export-led and import-led economies would remain unchanged.

3.2 Is the CBA a Superior Alternative?

Although CBA resolves the issue of allocating the emissions embedded in international trade and the literature concurs on its benefits over TBA and PBA, it is not universally accepted that the benefits have outpaced its drawbacks. Its comparative advantages and its setbacks are discussed below.

The different emissions resulting from PBA and CBA could reverse the outcomes of policies designed to reduce emissions because PBA emissions are not capable of capturing the ultimate driver of increasing GHG emissions, i.e. consumption growth (Fan et al. 2016; Barret et al. 2013; Druckman et al. 2008; Storm and Mir 2016). Druckman et al. (2008) stress that according to CBA, UK emissions will continue to rise despite TBA showing otherwise. They remark that any achievements in reducing emissions in the UK would be negative if CBA is used. Storm and Mir (2016) demonstrate that although Annex I countries show a trend of carbon emission decoupling or reducing their carbon intensity despite GDP growth, the reduction trends are merely the result of PBA calculations. Their CBA calculations show no such signs. They further postulate that the GHG reduction shown in PBA is the result of carbon leakage, the relocation of the production base outside of Annex I countries, and not the result of effective climate change mitigation policies.

By bringing together the contribution of the service sector and highlighting that of key traded products, the CBA-NEI enables policymakers to tackle polluters differently, but arguably not always *more* effectively. The conventional TBA-NEI places the spotlight on energy production, energy-intensive sectors, and transportation, whereas the CBA-NEI incorporates the service sector and consumption of manufactured products such as cars, electronic appliances, food, and textiles (Barret et al. 2013). The emissions from consumption-led growth can neither be captured by TBA nor PBA. Consequently, climate change mitigation policies based on the CBA-NEI focus on reductions of GHG emissions from whole value chains (from raw materials, production processes, transportation, consumption and disposal) of food and apparel products, or the tourism industry such as hotels and aviation, rather than solely focusing on power stations and transportation systems as in the current practice.

In contrast, a rather cautious strand of literature argues that the different policy responses resulting from CBA emissions might not turn out to be more effective (summarized in Afionis et al. 2016) due to the fact that CBA does not differentiate between products produced from low- and high-carbon intensity provided that they are finally consumed in the same place. Which is to say, export-induced emissions are passed on to final consumption countries. However, exporting countries could completely ignore reducing the emissions from their production processes and total emissions could still be lessened on the condition consumer countries implement their own mitigation policies effectively. Simply put, CBA emissions do not always correlate directly with policy responses and changes in emissions (summarized in Afionis et al. 2016). In this sense, this argument bases itself on the assumption of having a CBA-NEI for the purpose of measuring internationally traded goods, and hence monitoring emissions on the global scale only.

An international climate regime grounded in CBA-NEIs could address concerns on international trade competitiveness and also moderate the climate justice debate revolving around historical responsibility (Afionis et al. 2016). Through adoption of the CBA accounting scheme, the literature argues that the playing field could be levelled for policy measures such as border carbon adjustment could be imposed across a product's supply chain.

CBA suffers from its methodological data challenges and a double degree of uncertainty, consequently preventing it from being internationally accepted as an agreed methodology at the UNFCCC. The methodological discussion to draw clear boundaries between each sub-sector including its associated activities (of GHG emissions) would become highly debatable and arriving to a consensus would not be possible at any level, either global or country.

Afionis et al. (2016) assert that due to data availability at country level and coupled with the assumptions made regarding missing values during conversion from TBA to CBA-NEIs, CBA-NEIs would come with a double degree of uncertainty (Afionis et al. 2016). These drawbacks impede political acceptance within the international community at the UNFCCC. One proposed solution is that of a mechanism that discloses the carbon emissions of all internationally traded goods. Afionis et al. (2016) posit that such an international carbon disclosure system itself might pose more harm than good in the implementation of CBA.

Lastly, Kander et al. (2015) concluded that CBA does not fulfil criteria for being an ideal principle for NEI. Nonetheless, according to their study (Kander et al. 2015), neither TBA nor PBA fits to be an ideal principle for NEI.

In conclusion, although the CBA has been debated for in the literature, it does not demonstrate to be a superior alternative. Yet this conclusion might not fully capitalize on the potential of CBA. The author asserts that there is a benefit of having CBA as a complement to the obligations of current inventory countries in the UNFCCC—because policymakers could capture the different picture of their country's emissions.

4 Possible Institutional Change

4.1 At the UNFCCC Level

Changing institutions is costly and once in place an existing institutional framework will resist to change (North 1990). Institutional change involves multi-actors, can be incremental, and is also path-dependent (Purkus 2016). The historical process that has shaped current institutions influences and may constrain future opportunities for change or path dependence (North 1990). To put the concept into perspective, a current institution may have formed as a result of the bargaining power of different actors, and institutional change would eventually lead to redistributed costs and benefits among those actors. Therefore, as long as the coalitions or groups of institutional entrepreneurs advocating change are unable to defeat or compensate those who can block or slow down the proposed change; in particular, if the proposed change might devalue their assets, the proposed change might not occur (Brousseaua et al. 2011).

In addition to maintaining the status quo, institutions are resistant to change if they are self-reinforcing—meaning they produce positive feedback, increase returns, and complement one another. These positive feedback mechanisms build institutional resilience and result in a high degree of path dependence for institutions (Andrew-Speed 2016; Purkus 2016).

Nonetheless, path dependence means more than the mere recognition that history matters (Woerdman 2004). It refers to situations in which an institutional change process depends on earlier choices and events (Woerdman 2004). It also means that the institutional change process would involve such impactful self-reinforcing dynamics that impede, or even resist, the evolution or change to a more efficient alternative.

Applying the theory of institutional change to the NEI's context, in order for the climate regime to arrive at the decision to replace the TBA with CBA or even introduce CBA to be used as an official secondary NEI, there have to be changed in power-distribution among parties on this issue. In other words, the majority of parties have to agree and this could be a herculean task. In addition, it depends on the self-reinforcing dynamics of the TBA's advocacy coalitions.

The underlying factors impeding this power-distribution change are discussed as follows: Firstly, all parties have to hold full understanding of the CBA and the benefits it would bring to the table. Thus far, discussions at the COP have not gone in this direction. Partially it is due to the fact that none of PBA and CBA has proven to be a superior alternative as discussed in the previous section. In particular, more research on CBA-NEIs in non-Annex I countries, with the exception of China, could an effective tool to cultivate better understanding of CBA.

Secondly and more importantly, at the climate regime level, there has been no apparent agent advocating for change to counter the power balance of TBA advocacy coalitions. The IPCC has operated the National Greenhouse Gas Inventories Programme since 1991[6] to develop and improve upon the IPCC guidelines. The Programme also provides technical capacity building and training programmes for experts from parties. Additionally, at this point, non-Annex I countries have started to build up local advocacy coalitions of the TBA. In Thailand alone, there are quite a number of experts, the majority of who are in universities, and research organizations—who categorize themselves as IPCC guidelines' sectoral experts. These national experts have loosely formed an advocacy coalition together with international coalitions such as IPCC experts—working in the National Greenhouse Gas Inventories Programme—who have worked collaboratively over these past two decades to create self-reinforced mechanisms leading to strong network elasticity that also is resistant to change.

Lastly, the emissions from international bunker fuels have been reported by International Civil Aviation Organization (ICAO) and International Maritime Organization (IMO)[7] implying that the parties do not need to report to UNFCCC on the emissions from these two sources. Therefore, PBA-NEI from countries is not required.

Because of the fact that the climate change regime demonstrates no change that would trigger change of its rule on the use of the IPCC guideline on NEI accounting

[6]From http://www.ipcc-nggip.iges.or.jp, last accessed 24 June 2017.

[7]From http://unfccc.int/methods/emissions_from_intl_transport/items/1057.php.

principle, and also the fact that IPCC has been improving the new guidelines aiming at issuing them in 2019,[8] it is assured that the TBA continues to be *the* accounting principle within years to come. Therefore, in accordance with the literature, completely replacing the TBA with PBA or CBA is not possible, and for the COP to adopt CBA as an official secondary NEI is also not possible.

It would still be possible, although having a minuscule chance and not within the near future, that there is a change in the power balance in climate regime on this issue. This change could be driven either internally, such as change from interactions among parties or externally by an agent advocating for the adoption of CBA, or by a policy entrepreneur advocating for the CBA.

4.2 At a Country Level—A Case of Thailand

In contrast to Afionis et al. (2016), the author argues that while displacement is not possible, *layering* for institutional change is possible at a country level. This segment illuminates conditions for the adoption of PBA and CBA starting at a country level, in the light of SDG reporting. This segment starts with laying down the NEI context in Thailand, followed by the conditions under which lead to the adoption of the PBA and the CBA.

Thailand climate change policymaking context

The National Committee on Climate Change Policy (NCCC) established in 2007, chaired by the Prime Minister, and co-chaired by the Minister of Ministry of Natural Resources and Environment (MONRE), and Minister of Foreign Affairs, was mandated to primarily be responsible for (i) formulating overall climate change policy and strategy; (ii) determining national position towards the international negotiations under the UNFCCC and any relevant international agreements; and (iii) monitoring and evaluating implementation results of government agencies as stated in the national policy and strategy. To support the work of the NCCC, three sub-committees, namely Integrated Policy and Plan Sub-Committee, Technical and GHG Database Sub-Committee, and Coordination and Negotiation Sub-Committee, have been established. The NCCC is composed of members from 16 ministries, one local authority, and several experts (Thailand's First Biannual Update Report 2015).

Structurally as an inter-ministerial arrangement, the decision-making process of the NCCC could potentially be influenced by the Minister of MONRE. Because the Minister of MONRE is named among the top three most influential agents in the environmental policymaking process in the country along with the cabinet, and the National Environment Board (Prukkanone and Wang 2016).

[8]From http://www.ipcc-nggip.iges.or.jp/meeting/meeting.htm, last accessed 24 June 2017.

Governing the NEI

According to Thailand's First Biannual Update Report (2015), Thailand's NEI is managed and maintained under the Climate Change Coordination Unit, Office of National Environment Policy and Planning (ONEP), under the MONRE. The technical and GHG database sub-committee under the NCCC is in charge of preparing the National Communications and Biennial Update Reports under which the NEI is included. Five Sectorial Working Groups have been set up to review activity data and their relevant emission factors used to calculate sectorial GHG emissions and removal as well as to provide recommendations on the Measurement, Reporting, and Verification (MRV) system best suited to the country. The ONEP's Climate Change Coordination Unit (CCCU) coordinates the five sectorial working groups and serves as the secretariat of the technical and GHG database sub-committee.

Even with the sub-committee on the technical and GHG database in place, obtaining inter-ministerial data has been ineffective in a couple of ways: First, the accessibility issue remains difficult for sectoral experts who are contracted by ONEP's CCCU to perform sectoral GHG emission accounting. These experts have to contact associated sectoral-governmental agencies by themselves to obtain the necessary data. Those associated units who are the custodians of the data are not those same units who sit in the sub-committee representing their units. Even some representatives assigned to sit in the sub-committee are from the custodian agencies not authorized to give out data. This misrepresentation and lack of authority have obstructed the accessibility of the data by the experts. A recent official memorandum of understanding between the ONEP and Department of Industrial Works under Ministry of Industry was signed in late 2016 to facilitate inter-ministerial data accession on NEI, despite the fact that the sub-committee was set up in 2007. The need of the MOU as an enforcement tool demonstrates the pertaining issue of obtaining the data.

Lastly, even when they exist and are accessible, the data are incompatible with IPCC's requirements, and as a result, the data need some configurations to make them compatible or in a worse case scenario, they cannot be used at all.

PBA- and CBA-NEI as a secondary NEI

Results of the synthesis focus indicate that *layering* of additional principles is possible under the following conditions:

(a) Activity data at sub-sectoral level will need to be gathered for comprehensive PBA and CBA-NEIs. At the moment, activity data at the sub-sectoral level are still missing. It is highly recommended that prior to constructing PBA and CBA-NEIs, a nation-wide survey needs to be conducted to gather the necessary data. Both PBA and CBA principles require activity data at a sub-sectoral level. In addition, to set clear boundaries of the activity data at sub-sectoral level as required by PBA, the guidelines might not be straightforward, due to the fact that industries are coded by type of industries not by activities. For instance, two factories with the same code, producing the same product, but with different production activities, will be grouped in the same GHG category;

whereas, the GHG emissions of both factories can differ due to different production activities. Therefore, activity data at economic sub-sectoral level will be a huge issue to settle in extending the TBA to PBA.

(b) The governing structure of NEI should be restructured to better support data compatibility across sectors and additional data sets for PBA and CBA. The restructuring would include increased technical and resource capacity, as well as mandates of ONEP's CCCU. The CCCU should function as a gate-keeper of NEI not just a secretariat as it is now in its current role. Being the gate-keeper necessitates the CCCU to be housed with at least one technical expert per sector to efficiently monitor and control the quality of the NEI. With the current practice, there is no gate-keeper of the NEI, making it difficult to consider adopting additional principles.

However, a stumbling block for these conditions to occur lays on the fact that climate change mitigation policy has not been formulated based on national emission profile. For Thailand, the climate change issue has been perceivably dropped on the country as a part of being a party to the UNFCCC. The act of constructing the NEI and reporting to the UNFCCC has been treated as an obligation to the climate change regime, rather than a tool to formulate policies. Hence, in particular the policies to reduce GHG emissions are not based on the NEI, they are mostly based on other priorities. With this underlying notion, the question arises of what would be the benefits of having additional NEI when they will not be used for formulating a policy. Responding to this question in extending the NEI to PBA would be beneficial for the country to report on SDG's Target 9.4 on carbon intensity per sectoral value added, and having CBA would prepare the country for SDG's Targets 8.4 and 12.2 which are linked to the material footprints of the nation.

4.3 An Agent Drive for Change

There is a need for policy entrepreneurs to drive for the layering effort of both the national and international. Obviously, neither the UNFCCC nor IPCC could be such an agent. However, the existing experts on TBA who are mostly working in universities and climate change research institutes could be built up as policy entrepreneurs advocating for PBA and CBA-NEIs. This means to perform layering on existing TBA, both international and national experts advocating for PBA and CBA might be a practical way to bring about the change from countries. Rather than being obligated by the UNFCCC, countries—triggered by these policy entrepreneurs—could start implementing PBA and CBA by themselves. After all, institutions usually do evolve gradually. Therefore, it is not a question of why an institution does not evolve. Instead, it is a question of how to foster an institutional change to occur gradually (Mahoney and Thelen 2010).

5 Recommendations

Firstly, similar research should be conducted in non-Annex 1 countries, in particular to examine to what extent policymakers understand the three principles. Perhaps, a global online questionnaire could be established to address this issue.

Secondly, a NEI based on PBA and CBA for countries who do not have them yet in place, such as Thailand, should be constructed in preparation for the SDG reporting as well as for the policymakers' own benefit rather than an instruction from the climate regime. These additional NEIs would show the country's emissions that differ from the TBA; hence, having these two NEIs in place would be an effective tool to educate policymakers of the different emission profiles. There are tools ready to covert and make all three NEIs compatible and comparable (Fan et al. 2016), while the sub-sectoral activity data will still need to be surveyed. Most policymakers, even though they are familiar with the IPCC guidelines and have worked on GHG inventory for years and are aware that other principles exist, do not realize that each PBA and CBA-NEI would result in different emission profiles to the currently used TBA. Additionally, the main benefit of CBA-NEI is that policymakers could formulate GHG reduction policies and measure the aim to change behaviour because the CBA-NEI reveals emissions associated with the consumption profile of a country, whereas the TBA and PBA offer the technology-based emissions.

Thirdly, building up agents for change in Thailand and other countries advocating a secondary NEI would be an effective stepping stone. It may be worthwhile to turn those existing policy entrepreneurs on TBA to be agents for change advocating for the PBA and CBA.

6 Conclusion

Before the paper is concluded, limitations that the paper contains may be announced as follows. While the paper focuses on Thailand, its findings could be applicable to some other countries in Southeast Asia, namely Malaysia, Vietnam, Philippines, and Indonesia, due to similar level of capacities on constructing NEIs. These countries are reported to house a good number of national experts on TBA. However, the findings may not be applicable to Myanmar, Lao PDR, and Cambodia because of insufficient local capacity to construct robust TBA-NEI. An extension of the findings to other countries beyond Southeast Asia is not recommended due to limited knowledge of the author. A similar study to this paper is otherwise encouraged.

This paper has demonstrated the differences and benefits of each accounting principle and encourages countries to implement PBA, and CBA-NEIs accordingly, so that their policymakers can have a complete view of the country's emission profiles. But because of the fact that policymakers in countries may not have the full understanding of these benefits, policy entrepreneurs advocating for these additional principles are needed.

Most importantly, through the lens of institutions and institutional change and using Thailand's case, the paper concludes that the layering of the PBA and CBA on the TBA is recommended and could be achievable; however, not the replacement of due to the fact that through implementation of IPCC guidelines, countries have built the network of TBA experts on the ground which is entrenched, therefore introducing and advocating PBA and CBA-NEIs through this existing network by enhancing their capacity might be an effective approach to gradually evolve into the CBA.

Lastly and to conclude, climate scientists and research communities should continue to research for a more effective principle to count the hot air. More research on CBA-NEI using the EEBT, instead of the MRIO, might solve the issue of counting the emissions at the final consumer countries as the EEBT counts the emissions of imported goods at the country of import. This way the country could influence the action in response to the emissions embodied in the imported goods as well as a lower carbon-intensive production for exports being credited. Hence, the CBA-NEI using the EEBT may fulfil conditions suitable to be an ideal NEI.

References

Afionis, St., Sakai, M., Scott, K., Barret, J., Gouldson, A., & Stéphane, H., et al. (Eds.). (2016). Consumption-based carbon accounting: does it have a future? *WIREs Clim Change*. https://doi.org/10.1002/wcc.438.

Alston, J. L. (2008). The "case" for case studies in the new institutional economics. In E. Brousseau & J.-M. Glachant (Eds.), *New institutional economics: A guidebook* (pp. 103–121). Cambridge: Cambridge University Press.

Andrews-Speed, P. (2016). Applying institutional theory to the low-carbon energy transition. *Energy Research & Social Science, 13,* 216–225.

Barret, J., et al. (2013). Consumption-based GHG emission accounting: A UK case study. *Climate Policy, 13*(4), 451–470.

Bastinanoni, S., Pulselli, F. M., & Tiezzi, E. (2004). The problem of assigning responsibility for greenhouse gas emissions. *Journal of Ecological Economics, 49,* 253–257.

Brousseaua, E., Garrousteb, P., & Raynaud, E. (2011). Institutional changes: Alternative theories and consequences for institutional design. *Journal of Economic Behaviour & Organization, 79,* 3–19.

Druckman, A., Bradley, P., Papathanasopoulou, E., & Jackson, T. (2008). Measuring progress towards carbon reduction in the UK. *Journal of Ecological Economics, 66,* 594–604.

EUROSTAT. (2015). Manual for Air Emissions Account [Online]. Available from https://www.unece.org/fileadmin/DAM/stats/documents/ece/ces/ge.33/2016/mtg1/Manual_Air_Emission_Accounts.pdf. Last accessed December 6, 2016.

Fan, J.-L., Hou, Y.-B., Qian Wang, Q., Wang, C., & Weian, Y.-M. (2016). Exploring the characteristics of production-based and consumption-based carbon emissions of major economies: A multiple-dimension comparison. *Journal of Applied Energy*. https://doi.org/10.1016/j.apenergy.2016.06.076. (Article in Press).

Intergovernmental Panel on Climate Change (IPCC). (2016). Accessed December 7, 2016 http://www.ipcc-nggip.iges.or.jp/public/ http://www.ipcc-nggip.iges.or.jp/public/gl/pdfiles/invs2a.pdf http://www.ipcc-nggip.iges.or.jp/public/2006gl/pdf/1_Volume1/V1_1_Ch1_Introduction.pdf.

Kander, A., Jibborn, M., Moran, D. D., & Wiedmann, T. (2015). National greenhouse-gas accounting for effective climate policy on international trade. *Journal of Nature Climate Change, 5,* 431–435.

Lane, L., & Montgomery, W. D. (2014). An institutional critique of new climate scenarios. *Journal of Climatic Change, 122,* 447–458.

Mahoney, J., & Thelen, K. (2010). A theory of gradual institutional change. In J. Mahoney & K. Thelen (Eds.), *Explaining institutional change. Ambiguity, agency and power*. Cambridge: Cambridge University Press.

Majchrzak, A. (1984). *Methods for policy research applied social research study*. Sage.

North, D. C. (1990). *Institutions, institutional change and economic performance*. Cambridge: Cambridge University Press.

North, D. C. (1995). The new institutional economics and third world development. In J. Harris, J. Hunter, & C. M. Lewis (Eds.), *The new institutional economics and third world development*. London: Routledge.

Peters, G. P. (2008). From production-based to consumption-based national emission inventories. *Journal of Ecological Economics, 65,* 13–23.

Peters, G. P., Andrew, R. M., & Karstensen, J. (2016). Global environmental footprints—A guide to estimating, interpreting, and using consumption-based accounts of resource use and environmental impacts [Online]. Available from http://folk.uio.no/roberan/docs/Peters2016_NordicFootprints.pdf. Last accessed December 6, 2016.

Peters, G. P., & Hertwich, G. E. (2008). Post-Kyoto greenhouse gas inventories: Production versus consumption. *Journal of Climate Change, 86,* 51–66.

Prukkanone, K., & Wang, G. (2016). Stakeholder analysis of environmental policymaking process in Thailand. *Journal of Public Administration and Governance, 6*(2).

Purkus, A. (2016). *Concepts and instruments for a rational bioenergy policy: A new institutional economics approach*. Switzerland: Springer International.

Rypdal, K., & Winiwarter, W. (2001). Uncertainties in greenhouse gas emission inventories—evaluation, comparability and implications. *Journal of Environmental Science and Policy, 4,* 107–116.

Storm, S., & Mir, G.-U.-R. (2016). Carbon emissions and economic growth: Production-based versus Consumption-based evidence on Decoupling. Working Paper No 41, Institute for New Economic Thinking.

Thailand's First Biannual Update Report. (2015). Thailand's First Biannual Update Report under United Nations Framework on Climate Change Convention, December, from http://unfccc.int/resource/docs/natc/thabur1.pdf. Accessed June 2017.

Wiedmann, T. (2009). A first empirical comparison of energy Footprints embodied in trade—MRIO versus PLUM. *Journal of Ecological Economics, 68*(7), 1975–1990.

Woerdman, E. (2004). *The institutional economics of market-based climate policy*. Amsterdam: Elsevier.

Sooksiri Chamsuk is a doctoral student at the Institute for Policy Research, University of Bath (UK), on a part-time basis. She works full time for United Nations Industrial Development Organization (UNIDO) as a Programme Officer based in Bangkok, Thailand. She has more than fifteen years of experience working on sustainable development with specific focus on energy, climate change, and chemical management in Southeast Asia. She holds a Master's in Energy Technology for Sustainable Development, University of Salford (UK), and a bachelor degree in Mechanical Engineering, Chiangmai University, Thailand.

Climate Change "Conditionality": The Case for Bundling the Fate of International Trade Legal Obligations and Climate-Change-Relevant Legal Obligations

Eduardo Alvarez Armas

Abstract This paper intends to make a contribution to the global debate on climate change mitigation by suggesting the transposition of a peculiar human rights-enhancing notion (developed by the European Institutions over the past three decades) towards the intersection between international trade law and climate change issues. The said notion, "conditionality", is a practice which entails that the obligations assumed by the EU in the international cooperation and development agreements it signs with developing countries are conditional on those recipients' continuous and clear respect of basic human rights. This paper's hypothesis is that if obligations in cooperation and development agreements can be made conditional on the respect by other parties of human rights, then obligations in international trade agreements can be made conditional on the respect by other parties of their respective climate change obligations. In other words, "climate change conditionality" would mean that there would be no free-trade benefits without effective fulfilment of international climate-change-related obligations. The feasibility of this idea will be assessed via the textual analysis of the current wording and content of environmental/climate-change-related elements already present in a sample of relatively recent and/or upcoming free-trade agreements. This exercise intends to ascertain whether elements already present therein may be used to build climate change conditionality upon them.

Keywords Climate change conditionality · International trade law
Human rights conditionality

E. Alvarez Armas (✉)
Centre for Commercial Law Studies, Queen Mary University of London,
67-69 Lincoln's Inn Fields, London WC2A 3JB, UK
e-mail: e.alvarezarmas@qmul.ac.uk

© Springer International Publishing AG, part of Springer Nature 2019
W. Leal Filho and R. Leal-Arcas (eds.), *University Initiatives in Climate Change Mitigation and Adaptation*, https://doi.org/10.1007/978-3-319-89590-1_21

1 Introduction

The following pages will introduce the contribution that international trade law may make to global decarbonisation, as inspired by human rights practices of the European Union. As disparate as these topics and realms may seem at first sight, there is a significant potential in transposing a peculiar human rights-enhancing notion, developed by the European Institutions over the past three decades, towards the intersection between international trade law and climate change.

In a nutshell, the said notion, "conditionality", is a practice that involves linking the effective fulfilment of European cooperation and development engagements towards third parties, on the one hand, with the respect of basic human rights by those third parties, on the other hand. In other words, conditionality entails that the obligations assumed by the European Union in the international cooperation and development agreements it signs with developing countries are conditional on those recipients' continuous and clear respect of basic human rights (Velluti 2016; Hachez 2015).

This paper's hypothesis is that if obligations in cooperation and development agreements can be made conditional on the respect by other parties of human rights, then obligations in international trade agreements can be made conditional on the respect by other parties of their respective climate change obligations. In other words, international trade law may be used to help decarbonise the economy by bundling the fate of free-trade agreements—hereinafter FTAs—and international climate change obligations: the breach of the latter would give rise to legal consequences under the dispute settlement mechanisms of the former (i.e. no free-trade benefits without effective fulfilment of international climate-change-related obligations).

The feasibility of this idea will be assessed via the textual analysis of the current wording and content of environmental/climate-change-related elements already present in a sample of relatively recent and/or upcoming FTAs. This exercise intends to ascertain whether elements already present in FTAs may be used to build climate change conditionality upon them. However, as a complement to pure textual analysis, a brief reference will be made to similar academic endeavours by other authors, which have resorted to either resembling or alternative methodologies. Thus, the rather lengthy textual analysis (2) and the brief academic reference (3) will be completed by a synthesis (4) and some conclusions (5).

2 Empirical Assessment: "Environmental" Language and Content of Sample FTAs

The first subsection will be descriptive: a review of the relevant language and content of a selection of 14 FTAs. When analysing their environmental and/or climate-change-relevant substantive content, and its potential articulation with dispute settlement mechanisms, at least five different configurations can be identified therein. Let us assess them in sequence.

First configuration—Enforceability, but weak substantive content: The TPP and the ASEAN agreements with China and India

The first possible configuration to be found within the sample selected is embodied in three different agreements within the sample: (i) the "*Trans-Pacific Partnership*" (TPP), a trade agreement originally signed by twelve Asia-Pacific nations on 4 February 2016; (ii) the "*Framework Agreement on Comprehensive Economic Cooperation between the Association of South East Asian Nations and the People's Republic of China*" (ASEAN–China Agreement), an international agreement which created a free-trade area between the members of the ASEAN and China, and which entered into force on 1 January 2010; and (iii) the "*Framework Agreement on Comprehensive Economic Cooperation Between the Republic of India and the Association of Southeast Asian Nations*" (ASEAN–India Agreement), another international agreement which also created a free-trade area between the members of the ASEAN and India, and which equally entered into force on 1 January 2010.

The first agreement in the sample, the TPP is not yet in force, and its future has been called into question since the USA decided to withdraw from it in 2017 (CNN 2017). Interestingly for this paper's purposes, the TPP contains a chapter specifically entitled "environment" (Chapter 20). This chapter envisages in Article 20.23 resort to Dispute Resolution if other means to solve controversies arising therefrom have failed. This includes, notably, resort to Consultations (Article 28.5) and possibly to the establishment of a Panel (Article 28.7). Following Article 28.12, the function of a Panel "*is to make an objective assessment of the matter before it,* […] *and to make the findings, determinations and recommendations as are called for in its terms of reference and necessary for the resolution of the dispute*". The Panel will prepare an initial report (Article 28.17), and, where appropriate, a final report (Article 28.18), whose implementation will be assessed (Article 28.19). In the event of non-implementation, the non-complying party may either need to provide compensation, or may either face a suspension of the benefits granted under the TPP (or, under certain conditions, may have to pay a monetary assessment) (Article 28.20). This regime would, in principle, provide significant chances of international enforceability to the obligations contained in Chapter 20 of the TPP (Hillman 2016). However, while the said chapter does contain an article entitled "*Transition to a Low Emissions and Resilient Economy*" (Article 20.15), whose content is relevant for the purposes of this paper, the article's language is very weak, making it hard to substantiate any solid claim:

> *1. The Parties* ***acknowledge*** *that transition to a low emissions economy requires collective action.*
>
> *2. The Parties* ***recognise*** *that each Party's actions to transition to a low emissions economy should reflect domestic circumstances and capabilities and, consistent with Article 20.12 (Cooperation Frameworks), Parties shall* ***cooperate*** *to address matters of joint or common*

> *interest.* [...] *Further, the Parties shall, as appropriate,* ***engage in cooperative and capacity-building activities*** *related to transitioning to a low emissions economy.* [Emphasis added]

Such weak substantive content makes irrelevant the fact that Article 20.15 comes within the scope of the Dispute Settlement scheme provided for in the Treaty.

A similar, albeit slightly different description can be made of the ASEAN–China and the ASEAN–India Agreements, the second and third agreements mentioned above. Since their structure and content are almost identical, they may be dealt with simultaneously: while the respective contents of these agreements are meant to be enforceable under additional dispute settlement agreements between the parties, their actual environmentally relevant contents are restrained to simply providing for cooperation on environmental issues, as an area of "*economic cooperation*". This renders the availability of enforcement mechanisms and redress, again, irrelevant.

Second configuration—Ambiguous substantive content, and non-enforceability: Selected EU FTAs

The second possible configuration that may be found within the sample of treaties is embodied in the following five EU-promoted trade agreements: the "*free-trade agreement between the European Union and its Member States, of the one part, and the Republic of Korea, of the other part*" (EU–South Korea FTA; in force as of 13 December 2015); the "*Trade Agreement between the European Union and its Member States, of the one part, and Colombia and Peru, of the other part*" (EU–Colombia–Peru TA; provisionally applied as of 1 March 2013 in respect of Peru, and as of 1 August 2013 in respect of Colombia); the "*Association Agreement between the European Union and its Member States, of the one part, and Ukraine, of the other part*" (in force as of 1 September 2017); the "*Comprehensive Economic and Trade Agreement (CETA) between Canada, of the one part, and the European Union and its Member States, of the other part*" (CETA; Not yet in force, signed on 30 October 2016); and the "*Japan–EU Economic Partnership Agreement*" (JEEPA; not yet signed). Overall, the environmental language and content of the post-2008-financial-crisis EU FTAs seem to be relatively stable (with the notable exception, as it will be seen below, of the EU–Colombia–Peru TA), and, therefore, they will be assessed simultaneously. In terms of substantive content, generally speaking, these agreements tend to contain a chapter on "*Trade and sustainable development*" or "*Trade and environment*", within which an article on "*Multilateral environmental agreements*" may be found. The content of the said article has a constant basis, upon which other elements are pinned. The said basis is constituted by the following language:

> *1. The Parties* ***recognise the value*** *of international environmental governance and agreements as a response of the international community to global or regional environmental problems.*
>
> *2. The Parties* ***reaffirm their commitments*** *to the effective* ***implementation*** *in their laws and practices of the multilateral environmental agreements to which they are party.*

This common basis is sometimes accompanied by further paragraphs which, despite depicting progressive intentions, are unable to lead to ambitious environmental outcomes. As brief examples, in the Agreements with South Korea and Japan, the parties mention the United Nations Framework Convention on Climate Change and the Kyoto Protocol (in the case of the JEEPA, the Paris Agreement), but only to state that they "*reaffirm their commitment to reaching* [the Agreements'] *ultimate objective*". The Agreement with Ukraine establishes further means through which the parties "*shall cooperate*"; as does CETA, which adds that the parties "*commit to consult*" and "*exchang*[e] *information*". Equally, the first paragraph above is sometimes accompanied by additional elements which remain within the same vague narrative (the parties "*commit to consulting and cooperating*", "*stress the need to enhance*", etc.).

However, beyond these considerations, which would tend to sustain the conclusion that the substantive content of the concerned Agreements is purely and simply weak, a qualification may be made: the second paragraph is ambiguous in its content. The paragraph begins with a typically "weak" wording ("*The Parties reaffirm their commitments*") but, instead of referring to weak policy objectives, it refers to the "*effective implementation in their laws and practices* [...] *of the multilateral environmental agreements to which they are party*". It, therefore, could be argued that, despite its beginning, the paragraph actually links, in a solid manner, the trade agreement to the relevant multilateral environmental agreements (any relevant agreement to which they are party, without limitation): in other words, it could be argued that a lack of "*effective implementation in their laws and practices*" of the relevant environmental agreements entails a breach of paragraph 2.

This potential interpretation is further reinforced by the specific wording of the relevant provisions of the EU–Colombia–Peru FTA. The second, third and fourth paragraphs of Article 270 ("*Multilateral Environmental Standards and Agreements*"), included in Title IX ("*Trade and sustainable development*") of the said FTA, read as follows:

> *2. The Parties* ***reaffirm their commitment to effectively implement*** *in their laws and practices the following multilateral environmental agreements: the Montreal Protocol on Substances that Deplete the Ozone Layer adopted on 16 September of 1987, the Basel Convention on the Control of Transboundary Movements of Hazardous Wastes and their Disposal adopted on 22 March 1989, the Stockholm Convention on Persistent Organic Pollutants adopted on 22 May 2001, the Convention on International Trade in Endangered Species of Wild Fauna and Flora signed on 3 March 1973 (hereinafter referred to as "CITES"), the CBD, the Cartagena Protocol on Biosafety to the CBD adopted on 29 January 2000, the Kyoto Protocol to the United Nations Framework Convention on Climate Change adopted on 11 December 1997 (hereinafter referred to as "Kyoto Protocol") and the Rotterdam Convention on the Prior Informed Consent Procedure for Certain Hazardous Chemicals and Pesticides in International Trade adopted on 10 September 1998.*
>
> *3. The Trade Committee* ***may recommend the extension of the application*** *of paragraph 2 to other multilateral environmental agreements* [...].

> *4. Nothing in this Agreement shall limit the right of a Party to* ***adopt or maintain measures to implement*** *the agreements referred to in paragraph 2.* [...] [Emphasis added, original footnotes omitted]

Paragraph 2 of the article contains a list with a limited number of precisely identified agreements. At its present state, this provision does not seem very prone to providing coverage to climate change instruments (Only one of the instruments listed in paragraph 2, the Kyoto Protocol, may be characterised as a climate change instrument. Beyond this, in the EU–Colombia–Peru FTA, climate change agreements are the object of a separate, notably weak provision, Article 275.). Notwithstanding this fact, it may be argued that it does not contain vague policy objectives and statements, but clear, precise and solid obligations. If this is the case, the above-referred parallel provisions in other EU FTAs may be interpreted accordingly, as being open-ended clauses (hence, not restrained to a limited list of treaties), containing the same kind of clear, precise and solid obligations.

However, in order for this ambiguous possibility to have any practical impact, the said provisions would need to be enforceable under the relevant "*Dispute Settlement*" provisions in the respective EU FTAs, which does not seem to be the case. While, in most cases, the dispute settlement chapters of the selected FTAs are assorted with provisions which state that their scope is general and covers the entirety of the said agreements, these provisions operate only "*unless otherwise provided*". The environmentally relevant chapters, therefore, exclude resort to dispute settlement under various formulations. In general, they only provide for soft and conciliatory mechanisms, such as "*Government consultations*", or the establishment of "*panels*" or "*groups*" of experts, who draft "*initial*" and, "*final*" reports, on how to solve issues. The parties then "*shall take into account the final report*", or "*inform the* [relevant entity] *of its intentions as regards the recommendations*", or "*discuss actions or measures to resolve the matter in question, taking into account the* [...] *report and suggestions*". The parties' courses of action are often "*monitor*[ed]", but no consequence is provided for.

Third configuration—Solid substantive content, but non-enforceability: The EU–Singapore FTA

A third possible configuration is reflected in the text of the "*free-trade agreement between the European Union and the Republic of Singapore*" (EU–Singapore FTA; not yet signed). Chapter 13 ("*Trade and sustainable development*") contains in Section C ("*Trade and Sustainable Development—Environmental Aspects*") an Article on "*Multilateral Environmental Standards and Agreements*" (Article 13.6) which bears a structure which greatly resembles the one described in the analysis of the previous configuration (i.e. "*ambiguous substantive content but weak enforceability*"). However, in this occasion there is no possible ambiguity:

> *2. The Parties shall* ***effectively implement*** *in their respective laws, regulations or other measures and practices in their territories,* ***the multilateral environmental agreements to which they are party****.* [Emphasis added; original footnote omitted]

The wording of the second paragraph of this Article is clearly assertive, without any sort of resort to the typical vague language found in weak policy statements. It can hardly be argued that it does not contain clear, precise and solid obligations. A priori, its only fault, for the purposes of this paper, is that it clearly does not cover climate change obligations, for climate change is the object of the third paragraph of the Article, which is undoubtedly weak in its content:

> *3. The Parties **reaffirm their commitment** to reaching the ultimate objective of the UN Framework Convention on Climate Change (hereinafter referred to as "UNFCCC"), and of its Kyoto Protocol in a manner consistent with the principles and provisions of the UNFCCC. They **commit to work** together to strengthen the multilateral, rules-based regime under the UNFCCC building on the UNFCCC's agreed decisions, and to support efforts to develop a post-2020 international climate change agreement under the UNFCCC applicable to all parties.* [Emphasis added]

In any case, again, as was the case in the second model, for the second paragraph of this provision to have any practical impact, it would need to be enforceable under the relevant "*Dispute Settlement*" chapter. And, again, this does not seem to be the case.

Fourth configuration—Enforceability, but ambiguous substantive content: The TTIP

The "*Transatlantic Trade Investment and Partnership*" (TTIP), that the European Union and the United States began negotiating between 7 and 12 July 2013, is at the roots of the fourth configuration detected in the sample of Treaties. Notwithstanding the fact that a significant amount of uncertainty swirls around its future, draft documents on this envisaged agreement, reflecting different stages in the negotiations, will serve as the basis for the analysis. The first of such documents is the EU's "*initial proposal for legal text on* "*Trade and Sustainable Development*" drafted to serve during the negotiating round with the USA that took place between 19 and 23 October 2015 (EU Commission 2015). Within "*Section III*" (named "*Trade and Sustainable Development—Environmental aspects*"), Article 10 ("*Multilateral environmental governance and rules*") contains the following language:

> *1. The Parties **recognise** the value of global environmental governance and rules, including Multilateral Environmental Agreements, to tackle environmental challenges of common concern and stress the need to enhance the mutual supportiveness between trade and environment policies, rules and measures.*
>
> *2. Each Party **reaffirms its commitment** to effectively **implement** in its domestic laws and practices the Multilateral Environmental Agreements to which it is a party.*
>
> *3. The Parties should **continue to strive** towards further ratification of Multilateral Environmental Agreements and cooperate in this regard, including through **exchanging information** on advancement and **supporting each other**'s full participation in, or membership to, multilateral environmental agreements, international bodies, and processes.*
>
> *4. The Parties **commit to consult and cooperate** with each other as appropriate in Multilateral Environmental Agreements and other global environmental fora, in particular trade-related environmental issues.*

5. The Parties ***acknowledge*** *that nothing in the Agreement should prevent either Party from adopting or maintaining measures to implement the Multilateral Environmental Agreements to which it is a party,* [...]. [Emphasis added; original footnotes omitted]

In general terms, the wording of Article 10 roughly corresponds with that of equivalent substantive provisions in the EU FTAs that were classified as having an "*ambiguous substantive content*" but lacking enforceability (i.e. the second configuration above). However, what distinguishes this early TTIP document from the above-commented EU FTAs is precisely the fact that, following an EU's "*initial proposal for legal text on "Dispute Settlement (Government to Government)"*" the TTIP's relevant provisions would be enforceable. Article 2 ("*Scope of application*") in Section 1 ("*Objective and Scope*") establishes that the dispute settlement chapter applies to "*any dispute concerning the interpretation and application of the provisions of this Agreement, except as otherwise provided*".

This assessment, made on the basis of a rather early-stage document, may be altered if the content of a second document, stemming from a later round of the negotiations (11–15 July 2016), gets to be inserted into the relevant chapter of the TTIP. The EU's "*initial proposal for a legal text on climate which would be included* [sic] *to the "Trade and Sustainable Development" chapter in TTIP*" (EU Commission 2016) contains lengthy language on "*Trade favouring low-emission and climate-resilient development*" and "*Protection of the Ozone Layer and Measures Related to Hydrofluorocarbons*". For the sake of brevity (especially taking into account the above-referred uncertainty surrounding the negotiations) suffice it to say that most of the content of the draft provisions can be easily qualified as "weak", for it reflects the same kind of language as other provisions transcribed in previous pages. However, each of the two unnumbered draft provisions features a paragraph bearing more solid content:

4. [...] *the Parties shall:* [...] *b)* ***effectively implement*** *the WTO Environmental Goods Agreement (EGA) and in this context cooperate to reduce or, as appropriate, eliminate non-tariff barriers related to environmental goods and services;* [Emphasis added]

2. [...] *each Party* ***shall take measures*** *to control the production and consumption of, and trade in, substances within the scope of the Montreal Protocol on Substances that Deplete the Ozone Layer, including any future amendments thereto.* [Emphasis added]

Nevertheless, as none of these paragraphs with "solid" content refers to any of the core climate change agreements, they may only be collaterally relevant, if anything, for the purposes of this section.

The potential insertion of this language into the TTIP would make it unlikely that climate change be covered by the above-referred Article 10. This would mutate this fourth configuration, thus making it partially resembling to the first configuration described above (namely, "*enforceability, but weak substantive content*"). This fourth configuration would, nevertheless, remain a distinct one, for despite the provision's general weakness, it does contain solid obligations as regards collaterally relevant agreements, which is not the case regarding its counterpart in the TPP.

In any case, irrespective of the "ambiguity" in Article 10, and irrespective of the potential impact of the insertion of the unnumbered provisions just discussed, a major hindrance to the enforceability dimension of the TTIP, which is also shared by the TPP, would always remain: putting aside the actual architecture and specific features that a final TTIP may possibly come to display, what is really fundamental is the fact that the USA has not ratified the Kyoto Protocol and is possibly going to withdraw from the Paris Agreement on Climate Change (New York Times 2017). Consequently, there would be no relevant climate change obligation that could potentially come to be enforced via this trade agreement.

As interesting for these papers' purposes as they are, these developments on the TTIP negotiations, as it has been already highlighted, do not provide a sufficient degree of certainty. Therefore, discussion will stop at this point and switch to the last treaty configuration detected in the sample.

Fifth configuration—Solid substantive content, and enforceability, but restrictive scope: Selected US FTAs

The fifth and final configuration found in the sample of treaties chosen can be found in three US international economic agreements: the "*free-trade agreement between the United States of America and the Republic of Korea*" (KORUS; entered into force on 15 March 2012), the "*United States–Peru Trade Promotion Agreement*" (US–Peru TPA; entered into force on 1 February 2009) and the "*United States–Colombia Trade Promotion Agreement*" (US–Colombia TPA; entered into force on 15 May 2012). As it was the case with the EU FTAs, these agreements bear a great resemblance in their wordings, for the purposes of this paper. Therefore, the following commentary will focus on the US–Peru TPA, but all observations are applicable, *mutatis mutandis*, to the other two legal instruments. Chapter 18 of the US–Peru TPA ("*Environment*") contains an article on "*Environmental Agreements*" (Article 18.2) which reads as follows:

> *A Party* ***shall adopt, maintain, and implement*** *laws, regulations, and all other measures* ***to fulfill its obligations*** *under the multilateral environmental agreements listed in Annex 18.2 ("covered agreements").* [Emphasis added]

The restrictive list of "*covered agreements*" referred to (which is identical in all three treaties) is composed of seven international agreements, amongst which the "*Convention on International Trade in Endangered Species of Wild Fauna and Flora*" (CITES), the "*Montreal Protocol on Substances that Deplete the Ozone Layer*" and the "*Protocol of 1978 Relating to the International Convention for the Prevention of Pollution from Ships, 1973*" (MARPOL), as well as other less significant instruments. The relationship between, on the one hand, the US–Peru TPA and any of the targeted multilateral environmental agreements is further shaped by Article 18.13 ("*Relationship to Environmental Agreements*"):

> *1. The Parties* ***recognize*** *that multilateral environmental agreements to which they are all party, play an important role globally and domestically in protecting the environment and that their respective implementation of these agreements is critical to achieving the*

> *environmental objectives thereof. The Parties further* ***recognize*** *that this Chapter and the ECA can contribute to realizing the goals of those agreements.* [...]
>
> *2. To this end, the Parties shall* ***consult****, as appropriate, with respect to negotiations on environmental issues of mutual interest*
>
> *3. Each Party* ***recognizes the importance*** *to it of the multilateral environmental agreements to which it is a party*
>
> *4. In the event of any inconsistency between a Party's obligations under this Agreement and a covered agreement, the Party shall* ***seek to balance*** *its obligations under both agreements, but this shall not preclude the Party from taking a particular measure to comply with its obligations under the covered agreement,* [...]. [Emphasis added]

Putting aside the fact that Article 18.13 resorts to weak and vague language (possibly in order not to force the pre-eminence of environmental values over economic ones), the main article discussed, namely Article 18.2, is precise, unconditional and assertive, and neatly establishes clear obligations which, in this case, are actually enforceable under the relevant dispute settlement provisions. In this sense, Article 18.12 ("*Environmental Consultations and Panel Procedure*") establishes, amongst other aspects, in paragraph 6, that a complaining party may "*as provided in Chapter 21 (Dispute Settlement), thereafter have recourse to the other provisions of that Chapter*". Specifically, by establishing how "*a panel convened under* [the] *Dispute Settlement* [chapter]" is meant to make its "*findings*", paragraph 8 of the same article provides guidance as regards the operation of the dispute settlement provisions in respect of Article 18.2 ("*Environmental Agreements*"). Within the provisions on dispute settlement of the US–Peru TPA, Article 21.16 ("*Non-Implementation—Suspension of Benefits*") establishes that an infringing party may have to provide compensation or may, under certain conditions, see her rights under the Treaty suspended.

However, despite the Agreement's clear enforceability (and solid substantive content), there are, at least, two restrictions in the scope of the relevant provisions, which render this fifth Treaty configuration unfit for the purposes of this paper. The first one arises from the fact that, as already mentioned, the Treaties feature a restrictive list of "*covered agreements*" which does not contain any climate-change-relevant international instrument. The second one, which is just equally important, arises from a footnote placed immediately after the title of the "*Environmental Agreements*" provision in each of the Treaties. The one in Article 18.2 of the US–Peru TPA reads as follows:

> *1. To establish a violation of Article 18.2 a Party must demonstrate that the other Party has* ***failed*** *to adopt, maintain, or implement laws, regulations, or other measures* ***to fulfill an obligation*** *under a covered agreement* ***in a manner affecting trade or investment between the Parties***. [Emphasis added]

This entails that only a specific kind of breach of the "*covered agreements*" may potentially lead to legal consequences under the dispute settlement provisions of the relevant Treaties. As it may be understood, therefore, the potential of this fifth Treaty configuration is severely hindered. While it is still too early to say, it is likely that the potential recast of the North American Free Trade Agreement (NAFTA), whose renegotiation rounds begun during Summer 2017, may end up belonging to this category. Following a "*Summary of Objectives for the NAFTA Renegotiation*" released by the office of the United States Trade Representative on 17 July 2017, the United States intends to insert provisions on environmental issues into the text of the NAFTA (USTR 2017). Specifically, the very little guidance provided by the said document possibly hints at the replication of the structural configuration described above, including its restrain to a limited list of "*covered agreements*".

Having concluded the description of the fifth configuration, let us briefly consider other similar academic accounts which have resorted to other methodologies.

3 Other Academic Accounts

The textual analysis in previous pages needs to be complemented by a brief reference to other academic assessments, which have been performed along similar lines (resembling focuses, either under different or slightly varying methodologies) in recent years. These studies provide a wider context to the referred textual analysis and further enlighten the potential contribution of FTAs to international environmental and climate-change-related issues.

A first account to be mentioned was performed by S. Jinnah and E. Morgera in 2013. It comprised a sophisticated analysis, based on coding, of Environmental provisions in American and EU Free Trade Agreements (Jinnah and Morgera 2013). This interesting and timely work does not focus, however, on climate-change-relevant provisions, which are only very briefly and generically addressed. Nevertheless, this piece is extremely valuable, inasmuch as it shows (although not explicitly stated) the potential that future climate-relevant provisions could come to bear, if, in policy terms, an approach similar to the one taken in respect of generic environmental provisions was adopted.

A second account which runs along similar lines and deserves being highlighted is the very recent work performed by J. F. Morin and R. Gauthier Nadeau on what they label as "*environmental gems*" in trade agreements: in their opinion, several very innovative environmentally relevant provisions found in a rather limited number of trade agreements are good examples of "*best practices*" that should actually be amplified and replicated in future agreements. Most of these "legal one-hit-wonders" are not directly climate-change-relevant (there is only a brief reference to "*two Stabilisation and Association Agreements of the European Union* [calling] *for the ratification of the Kyoto Protocol on Climate Change*"), yet, again, they point to the huge potential that a similar approach in climate change terms would have (Morin and Nadeau 2017).

Finally, N. A. J. Croquet has performed an in-depth analysis of "The climate change norms under the EU–Korea Free Trade Agreement" which rightly points to the need of assessing, in every case, what is the actual value of the climate-change-relevant provisions that may be found in trade agreements (Croquet 2015). When discussing this point as regards the EU–Korea FTA, Croquet, following assessments by A. Boyle, and by K. Abbott and D. Snidal, places this question within the academic soft law–hard law continuum. It is under these coordinates that he asserts, following the referred authors, that "*given that the EU–Korea FTA constitutes hard law vis-à-vis the contracting parties*", climate change provisions are to be considered from the standpoint of two additional elements: "*their degree of precision*" and "*their enforcement/dispute settlement mechanism*". Consequently, Croquet goes on to consider: (i) that what we have labelled as "*ambiguous provisions*" in previous pages are actually "*renvoi clauses*", i.e. "*confirmatory clauses that urge the Contracting parties to live up to their pre-existing climate change treaty commitments*"; (ii) that in the EU–Korea FTA case the renvoi clause is unambiguous and reflects an unequivocal obligation to "*effectively implement*" previously assumed climate-change-relevant obligations; (iii) "*that these pre-existing treaty commitments now fall under the Chapter 13 dispute settlement mechanism*". As relevant and enlightening as this account is, it is not possible to fully subscribe it, nevertheless: while it may be possible to agree with (ii), as Croquet quotes documentation by the "WTO Committee on Regional Trade Agreements", it is not possible to fully agree with (iii). What Croquet labels as "dispute settlement" is a set of provisions within Chapter 13 ("Trade and sustainable development") which is independent from the "Dispute settlement" Chapter (Chapter 14) and which, ultimately, is not accompanied by any sort of remedy.

In the light of all of these considerations, let us synthesise the findings of previous pages.

4 Synthesis: The Case for Bundling Trade and Environmental Agreements

Textual analysis shows that, as the law stands nowadays, there are, at least, four sorts of obstacles to the hypothesis presented in this paper:

1. Weak and/or vague language of the relevant "primary" obligation (e.g. the TPP): whenever a given climate-change-relevant provision is covered by the dispute settlement mechanism, the actual content of the relevant obligation is weak, and thus rather useless.
2. Lack of coverage by the relevant dispute settlement provisions (e.g. the EU–Singapore FTA): reverse problem; while the content/language of the relevant obligations is strong, the provision is not covered by the dispute settlement mechanism.
3. Severe limitations in the scope of the relevant provisions (e.g. the KORUS): whenever both the content/language of the primary obligations is unequivocal and the enforceability dimension thereof is strong, the relevant provision only covers a selected number of environmental agreements and/or is only triggered under certain narrow conditions.

4. Lack of environmental commitment by at least one of the parties (e.g. the TTIP): no bundled climate change enforcement would be possible because the relevant party has not ratified any meaningful international climate change obligation.

Depending on the reading that may be made of the provisions labelled above as "ambiguous", a fifth point could be added: if the said provisions are to be considered as actually "ambiguous", an additional difficulty would arise from the fact that the precise content of the relevant obligation is not, in principle, easy to ascertain due to the combination of seemingly "weak" and "solid" elements in its wording. If, however, the said provisions are to be understood, following Croquet's analysis, as being actually clear, no addition needs to be made.

Notwithstanding all of these facts, in normative terms (*de lege ferenda*), an effective way for the European Union to move forward regarding the enforcement of international climate change mitigation obligations would be to bundle the fate of climate-change-relevant obligations with that of trade agreements. Since the relevant elements for "bundling" are already disseminated in various FTAs (some feature solid climate change linkage language, others feature coverage by the relevant dispute settlement mechanism), lawmakers would need to simultaneously include both sorts of elements in future FTAs to obtain the desired result:

1. In substantive terms, possibly the clearest wording found in the analysed agreements is the one featured in Article 13.6 of the EU–Singapore FTA:

> *2. The Parties shall* ***effectively implement*** *in their respective laws, regulations or other measures and practices in their territories,* ***the multilateral environmental agreements to which they are party***. [Emphasis added; original footnote omitted]

It is not a closed or limited list of agreements, but a general clause, and leaves no doubt regarding the kind and extent of obligation being assumed.

2. In enforceability terms, it will suffice to follow the example of the TPP and the US FTAs: in other words, the relevant provision needs to be covered by the Dispute Settlement Chapter of the relevant agreement, which needs to ultimately provide access to remedies such as compensation and suspension of benefits in case of "non-implementation" of panel reports.

5 Conclusions

International trade law may indeed make a substantial contribution to global decarbonisation if "climate change conditionality" becomes an established practice. The textual analysis performed in this paper has showed that there is an interesting potential in this sense, and it would only require using a correct combination of already-existing provisions and language when drafting future treaties and agreements, and amending past ones.

Interestingly, a "partially progressive" trend in environmental terms is possibly starting: while CETA and JEEPA face, cumulatively, ambiguous content and un-enforceability, the EU–Singapore FTA only suffers from un-enforceability, but has clearly solid content. Climate change conditionality could perfectly fit into this trend, provided that political will actually concurred.

Therefore, in a nutshell, the means to bundle the fate of international trade benefits to effective compliance with climate change obligations are available. The European Union, moreover, is placed in a privileged position to act in this sense, as it did in the past with human rights conditionality. Again, it would only take political will.

References

CNN. (2017). TPP unravels: Where the 11 other countries go from here. http://edition.cnn.com/2017/01/24/asia/tpp-other-11-countries-what-next/index.html. Last accessed August 15, 2017.

Croquet, N. A. J. (2015). The climate change norms under the EU-Korea free trade agreement: Between soft and hard law. In J. Wouters, A. Marx, D. Geraerts, & B. Natens (Eds.), *Global governance through trade EU policies and approaches. Leuven Global Governance series* (pp. 124–157). Cheltenham: Edward Elgar Publishing.

EU Commission. (2015). EU textual proposal—Trade and sustainable development—TTIP Negotiations. http://trade.ec.europa.eu/doclib/docs/2015/november/tradoc_153923.pdf. Last accessed August 22, 2017.

EU Commission. (2016). European Union's initial proposal for a legal text on climate—TTIP Negotiations. http://trade.ec.europa.eu/doclib/docs/2016/july/tradoc_154800.pdf. Last accessed August 22, 2017.

Hachez, N. (2015). 'Essential Elements' clauses in EU Trade Agreements Making Trade Work in a Way that Helps Human Rights? Leuven Centre for Global Governance Studies, Working Paper 158, pp. 1–24.

Hillman, J. (2016). Dispute settlement mechanism. In C. Cimino-Isaacs & J. J. Schoot (Eds.), *Trans-Pacific partnership: An assessment.* Washington D.C: Peterson Institute for International Economics.

Jinnah, S., & Morgera, E. (2013). Environmental provisions in American and EU free trade agreements: A preliminary comparison and research agenda. *RECIEL, 22*(13), 324–339.

Morin, J. F., & Nadeau, R. G. (2017). Environmental gems in trade agreements—Little-known clauses for progressive trade agreements. CIGI Papers 148, pp. 1–8.

New York Times. (2017). Trump Will Withdraw U.S. From Paris Climate Agreement, https://www.nytimes.com/2017/06/01/climate/trump-paris-climate-agreement.html?mcubz=1. Last accessed August 22, 2017.

USTR. (2017). Summary of Objectives for the NAFTA Renegotiation. https://ustr.gov/sites/default/files/files/Press/Releases/NAFTAObjectives.pdf. Last accessed August 30, 2017.

Velluti, S. (2016). The promotion and integration of human rights in EU external trade relations. *Utrecht Journal of International and European Law, 32*(83), 41–68.

Dr. Eduardo Alvarez Armas obtained his "Licenciatura en Derecho" (LL.B) *Summa Cum Laude* at the Universidad de La Laguna (Tenerife) in 2006. Upon graduation, he began working for J&A Garrigues, one of Spain's leading law firms, in their civil and commercial litigation department. In 2008, he moved to Belgium where he received a Masters in European Law at the College of Europe (Bruges—2009) and a Masters in International Law at the Université catholique de Louvain (2010—*Summa Cum Laude*). Between 2010 and 2015, he worked as a Research Fellow in legal sciences for the Belgian National Fund for Scientific Research (FNRS), teaching and researching at the Université catholique de Louvain on private international law, international environmental law, global governance and the law of the European Union. After obtaining an LL.M in American Law (*Honours*) at Stanford Law School in 2016, he defended his Ph.D. dissertation in 2017 (Double Ph.D. diploma, Université catholique de Louvain and Universidad de Granada—*Summa Cum Laude*). He is currently a Research Associate at Queen Mary University of London.

Peruvian Amazonian Natives and Climate Change: Minorities Facing a Global Problem

Guillermo Mandelli Constant, Erika Sänger Do Amaral, Alberto Paucar-Caceres, Silvia Quispe Prieto and Manuel Caipa Ramos

Abstract Indigenous people knowledge is classified as comprehension of the skills, practices and ideologies of cultures that have had long histories of harmonious living with their natural surroundings. This knowledge forms important contributions to the world's adaptive capabilities and is a contributor to the accomplishments of the 2030 Paris Agreement. Granting land rights to indigenous people societies has been proven to significantly curb deforestation and thus is a low-cost solution to reducing carbon emissions. Considering the importance of indigenous people knowledge, and the role the Amazon rainforest and its original inhabitants play in climate change, a three-month Participatory Action Research approach was embarked upon in various communities of the Peruvian Amazon. This paper aims to highlight how the Peruvian government and external agents have interacted with the natives throughout history and emphasises the outcomes of these encounters in regard to the protection of indigenous people knowledge, the rainforest and climate change. Due to the short time span of the field research, the indigenous people communities where the primary knowledge was obtained cannot be representative of the whole Peruvian Amazon; however, together with the literature review and further research, the stance of both the Peruvian government and the indigenous people populations can be seen. While evidence shows that the

G. M. Constant
Independent Researcher, Manchester, UK
e-mail: guillermomandelli@hotmail.com

E. S. Do Amaral · A. Paucar-Caceres (✉)
Manchester Metropolitan University Business School, All Saints Campus, Oxford Road, Manchester M15 6BH, UK
e-mail: a.paucar@mmu.ac.uk

E. S. Do Amaral
e-mail: sanger.erika@gmail.com

S. Q. Prieto · M. C. Ramos
Universidad Nacional Jorge Basadre Grohman, Tacna, Peru
e-mail: silviaq24@hotmail.com

M. C. Ramos
e-mail: mcaipa26@hotmail.com

© Springer International Publishing AG, part of Springer Nature 2019

W. Leal Filho and R. Leal-Arcas (eds.), *University Initiatives in Climate Change Mitigation and Adaptation*, https://doi.org/10.1007/978-3-319-89590-1_22

world has never been better geared towards legally granting the indigenous people land rights, the results in this paper seem to indicate that Peru's progress towards securing tenure rights and upholding the sustainable development goals highlighted in the Conference of Parties and United Nations Framework Convention on Climate Change is not being met, and consequently, the knowledge, rights and lives of the natives are being overlooked, causing a ripple effect on the rest of the world.

Keywords Indigenous people knowledge · Peruvian Amazon · Deforestation Climate change · Government institutions

1 Introduction

Humans are known for their adaptability, and historically, societies have always adapted to socio-ecological changes. However, this accelerated climate change we are experiencing will require higher levels of adaptation then ever before (Biesbroek 2014). Climate change adaption does not come without its obstacles, and from an economic point of view, implementation of solutions has been hindered by unattractive barriers to entry, especially for developing nations such as Peru.

With global progress in 'acknowledging' the climate problem, there still lies a complex pattern of discrimination when looking towards the indigenous people as a possible solution to the issue. Evidence shows that granting land rights to the indigenous people can reduce carbon emissions by 30%, making it the lowest cost strategy in reducing global emissions (CGD 2014; WRI 2014). Forest and landownership in the hands of natives have proven to not only conserve, but also enhance landscapes resulting in a higher number of people benefiting as opposed to state or privately owned land (RRI 2017).

A recent study using community-level longitudinal data resulting from high-resolution satellite images has noted that within one year of titling land to natives there was an 81% decrease in the deforestation rate and a 56% decrease in the following year (Blackman et al. 2017). Peru holds the third largest area of tropical rainforest after Brazil and Democratic Republic of Congo so they play an extremely important role in climate change mitigation.

In recent years, Peru has demonstrated concern for protecting the environment by setting ambitious sustainability goals and implementing environmental policies at a federal level. However, the Peruvian track record for fulfilling such objectives at state and local levels has not always been successful, with many cases of laws contradicting environmental agendas (Catanoso 2016; Collyns 2018).

Although Peru is not a big polluter accounting for only 0.3% of global emissions per capita, it is classified as a 'particularly vulnerable' country to the effects of climate change. The increase in Peru's anthropogenic activities is further intensifying their susceptibility (UNFCCC 2015a, b).

Our fieldwork seems to indicate that Peru's external environmental agenda often does not correlate with the internal realities, consequent of its desired rapid economic growth. Due to its geographic remoteness and sheer size, keeping an eye on illegal activities that lead to deforestation, contamination and ethnocide is a difficult but not impossible task for the Peruvian government. This paper explores the relationship the Amazon rainforest and its inhabitants have had with resource extraction, external agents, government institutions and global climate change, looking closely at how history and current affairs have affected the ability for the natives to continue protecting their territories, thus aiding the 'global villages' environmental issue.

2 Literature Review

In order to contextualise our research, we have selected some key points of information to be included in the literature review. Firstly, we found it necessary to set the scene with Peru, highlighting its diverse demographics, vast biodiversity, economic agenda and their attitude towards their indigenous people and environment. We wanted to emphasise what is indigenous people knowledge and why has it become such a significant topic in regard to world climate science and adaption. Furthermore, what can we do to promote it and protect it? We look at how international institutions are coming together in order to address climate change through global projects, and how new environmental agendas unanimously agreed upon by the world seem to be addressing the lives of those most vulnerable.

2.1 Peru Overview

Peru has a population of over 31 million inhabitants, 50% of which are considered indigenous people (UNESCO 2017a, b). It has the second largest section of Amazon rainforest (13%) after Brazil (60%) making it one of the most biodiverse countries on the planet and boasts some of the highest levels of animal species known to man (Butler 2017). Although rainforest covers 50% of Perus' territory, only 12% of its population live there. Within the 2–3 million people who live in the Amazon basin, approximately 500,000 are indigenous people constituting 2% of the countries' population (INEI 2018).

Peru has experienced high levels of GDP growth over the past decade (World Bank 2018). A combination of macroeconomic stability and market liberalisation pushed by a series of probusiness presidents has allowed Peru to benefit from exports of their rich resource deposits. By establishing a series of free trade agreement (FTA), they have been able to profit from the past decade's commodity boom (Mapstone 2010).

Externally, Peru's growth has been impressive and improvements in quality of life have been felt by some of the coastal populations; however, when we look at the indigenous people populations of the Andes and more specifically the Amazon, this economic growth model is not sustainable and the long-term exploitation of natural resources has wounded Peru's forests and has caused uproar in the indigenous people communities (UNESCO 2017a, b; UNEP 2016).

The impacts to the indigenous people communities have been horrific; resource extraction has left contaminated soil, water, fish and game and has led to an untold number of direct and indirect problems such as destruction of habitat, disease epidemics, miscarriages, skin diseases, elevated cancer rates and deaths. One case in northern Peru has seen companies such as American Giant Occidental, Petroperu, the Netherlands registered Pluspetrol, China National Petroleum Corporation and Canadian subsidiary Frontera Energy, all extracting oil and leaving the region and its peoples in unliveable conditions. The total estimate to clean up the mess is $1bn, but no one is assuming the responsibility (Hill 2017).

Government institutions, such as the 'Servicio Nacional de Areas Protegidas' (SERNAP), together with local indigenous people groups, NGO's and associations such as the 'Interethnic Association for the Development of the Peruvian Rainforest' (AIDESEP) work to create protected areas in order to defend wildlife and indigenous people communities. However, once these areas are formed, they become government-owned and Peru's economic political agenda is reflected, with 72–84% of Peru's Amazon territory allocated to oil and gas exploration, according to satellite images and a report released by RAISG in 2012.

2.2 *UNESCO Links—Local and Indigenous People Knowledge Systems*

2.2.1 The Importance of Indigenous People Knowledge

For the last 25 years, indigenous people knowledge (IK) has been recognised as playing a vital role in environmental management and development, particularly with regard to combating climate change (see Appendix 1). Over the same period, nations across the world together with different scientific and economic research institutions have recognised the need to incorporate indigenous people knowledge systems and practices with existing scientific understanding as a major resource for adapting to climate change and increasing adaptation capabilities.

The worlds' heritage and tradition are associated with IK and perhaps give the illusion of it being inflexible. Actually, as is science, it is constantly reassessed, reshaped and expanded, with each generation reinterpreting the knowledge and adapting it to the changes around them. Thus, making it a very dynamic set of information rendering the natives is very adaptable. For the indigenous people, it can be seen as the human laboratory of trial and error. The indigenous people have always been subject to environmental changes, due to their remote geographic locations, resource-based livelihoods and an absence of rights over resources, leaving them even more vulnerable to economic, social and environmental impacts. As such, their practices have always had to have an adaptive capability. This has led to an extensive understanding of sustainability—especially in the areas of planting and harvesting; transhumance or herd mitigation; and timing and locations of hunting—all of which are heavily dependent on detailed understanding of weather and climate. IK can directly contribute to climate science, with observations on a much wider temporal depth and a finer spatial scale, with knowledge being gathered for thousands of years.

From living so closely with nature and being dependant on it, the indigenous people are the best monitors of biodiversity and data collection. It has been proven that the indigenous people can enhance biodiversity by transforming landscapes, making them key players in research monitoring and raising awareness worldwide. IK recognises that the key to conservation is not in separating humans from nature, as the Western ideologies do; for indigenous people communities, this is unacceptable and renders landscapes useless if social systems and ecosystems are severed.

This diversity of knowledge comes from indigenous people languages and is a direct reflection of their worldviews. The things that different societies decide to name reveal the way in which they engage with the world. They contain highly specialised lexicons and taxonomies composing extremely detailed, high-resolution, conceptual frameworks for describing nature and subtle environmental changes and trends. Through weak government institutions, we see ethnic groups rapidly losing their language and with it a unique system of comprehending our natural world.

Synergies between Western science and IK cannot be made if scientists try to validate IK before accepting it. Assuming the Western systems are always correct, passing judgement on cultural roots causes an instant hindrance to the whole collaboration. IK and science both contribute different things as they are sometimes very contrasting: qualitative versus quantitative, holistic versus specialised; these approaches are highly complementary. For effective coproduction of knowledge, there needs to be recognition that both forms are legitimate. In order to create a viable platform, the building of effective dialogues, the breaking of cultural barriers and the establishment of mutual respect need to be implemented.

2.3 *International Institutions*

Climate change has been on international agendas for the last few decades with annual gatherings highlighting aims and objectives of each nation in order to achieve unifying commitments to the planet. International law does not enforce these agreements, and no one is penalised for not achieving their goals. Within these conferences, indigenous people rights and opinions are referred to, but often dismissed, and no real measures are taken to safeguard their knowledge or their existence (Glennie 2014). The participation of indigenous people in environmental conferences such as the COPs can be seen as a way of allowing access to power but not being able to influence it (Naidoo 2015). This goes in line with the 'recognition' of indigenous people input in the Paris Agreement as well as 'acknowledging' their key role in fighting climate change, and this choice of words is questioned by several indigenous people representatives for its real power of enforceability (UNFCCC 2015a, b; Goldtooth 2014). Despite in-depth research regarding their IK contributions to climate change and the correlation between granting indigenous people land rights reducing global emissions, it is business as usual for most nations (WRI 2014).

Peru is a proactive member of the United Nations Framework Convention on Climate Change (UNFCCC) and has set ambitious goals such as reducing emissions by 30% by 2030 and 0% net deforestation by 2021. They have also stipulated within their adaptation plan to 'encourage participation of indigenous people organisations in actions on climate change'. The indigenous people are highlighted as the most vulnerable to climate change along with rural communities (UNFCCC 2015a, b).

Development institutions, such as the United Nations and the World Bank, have joined forces to create initiatives like the REDD+ programme (reducing emissions from deforestation and forest degradation) to combat climate change.

Funded by the carbon market, this programme allows developing countries with key green areas of our planet to receive financial help as well as regulations and knowledge concerning sustainable forest management by creating carbon stocks (The REDD Desk 2018). By sectioning off parts of forest, the aim of the REDD+ programme is to protect large areas from deforestation, therefore creating a positive carbon offset. However, these green areas coexist within indigenous people societies' land. A traditional style of indigenous people farming is known as swidden farming; these landscapes are left untouched after they have been farmed for several years to encourage healthy regeneration. However, these areas are often seen as easy targets for governments and 'carbon cowboys' looking to cash in from the REDD program, by planting fast-growing and ecologically harmful plant species (Kashwan 2017; AIDESEP 2011).

3 Methodology

The role of the indigenous people in protecting the Amazon has been nominated as crucial to fight climate change (UNESCO 2017a, b; WRI 2014) Therefore, our rationale behind this research was to comprehend how these societies are positioned for such responsibility and the realities of their lives in this modern age. Importantly, within the country lines, their communities happen to fall into, how is their relationship with the government, and are external actors influencing their development. This particular situation led us to take a deductive research, investigating whether theory regarding indigenous people knowledge and rights is being upheld and implemented, in our case, is the Peruvian government acting accordingly as a credible institution. Furthermore, are the indigenous people given the adequate platforms to participate in the states' decisions or even to have a say in what happens to their ancestral land?

In order to understand the role of indigenous people communities in confronting climate change, we undertook a critical ethnographic methodology, empowering our informants by making their voice heard (Creswell 2007). The only way of truly understanding someone who lives in such a distinct way is to 'live as them and with them' (Whitehead 2005). This goes in line with the main principals of ethnography, in which researchers immerse themselves in a whole new and different culture and therefore acquire an in-depth perception through their informants' eyes (Corbin and Strauss 2008).

To broaden our results, we conducted three months of fieldwork in distinctive native communities across the Peruvian Amazon, in the areas of Madre de Dios and Ucayali. We were able to enter communities through different conservation projects and volunteering positions. At times, we were not welcomed for our work and some tried to charge us money for contributions of information; as this is an independent project, we were unable to provide funds. Instead, we provided friendly exchanges in forms of work, stories, tobacco, coca leaves and sometimes food. The results shown in this paper come from genuine people who believed in our cause and who did not request any monetary compensation.

As our journey continued, we became more accustomed to our surroundings and also gained trust from the locals and we were able to delve deeper and receive more interesting and more heartfelt answers to our questions. We were dynamic with our research gathering methods and also alternated time and place; for this reason, we were able to gauge a variety of our informants' feelings and our own observations more in-depth.

Two methods used were open interviews and participatory observation activities, which allow a much closer perspective of the communities. Another method practised during our fieldwork, of which is indeed deserving of explanation, was the actual fieldwork notes. These permitted us to not only describe our surroundings, the behaviours and interactions within the communities, but also correlate these descriptions with the data gathered in the interviews. This interpretation allowed us to verify whether the informants' responses were intrinsically linked to the methods employed.

Using a microphone and camera, we were able to register all our findings, and interviews were conducted across an array of people, from shamans to priests, community leaders to NGO representatives and local youth to the elders, enabling a deeper understanding of the communities' culture and behaviours. In addition, we found it important to attain information from external agents such as local governments, non-governmental organisations and private companies involved in the development of these communities; however, we found it hugely unsuccessful. This was due to a number of factors: contact numbers on official Websites were incorrect or disconnected, official office posts were empty, and our emails were never replied to. So, due to this limitation, we were not able to gather primary information from these external agents. Secondary data from past research, official institutional documents and environmental journalists were used to cover the perspective of these other groups.

As this research is based on the opinions and views of people, one must consider that the results are more qualitative than quantitative (Tinker 1990). Thus, when analysing the primary data gathered, a process of coding was performed. This was done in order to find any trends and build up a generalised opinion from the community members. Coding in this sense was used to distinguish general themes shared by different informants as correlations arose as the same questions were asked (Hay 2005). In order to provide a critical analysis of our findings, we used our field note observations together with the primary data obtained from our informants, while referencing pertinent academic research, which supports our findings and contributes to the coherency and accuracy of our results.

The following section provides the results gathered, with a critical analysis allowing us to paint a picture of the relationship the indigenous people of the Peruvian Amazon have with the outside world. As our primary data gathered were attained in Spanish, procedures of coding and categorising were exercised in Spanish and later translated into English to avoid any loss of meaning in our results. Only when information was given in native tongue did, we translate it before prior analysis.

4 Results and Findings

Due to the sensitivity of some of the information given, the identity of our informants is protected during our findings.

4.1 Brief History—The Arrival of Outsiders

From the community elders, we heard various ancestral tales of vast sedentary civilisations who sculpted their landscapes, cleared spaces for agriculture and cleverly managed rainforests for medicinal plants and ancient permaculture, improving their crops over thousands of years. These populous civilisations were reduced in number with the arrival of the European settlers apparently pushing the tribes further into the jungle, adopting a more nomadic life. Most significant perhaps were the effects of the industrial revolution on the Amazon populations, when the rubber boom took place (1850–1920) sending thousands of outsiders to the jungle in search of the rubber tree to produce vulcanised rubber mostly for their bicycle and car tires. These times are remembered as some of the darkest for the indigenous people, and the stories of ethnocide, torture and slavery are still echoing today. This was the natives first but certainly not the last encounter with resource extraction and the needs of the outside world.

When the news of the atrocities circulated the outside world, a wave of missionaries set off with the intention to protect the native people. The first contacts made by the missionaries in the Madre De Dios region caused widespread disease and death between the indigenous people, and they feared it was the spirits of those killed and tortured during the rubber boom plaguing them. Luring the indigenous people out with medicine for the European diseases they brought, the priests were able to form communities bringing together various different ethnic groups from within the jungle.

Combining distinct ethnic groups, moving them away from living under one roof, the *Maloca*, controlling who they married and suppressing their traditions caused a lot of tension and divisions between the groups resulting in loss of culture, especially language which some locals describe as 'ethnocide' (Gray 1996).

4.2 The Results of Church Intervention

It is necessary to know the past to understand the present situation and the conditions the indigenous people live in. Before the government ever made contact with the Amazon natives, the missionaries were there. Sixty years on how do the locals of this region feel about the church's presence and how are they remembered in the newer generations? Padre José Álvarez was one of the first missionaries in the Madre de Dios region venturing down the Ishiriwe River.

The opinions shown in Figs. 1 and 2 come from first-generation (post evangelical) locals about their memories of the changes and influences, giving us an insight into how they felt about the imposition of a new way of life by the Church. Speaking to more people, especially the elders, who felt the changes more drastically, felt very bitter about the events. As innocent as the intentions were, it was the church that disrupted these self-sufficient and culturally aware populations halting their natural development. Some teach them to cut trees and extract natural resources for monetary gains, introducing them to a new world in which they are at the bottom of the food chain—an open and deliberate acculturation (Gray 1997).

"Padre José Álvarez came with good intentions. He came to civilise our indigenous group and protect us from the rubber barons. It was a very demanding process, a little over the top. The church broke a lot of our traditions, even kicked our backsides sometimes. After Álvarez came more 'padrecitos' (little priests) but with other intentions, there were priests who came with tools like the two-man saw and started giving us things in exchange for trees. So everyone started cutting trees, we don't know how many we cut or if the trades were fair, we didn't know how to count money, sometimes for a few trees we would get a shotgun or a machete"

Fig. 1 Harakmbut local, ex-FENAMAD worker

"We see that in reality we have been tricked, they have stepped on our rights; this shouldn't be in the name of development. Firstly, with the missions implementing us with a religion without even knowing that we already had our own beliefs and traditions".

Fig. 2 Harakmbut elder, ex-community representative at FENAMAD

Many of the indigenous people groups we encountered spoke of loss of culture and tradition through the church and weak government institutions not protecting them; perhaps one of their biggest concerns was the loss of language. Languages are a direct expression of the indigenous people 'cosmovision', or worldview, and are a reflection of their natural environment. Being made to speak a non-native tongue can disassociate them with their own cultures and livelihood, thus heightening alienation and disorientation and therefore highlighting our finding of weakened ethnic groups (UNESCO 2017a, b).

This new way of life is accelerating a loss in culture for the natives, and as a result, thousands of years of adaptable indigenous people knowledge is being forgotten.

4.3 The Indigenous People Today and the Government

In today's reality, we saw that actual churchgoers were scarce and followers of the religion were minimal. However, we saw a different focus especially in the younger generation. They were not religious, nor were they fond of the jungle. All they were interested in was 'la plata'—the money. This demonstrates a cultural limbo in which the libraries of ancestral sustainable knowledge, the elders, are being ignored and economic interests are prevailing. External influences created by the church, government, NGOs, television, pop music and tourism have introduced the indigenous people to a capitalist driven mentality, but have left them living in a place with little opportunity for monetary gain, increasing the attractiveness of illegal activities.

In Peru, the percentage of 'poor' among the Amazon's native population reaches 60.5% in contrast to 15% in Spanish-speaking populations (Gushiken and Campos 2014). A conservation leader in one of the communities told us of a national census the government released. This census reached the Amazon, and some of the questions included: How many light bulbs do you have? How many times a week do you eat chicken? What material is your house made from? What modes of transport do you use? The same questions were asked to the residents of the capital Lima. Due to the asymmetric way of living the native Amazonians have, they are put in the 'seriously poor' bracket in terms of Peru's population. City and jungle living are two different worlds, and poverty cannot be measured in the same way. In the city, one needs money to be wealthy, and in the jungle, wealth is valued by a

"...development for a community isn't being financially comfortable, it's having good nutrition, clean water and to live in the best way without stress and pressure, no economy is needed."

Fig. 3 Harakmbut elder, ex-community representative at FENAMAD

healthy food source, good health and a healthy environment, as the following statement in Fig. 3 from an elder woman illustrates:

The Amazon indigenous people have been living without money for thousands of years and have lived off the land in harmony with the natural world around them in an exemplary form. Contributions of each indigenous person worldwide concerning the improvement of living conditions in the 'global village' have been a significant topic over the last 25 years, showing us the importance of creating real interculturalism to stop distributive injustice, economic inequalities and relationships of prestige and power (Tubino 2005). Yet while political institutions continue liberating all kinds of economic areas from ecological, ethical and legal work barriers, it is rare to see opportunities given to indigenous peoples encouraging or giving minimum conditions to put their wisdom into practice, allowing them to demonstrate their capability to advance into the future from an alternative holistic perspective (Hierro et al. 1998).

4.4 Are the Indigenous People Given Adequate Platforms to Participate in State Decisions, or Have a Say About What Happens on Their Ancestral Land?

In the jungle, the missionaries no longer lead the external development influences; it is now driven by dominant economic interests, backed by a government controlled by neoliberalist policies hunting the region for exportable goods aiming for territory jurisdiction. These economic forces provide the building blocks for progress-driven ideologies, leading to increasingly dehumanised objectifications of indigenous people as animal-like individuals whose destiny only leads to assimilation, displacement or death (Trigger et al. 1996). From the Peruvian governments' first contact with the Amazonian natives, there have been weak institutions protecting their rights and respecting their cultural heritage, especially in the areas of forming communities, a delicate process which was rushed, contributing to the weakeing of ethnic groups and exposing them to illegal activities. In Ecuador and Bolivia, each ethnic group formed their own community; with larger numbers, they have been able to embellish their cultures. Therefore, incorporation of local frameworks into state politics can be seen in both countries being the only ones in the world to grant nature a right in their constitutions. The Peruvian state has not incorporated indigenous people culture and not tried to integrate existing social forces into its orbit creating more of a divide further marginalising the original population of the country (Gray 1997).

In Fig. 4 this elder woman is referring to the leadership of Alan Garcia (1985–90, 2006–2011). Under his probusiness presidency, he 'waged a Cold War with the indigenous people (Aiello 2009). In 2007, he argued that indigenous people or communally owned land was inefficient use of natural resources referring to the issue as 'The dog in the manger' syndrome, coming from the famous Aesop's tale, with the indigenous people being the dog who does not give up its resources. In 2009, in the Amazonia province of Bagua, the indigenous people showed great unity and strength coming together in disagreement with government actions. Hundreds of police clashed with thousands of indigenous people, and according to UN Human Rights, report resulted in the deaths of 33 police officers, 10 local people dead and 200 locals injured. Clashes arose after two months of peaceful protests demanding the withdrawal of 'La Ley Forestal 1090' (OIT 2014)—the forest law, which allows for an easier process of extraction of resources bypassing the existing ILO C169 of prior consultation regarding new deals made putting their ancestral lands at danger of oil and mining exploitation. Alan Garcia responded to the events at Bagua with: 'indigenous people are not 1st class people' and '400,000 natives cannot hinder the development of 28 million Peruvians' (Viale and Bedoya 2010).

Figure 5 reflects a deal we came across out regarding a 40-year contract to 50% control on operations over 400,000 ha of protected forest reserve 'Amarakaeri' being granted to the oil company one year after it was created in 2003. The contract was made with the bare minimum of residents of the surrounding communities present.

As we can see, the government and its institutions are not directed to incorporate the indigenous people into their economic growth model, and quite the opposite, they are seen as a barrier to development and financial growth, so decisions to sell off their land to private companies are constantly being favoured. Over the decades, state interaction has led to an erosion of confidence in the government to provide any credible institutional commitments to the indigenous people (North 1991).

Alan Garcia enacted a series of laws and legislations to facilitate trade with the USA; within them were laws that directly affect the indigenous people; the objectives of these proposals are to make the whole idea of communal land disappear with legislations formed directly to get rid of these property rights. Peru's government only recognises 4% of Amazon to be communal land leaving the rest

"All the sudden because we didn't know how to read or write they told us we were the worst, it shouldn't be like this because we are wise we are good dreamers. We have been worked on psychologically saying we are ignorant, even the government saw we are 3rd class citizens and we don't let the country develop, for this reason they have been neglecting us as native communities. Due to this resistance the youth do not want to claim their heritage."

Fig. 4 Harakmbut elder, ex-community representative at FENAMAD

"Our fathers fought and fought for the 'Reserva Communal Amarakaeri' to be established in order to protect our ancestral land. Within one year the government had authorised Hunt Oil to begin exploration. Of course the little rights we have here as indigenous people they had to approach us first to get the 40-year deal. My fellow leaders were taken in helicopters to the capital where they have stayed for a week. The leaders had been bought out and when they returned the contract was signed, without anyone from the community knowing about it."

Fig. 5 Harakmbut community leader

up for grabs with up to 84% already allocated to oil and mining exploitation (RAISG 2012). This leaves the natives at serious risk of illegal activities from both MNCs and settlers who come from far and wide to exploit the quick cash available from illegal mining and illegal logging, pushing Peru's environmental commitments further away. Figure 6 illustrates a view on land rights issues the indigenous people face.

In 2006, the Alan Garcia administration modified a statute granting more authority to the Peruvian Agency of International Cooperation (APCI), enhancing governmental control over operations of NGOs. The mandate included 'supervising' of NGO projects, making sure they are in line with the states' own ideas of development and in this case free trade and resource extraction (Aiello 2009). Meaning the NGOs could have their legal status revoked if they stepped out of line with the state agenda. To some extent, perhaps this has made it so the NGOs have to plan more carefully so not to be shut down by the government. In 2009, Hunt Oil approached one of the communities and was able to provide $30,000 and the promise of jobs, to enter and begin their seismic testing in lot 76, the Amarakaeri Reserve. At a time when the indigenous peoples' official representatives FENAMAD were most needed, the oil company was able to bypass them and buy out some of the community with false promises of development. Leaving the town split in two: those desiring the short-term benefits of finite resource extraction and those fearing the long-term repercussions and longing to maintain the old way of life.

Even though Peru's actions have already gone against the UN Human Rights, they continue their mandate putting the lives of thousands of indigenous people at risk. Alan Garcia said during a speech in Lima 'The riches of Peru are for the Peruvians! Not just a group of people who were fortunate enough to be born there'.

"We have food security and health problems caused by the high levels of mercury due to illegal gold mining, if the land was ours maybe we could help stop it but the government only takes our land away they don't grant us land rights."

Fig. 6 Harakmbut conservation project leader

All of the abovementioned brings us our second main finding of very weak national institutions safeguarding the indigenous people of Peru.

Despite the government's acknowledgement of its serious vulnerability to climate change as stated in its UNFCCC report, and despite various international institutions proclaiming the importance of the indigenous people role in forest stewardship and sustainability goals, we see that the attitudes displayed towards the environment and those living there is not a priority.

4.5 Climate Change and Environmental Stewardship

The issue of oil and gold exploitation not only causes contamination of water, soil and food sources leading to health problems, but it is also a direct cause of deforestation along with infrastructure development such as roads and the illegal timber trade. The illegal wood market is a huge problem for Peru in which the state loses $250 million a year according to OSINFOR. Weak and corrupt institutions result in up to 80% illegal wood extraction (World Bank 2012). We can see that there are many indirect causes to deforestation such as landownership and use rights, governance and institutional capacity, corruption and criminality, human rights, access to justice and the criminalisation of protest, national development and land use policies, perverse incentives and weak state response (Valqui et al. 2015).

Protecting the Amazonian rainforest is critical in combatting global warming. The forests offset the effects of burning fossil fuels and help to regulate climate change patterns by absorbing massive amounts of carbon dioxide. In 2014, the IPCC released a report showing that as much as 30% of total carbon emissions mitigation would occur by halting and reversing tropical deforestation (CGD 2014). A report released by WRI and RRI in 2014 proved that strengthening community forest rights is a low-cost strategy and very successful model for deforestation resistance. It is seen that deforestation rate decreases by 2–3 times when indigenous people are present. It is worth mentioning how delicate the ecosystem of the Amazon actually is, with studies carried out in 2009, by leading climate change expert Dr. Vicky Pope, of the UK Met Office's Hadley Centre, projecting that even a slight increase in temperature (1–2 °C) could cause 1/3 of loss in flora. Figure 7 reflects these real changes felt by those most at risk—the indigenous people.

What option is best for planet earth regarding forest ownership? Is it governmental, private or communal? Our field research seems to indicate that, in

"Now on the subject of the environment and climate change, we feel it, with all the storms, heavy rains and flooding. The NGO's and even the government, what do they do? They don't change their system that is causing this, they just talk but we don't see any change."

Fig. 7 Harakmbut ex-FENAMAD worker

"External agents always come to 'teach us' about how to live and interact with the environment but they have their own western strategy of development that at times is not in accordance to our actual needs. The responsibility my father left me was to help the community while living in the community so to help nature we must live together with nature."

Fig. 8 Harakmbut conservation project leader

communal hands, the land thrives and provides much more in terms of long-term care of biodiversity and forest stewardship. This is achieved by preventing logging; roads; mining and other destructive forces. Peru's government are still internally promoting recent highway proposals, such as that connecting Iquitos to Saraniriza. Such projects in reality bring huge levels of deforestation and give access to pristine jungle to agribusiness and illegal mining. Externally, Peru's Environmental Ministry (MINAM) in the UNFCCC has set the goal of stopping all net deforestation by 2021; this was set in 2009, and rates have only increased (Zuazo 2017) (Fig. 8).

The idea of granting land rights to the indigenous people as a way of preventing climate change is a low-cost solution. Compared to other mitigation strategies, reducing deforestation and degradation is the most viable strategy in the land sector (Goodman and Martin 2014). Other projects such as the carbon market led by REDD+ pose higher costs and lead to disconnection from landscapes. In the west, the idea of conservation removes humans from nature, this idea of division for the indigenous people is considered unacceptable as ecosystems, and social systems must be intertwined for landscapes to be rendered useful (UNESCO 2017a, b).

4.6 Indigenous People Development Model

This brings us to our final, and foremost, finding: the indigenous people development model known as 'Buen Vivir'. Throughout our time in the Madre de Dios and Ucayali areas of Peru, when asking the indigenous people about the topic of development, the term 'Buen Vivir' seemed to correlate across the indigenous people groups we met: Harakmbut, Matsigenka and Shipibo. Sumac Kawsay is a term in Quechua that can be translated in Spanish as 'Buen Vivir' or in English as 'Good living'. The term describes 'a way of doing things that is community-centric, ecologically balanced and culturally sensitive', a stark contrast to the market-is-king model of capitalism (Balch 2013). This term has been used across different parts of Central and South America and has similar connotations to the natives. It is underpinned by the cosmovision that only by living in harmony with nature, the community as a whole can achieve a good living. It is these kind of mentalities and ideologies that the west seem to lack, perhaps the disconnection from nature and the ground beneath us has distanced us from the essence of our lives. Everyone wants to develop including the indigenous people, but at a natural pace that does not

"For the indigenous there's no such thing as economic development yet an intellectual development is far more important because you need knowledge to feel that you have capabilities. We want to create sources of work, as we no longer have natural wealth to depend on. We want our fellow indigenous to recognise the outside world but to maintain their native culture and create a combination of both knowledge's in order to protect the community in the future."

Fig. 9 Shipibo community leader

blatantly destroy our surroundings or undermine any living thing's existence, leaving an exhausted world for the next generations (Fig. 9).

5 Conclusions, Limitations and Future Prospects

Through our immersion, we have identified that the indigenous people possess a truly sustainable development model. For millennia, they have lived in harmony with their surroundings not merely taking from it but also complementing it. Scientific data and local evidence have proved that the indigenous people are the best stewards for the environment and they must be empowered.

Our results seem to indicate that the Peruvian government is not making good headway towards safeguarding indigenous people knowledge, nor making progress in the granting of land rights which, above-mentioned, has proven to reduce environmental destruction. Unfortunately, it appears to be quite the opposite. The relentless allocation of oil and gas concessions, and the complete disregard for the prior consent law, demonstrates the extent the government care about its indigenous people population, the environment and the global issue we are all facing, distancing itself more and more from the environmental commitments it made to the world.

Advances have been made in admitting to the problem and talking about it, but for all the agreements in COP conferences to actually progress, and for governments to make promises they can truly fulfil, environmental agreements need to resemble less of a gentleman's agreement and more of a solid framework which will compel countries to get on board with the global strategy. Free trade agreements are huge contributors to climate change. We suggest implementing ecocide and ecological crime within a new framework of enforceable international law, penalising companies and states. This may appear a drastic measure, but unless there is a shift in our global consumer mentality, then loosing money in court may be the only measure to save our environment.

The limitation of our work was the short time span in the field only allowing us only to visit a few communities in two regions, restricting our results to not be able to represent the whole Peruvian Amazon but to give us all a glimpse into the reality they live in and the struggles they face with their government, resource extraction and climate change.

In the last year, Peru has experienced some abnormally harsh weather conditions, from some of the worst droughts followed by some of the worst ever floods. Being such a vulnerable country to the effects of climate change means that the business-as-usual mentality is not sustainable. If Peru continues, its negative anthropogenic actions' extreme weather conditions will only worsen for them and the world.

Admittedly, it is a difficult task for a developing country to ignore the huge deposits of commodities lying beneath its territory; however, with its tropical forest reserves, Peru has a huge responsibility and potential to aid the global issue of climate change.

At the moment, Peru only uses a fraction of its renewable energy potential (0.65% wind and less than 1% solar) according to a 2011 report by CINYDE. New studies have shown renewables to be very attractive to developing countries as they can 'leapfrog' the fossil fuel dependency stage and go straight to clean energy (Kaufman 2015). With the ever-dropping price of renewable technologies and less infrastructural changes, this seems to be a feasible option for Peru.

Our 3-month research seems to indicate that the problem of deforestation in Peru comes from a lack of alternatives for the locals, perhaps looking towards the Amazon as a source of wind and solar energy (where deforestation has already taken place) could very well mitigate any further impact. Shifting investment away from oil and gas exploration is key, and creating government subsidies for clean technology would highly increase Peru's climate change mitigation potential, increase jobs and provide a new lease of life for those it has ignored for so long.

Acknowledgements This research was carried out as part of the programme financed by the European union Action 2 Mundus Partnerships, GRANT AGREEMENT NUMBER—2014-0870/001-001. The Euroinka (www.euroinka.eu) Project is lead by Manchester Metropolitan University, UK.

Trabajo hecho como parte del Programe Financiado por la Unión Europea-Action 2-Erasmus Mundus Partnerships, GRANT AGREEMENT NUMBER—2014-0870/001-001. El proyecto Euroinka (www.euroinka.eu) liderado por Manchester Metropolitan University, Reino Unido.

Special thanks to all the community leaders, shamans, NGO representatives, priests, government officials, conservationists, locals and all the genuine people who helped us along our journey. Many thanks go to the Universidad Nacional Jorge Basadre Grohmann in Tacna Peru for believing in the project and giving us the platform to share our story. Last but not least, Jessica Stoodley, Adam Rogerson, Tom Watts, Nilton Mandelli for your patience and honesty (Erika and Guillermo).

References

AIDESEP. (2011). The reality of REDD+ in Peru: Between theory and practice. https://www.forestpeoples.org/sites/fpp/files/publication/2011/11/reality-redd-peru-between-theory-and-practice-november-2011.pdf.

Aiello, K. (2009). Peru's cold war against indigenous. http://www.worldpress.org/Americas/3383.cfm.

Balch, O. (2013). Buen vivir: The social philosophy inspiring movements in South America. *The Guardian*. https://www.theguardian.com/sustainable-business/blog/buen-vivir-philosophy-south-america-eduardo-gudynas.

Biesbroek, G. R. (2014). Challenging barriers in the governance of climate change adaption. http://edepot.wur.nl/290520.

Blackman, A., et al. (2017). Titling indigenous communities protects forests in the Peruvian Amazon. http://www.pnas.org/content/114/16/4123.abstract#aff-1.

Butler, R. (2017). The Amazon rainforest: The World's largest rainforest. https://rainforests.mongabay.com/amazon/.

Catanoso, J. (2016). Peru's new environmental policies: What are they and will they work? https://pulitzercenter.org/reporting/perus-new-environmental-policies-what-are-they-and-will-they-work.

CGD, Center for Global Development. (2014). Ideas to action: Independent research for global prosperity. Why maintaining tropical forests Is essential and urgent for a stable climate—working paper 385. https://www.cgdev.org/publication/why-maintaining-tropical-forests-essential-and-urgentstable-climate-working-paper-385

Collyns, D. (2018). Peru passes law-allowing roads through pristine Amazon rainforest. https://www.theguardian.com/world/2018/jan/22/peru-passes-law-allowing-roads-through-pristine-amazon-rainforest.

Corbin, J. M., & Strauss, A. C. (2008). *Basics of qualitative research: Techniques and procedures for developing grounded theory* (3rd ed.). Thousand Oaks, CA: Sage.

Creswell, J. W. (2007). *Qualitative inquiry and research design: Choosing among five approaches* (2nd ed.). Thousand Oaks, CA: Sage.

Hierro, P. G., Hvalkof, S., & Gray, A. (1998). *Liberation through land rights in the Peruvian Amazon*. Copenhagen: International Work Group for Indigenous Affairs.

Glennie, J. (2014). Why are indigenous people left out of the sustainable development goals? https://www.theguardian.com/global-development/poverty-matters/2014/aug/14/indigenous-people-sustainable-development-goals.

Goldtooth, T. B. K. (2014). REDD and forests case verdict. https://therightsofnature.org/redd-forests-verdict/.

Goodman, R. C., & Martin, H. (2014). Why maintaining tropical forests is essential and urgent for a stable climate. https://www.cgdev.org/publication/why-maintaining-tropical-forests-essential-and-urgent-stable-climate-working-paper-385.

Gray, A. (1996). *Mythology, spirituality, and history*. New York: Berghahn.

Gray, A. (1997). *Indigenous rights and development*. Providence: Berghahn Books.

Gushiken, R., & Campos, M. Á. (2014). *Línea de base de brechas sociales por origen étnico en el Perú*. Lima: Ministerio de Cultura

Hay, I. (2005). *Qualitative research methods in human geography* (2nd ed.). Oxford: Oxford University Press.

Hill, D. (2017). $1bn to clean up the oil in Peru's northern Amazon. *The Guardian*. https://www.theguardian.com/environment/andes-to-the-amazon/2017/aug/03/us1-billion-oil-perus-amazon.

INEI. (2018). http://www.inei.gob.pe.

Kashwan, P. (2017). Bigotry against indigenous people means we're missing a trick on climate change. *The Guardian*. https://www.theguardian.com/working-in-development/2017/nov/15/bigotry-against-indigenous-people-means-were-missing-a-trick-on-climate-change.

Kaufman, C. (2015). "Leapfrogging": Can developing countries truly skip over fossil fuel reliance in favour of renewable energies? https://bpr.berkeley.edu/2015/11/28/leapfrogging-can-developing-countries-truly-skip-over-fossil-fuel-reliance-in-favor-of-renewable-energies/.

Mapstone, N. (2010). Peru's commodity-based boom hits the accelerator. *Financial Times*. https://www.ft.com/content/1d321438-f425-11df-886b-00144feab49a.

Naidoo, K. (2015). Cop-outs and denial: COP21—Climate emergency. *Vice*, (4). https://www.youtube.com/watch?v=lbXmSs9mPkQ&t=235s.

North, D. C. (1991). Institutions. *Journal of Economic Perspectives, 5*(1), 97–112.

OIT. (2014). Convenio Núm. 169 de la OIT sobre Pueblos Indígenas y Tribales. LIMA: OIT. http://www.ilo.org/wcmsp5/groups/public/—americas/—ro-lima/documents/publication/wcms_345065.pdf.

RAISG. (2012). Amazon oil and gas Concession satellite maps. https://www3.socioambiental.org/geo/RAISGMapaOnline/.

Rights and Resources Initiative. (2017). Securing community land rights: Priorities & opportunities to advance climate & sustainable development goals. http://rightsandresources.org/wp-content/uploads/2017/09/Stockholm-Prorities-and-Opportunities-Brief.pdf.

Tinker, I. (1990). *The making of a field: Advocates, practitioners, and scholars. Persistent inequalities: Women and World Development*. New York: Oxford.

Trigger, B., Macleod, M., & Adams, R. (1996). *The Cambridge history of native peoples of the Americas* (2nd ed.). Cambridge: Cambridge University Press.

Tubino, F. (2005). *Las prácticas discursivas sobre la interculturalidad en el Perú de hoy. Propuesta delineamientos para su tratamiento en el sistema educativo peruano*. Lima: Ministerio de Educación.

UNEP. (2016). Greening Peru's economy. http://web.unep.org/stories/story/greening-peru%E2%80%99s-economy.

UNESCO. (2017a). *Indigenous knowledge and practices in education in Latin America*. Santiago, Chile: UNESCO. http://unesdoc.unesco.org/images/0024/002477/247754e.pdf.

UNESCO. (2017b). Local knowledge, global goals. http://unesdoc.unesco.org/images/0025/002595/259599E.pdf.

UNFCCC. (2015a). Adoption of the Paris Agreement. In: *21st Conference of the Parties*. Paris: United Nations.

UNFCCC. (2015b). Intended nationally determined contribution (iNDC) from the Republic of Peru. http://www4.unfccc.int/ndcregistry/PublishedDocuments/Peru%20First/iNDC%20Perú%20english.pdf.

Valqui, M., Feather, C., & Espinosa, R. (2015). *Revealing the hidden*. Lima: AIDESEP and FPP. http://www.forestpeoples.org/sites/fpp/files/publication/2015/02/fppperureportenglishinternetfinalaug32015.pdf.

Viale, C., & Bedoya, F. (2010). Peru moves to implement greater indigenous rights. *RWI Research*. http://archive.resourcegovernance.org/news/peru-moves-implement-greater-indigenous-rights.

What is REDD+? | The REDD Desk. (2018). Theredddesk.org. https://theredddesk.org/what-redd.

Whitehead, T. (2005). *Basic classical ethnographic research methods*. Maryland: University of Maryland. http://www.cusag.umd.edu/documents/workingpapers/classicalethnomethods.pdf.

World Bank. (2018). Peru. http://www.worldbank.org/en/country/peru.

World Band Study. (2012). Justice for forests improving criminal justice efforts to combat illegal logging. http://siteresources.worldbank.org/EXTFINANCIALSECTOR/Resources/Illegal_Logging.pdf.

WRI. (2014). World resource institute 2014 Annual report. http://www.wri.org/annualreport/2014/

Zuazo, A. E. V. (2017). Peru lost more than 1 million hectares of Amazon forest over a period of 15 years. https://news.mongabay.com/2017/05/peru-lost-more-than-1-million-hectares-of-amazon-forest-over-a-period-of-15-years/.

If you have any concerns about our product,
[illegible]
productsafety@springernature.com
[illegible] Publisher [illegible]
the EU authorised representative is
Springer Nature Customer Service Center GmbH
Europaplatz 3, 69115 Heidelberg, Germany
Printed by Libri Plureos GmbH
in Hamburg, Germany

MIX
Papier aus verantwortungsvollen Quellen
Paper from responsible sources
FSC® C105338

If you have any concerns about our products,
you can contact us on
ProductSafety@springernature.com

In case Publisher is established outside the EU,
the EU authorized representative is:
Springer Nature Customer Service Center GmbH
Europaplatz 3, 69115 Heidelberg, Germany

Printed by Libri Plureos GmbH
in Hamburg, Germany